THE
OUTER
CIRCLE

Also by the Editors

THE OUTER CIRCLE

Women in the Scientific Community

Edited by

HARRIET ZUCKERMAN

JONATHAN R. COLE

JOHN T. BRUER

W · W · NORTON & COMPANY

NEW YORK · LONDON

THE TEXT OF THIS BOOK *is composed in Times Roman, with the display set in Centaur. Composition and manufacturing by the Maple Vail Book Manufacturing Group. Book design by Marjorie J. Flock.*

Y

Library of Congress Cataloging-in-Publication Data
The outer circle : women in the scientific community / edited by
    Harriet Zuckerman, Jonathan R. Cole, and John T. Bruer.
        p.   cm.
    Papers from 4 symposia held 1983–1986 at Stanford University,
        sponsored by the Josiah Macy, Jr. Foundation.
    Includes bibliographical references and index.
        1. Women in science—Congresses.   2. Women scientists—Congresses.
    I. Zuckerman, Harriet.   II. Cole, Jonathan R.   III. Bruer, John T.,
    1949–   .   IV. Josiah Macy, Jr. Foundation.
    Q130.087   1991
    305.43′5—dc20   90–40018

                        ISBN 0–393–02773–2

W.W. Norton & Company, Inc., 500 Fifth Avenue, New York, N.Y. 10110
    W.W. Norton & Company, Ltd., 10 Coptic Street, London WC1A 1PU

                        2 3 4 5 6 7 8 9 0

FOR JAMES HIRSCH
Scientist, Teacher, and Friend

Contents

I I I
WOMEN'S CAREERS
The Obstacle Course

I V
A THEORETICAL EXPLANATION

Preface

THIS VOLUME stems directly from the profound concern of James Hirsch, late President of the Josiah Macy Jr. Foundation, for the difficulties faced by many women scientists. Dr. Hirsch, a research scientist and a Dean of Graduate Studies at the Rockefeller University, was both sophisticated and experienced in the ways of the national scientific community. He wanted to know why universities and research institutions had failed to draw on the potentials of women scientists and why the scientific community so often failed to support their work. It was his idea to institute a symposium series that would examine the fate of women in science and would, he hoped, provide proposals for immediate action to be taken by the Macy Foundation.

To learn how the wide disparities in the careers of men and women scientists come about and are maintained, the Foundation sponsored a conference at Stanford University in January 1983. This meeting brought physical, biological, and social scientists together to explore their distinctive perspectives on these matters. The intense interest in the subject exhibited at the Stanford meeting led Hirsch and his Macy Foundation colleague, John Bruer, to ask Jonathan R. Cole and Harriet Zuckerman, both sociologists of science who had done research on women, to organize the symposium series. For their part, Cole and Zuckerman hoped that such a series would also contribute to advancing research on scientific careers. They agreed to organize a series of meetings which would focus specifically on developing a new and wide-ranging research agenda on the careers of women and men in science.

Four symposia were held between 1983 and 1986; the first, on the relation of marriage, family, and research performance; the second, on gender differences in rates of publication; the third, on gender discrimination in science; and the last, on research strategies and practices of contemporary scientists.

The governing idea was to bring together scholars and scientists from various disciplines. Some of them had studied the careers of women in science, while others had done research on cognate subjects. Other men and women scientists spoke from firsthand experience about the difficulties encountered by women scientists in the past and the present. Participants responded to working papers describing the currrent state of research on each subject and to an evolving plan for discussion. Throughout, the principal objective was to identify questions, ideas, theories, and methods which might contribute to a new research agenda on the subject of women in science.

Essays on subjects and problems noted in the symposia are collected here; and so are the empirical and theoretical essays by the editors which partly served as the basis for discussion. The volume also contains interviews with three women scientists, Andrea Dupree, Sandra Panem, and Salome Waelsch, who have reviewed and edited the verbatim transcripts. We are indebted to them for their collaboration and their candor in describing their scientific lives. We also thank two of our students: Annette Bernhardt, for her usual effective assistance in editing the lengthy interview transcripts, and Lisa Rosen, for the comprehensive index she prepared. Finally, Mary Cunnane and Rose Kernochan of W. W. Norton have been everything authors might wish of their editors. We are grateful to them.

Introduction

SINCE THE Scientific Revolution of the seventeenth century, women aspiring to work in science have encountered barrier after barrier. Among the limited numbers of women who succeeded, practically none were allowed to enter the inner circles of the emerging scientific community.[1] So antithetical were the social categories of women and science that, throughout the nineteenth century and well into the twentieth, it seemed only natural to call all practitioners of scientific inquiry "men of science." Long after the English philosopher of science, William Whewell, had coined the genderless term "scientist" in the early Victorian era, women scientists were still included in directories entitled "Men of Science."[2]

As late as the first decades of this century—and perhaps it is still true now—only one woman of science, Marie Curie, could be said to have won great fame. Having shared the Nobel Prize in physics with her husband, Pierre Curie, she captured the public imagination as the mother of research on radioactivity.[3] After Pierre Curie's death, she became the first woman to hold a chair at the Sorbonne and soon established herself as an independent scientific force. Her further research resulted in another Nobel Prize, this one in chemistry, making her the first scientist to be awarded the prize twice (she is still the only one to have received it both in physics and chemistry). She was not someone to be ignored by the topmost tier of her scientific contemporaries but, even so, she was never comfortably esconced in their inner circle. In fact, the Académie des Sciences excluded her from their ranks; she had missed selection by one vote.[4] Still, no other woman scientist of her time or before had ever been accorded such widespread peer recognition.

One of Marie Curie's near contemporaries, Sonya Kovalevsky—who has since been identified as "the greatest woman mathematician prior to the twentieth-century"—found doors closed to her throughout much of her life.[5] Barred from the study of mathematics in her native Russia, she entered into a marriage of convenience in order to pursue her education in

Germany. There, too, because she was a woman she was not admitted to the University of Berlin; but she studied privately with Karl Weierstrass, an exceedingly rigorous and critical mathematician at the university. Through his intervention, she was granted the doctorate at Göttingen—albeit *in absentia*—but she could find no academic post anywhere in Europe. Only years later, with the help of Sweden's most esteemed mathematician, Gösta Mittag-Leffler (another of Weierstrass's pupils) was she finally accorded a professorship at Stockholm.

Pioneering women scientists in the United States fared no better. The astronomer Maria Mitchell, born a half century before Curie, discovered a new comet in 1847 and calculated its precise position at the time of discovery. This feat brought her a medal from the King of Denmark and election to the American Academy of Arts and Sciences; she was the Academy's first woman member. But again, admission into the inner circle was grudging at best. The Academy's *Report* for 1848 records that she was elected "in spite of being a woman." It was only some decades later that she found an academic post—at Vassar College, one of the earliest institutions devoted to educating women.[6]

The American anatomist, Florence Sabin, was born just three years after Marie Curie. Her career was marked by notable achievements and, as Chapter 1 of this volume recounts, a succession of "firsts" in peer recognition. She was the first woman to serve on the faculty of the Johns Hopkins Medical School, the first to be elected president of the American Association of Anatomists, the first to be made a full member of the Rockefeller Institute, and, most significantly, in 1925, the first woman to be elected to the National Academy of Sciences since its founding in 1863.[7]

For all their great accomplishments, each of these four originative women of science remained poised, in her own time and her own space, somewhere between the outer and the inner circles of science.

Turning from such preeminent American women of science as Mitchell and Sabin to the larger population of women scientists in the nineteenth century, we find no reliable data on their number and distribution. Evidently, there were few of them.[8] As late as 1920, just 41 women earned doctorates in the physical and biological sciences, approximately 14 percent of the total.[9] As these numbers indicate, not many advanced degrees were conferred in these fields at all. For reasons that remain obscure,[10] the proportion of women earning degrees in the sciences shrank substantially in later decades so that by 1960 only 5 percent of new doctorates in the sciences were women. By 1970 the number of women taking degrees increased slightly (to about 7 percent) and then began to rise sharply.[11] In 1980 some 19 percent of such degrees went to women, and by 1988 that number had risen further to 22 percent.[12] There are new indications that the pace at which women are entering the sciences may be tapering off, but a return to the lean years of the 1950s and 1960s is unlikely.

Despite this growth in their numbers, women scientists have not acquired anything like parity with men in opportunities for doing research. Working scientists know, of course, that the concept of equal opportunity is only an abstraction. Concretely, equal opportunity for scientific work consists of equal access to research instrumentation, funds to pay for the research, laboratory space, time for work, appropriate collaborators, and good students. These components of opportunity in science have long been important, but now, in an age when big science has moved beyond physics into astronomy and the life sciences, access to these means of scientific production becomes ever more consequential in shaping life chances for significant scientific work.

It therefore comes as no surprise that, as the papers in this volume attest, practically all comparative studies find that, on average, women scientists have published from half to two thirds as many papers as their age peers among men. However as chapters in this volume also suggest, things have been changing, and especially so for young women of science.

Science remains a domain dominated by men, not only numerically but in the exercise of authority, power, and influence. Further, whatever the motives and attitudes of individual scientists, the practice of science is objectively competitive. How do these circumstances affect the careers of men and women in science? Are there differentials in the operation of the reward system of science for men and women? How does all this work in a profession marked by intense competition—whose practitioners claim prime emphasis on the quality of role performance? These and related questions about the comparative careers of women and men in science are at the focus of this collaborative four-part volume.

The chapters in Part I tell of the position of women in American science now and in the recent past. The second part examines the vexing question of the comparative research productivity of men and women scientists, while the third moves beyond science to treat professional women more generally. Part IV sets out a theory to account for disparities in the contributions of men and women scientists. Taken together, these varied papers describe the uncertain role women continue to play in science and the respects in which they remain in its outer circle.

The Outer Circle: The Position of Women in the Scientific Community

The first four chapters focus specifically on how women have recently fared in science. The first paper, by a sociologist, Harriet Zuckerman, systematically compares the status of men and women in American science and engineering. It examines three connected trends in the careers of contemporary scientists: first, persisting differences between men and women scientists in research performance and career attainments, rates of unemployment, organizational rank, salary, and honorific awards; second, a growing convergence in the career attainments of men and women scientists as a whole; and last, an increasing divergence between men and women scientists as they grow older such that the disparities in performance and standing between men and women age peers become greater at the end than they were at the beginning of their careers. The paper also examines the principal explanations that have been proposed for such observed differences and identifies gaps in our current knowledge of this subject.

Helen Astin, a psychologist, takes up related questions in the paper which follows. Do men and women differ in perceptions of their own work, the ownership of intellectual property, and the allocation of rewards? Drawing on her study of men and women authors of papers highly cited in the scientific literature, she finds that women seem less proprietary about their work and less clear about what motivates them to do it. They are more concerned with the responses of peers and are less likely to take for granted that their work will be rewarded by frequent citation. Like men, they emphasize the substantive importance of their contributions to science but feel that they have less control over how it will be judged by peers. Astin suggests that women may thus be more sensitive than men to external validation, a possible consequence of gender differences in early socialization and women's experience of discrimination. She speculates that women's greater concern with external response may, in turn, affect their motivation to engage in further research and publication. Her study raises questions about gender differences in perceptions of research contributions among scientist-authors whose research is seldom cited and about cross-gender evaluations of research. Do men tend to attribute women's success to situational factors rather than to talent? In contrast, do women tend to attribute men's success to talent rather than to a more favorable cognitive and social environment? Such questions of attribution tie directly into psychological research on attribution theory and gender.

Concluding Part I are three interviews with distinguished women scientists: the first, with Salome Waelsch, who began her research in genetics in the early 1930s; the second, with Andrea Dupree, an astronomer who began research 30 years later; and the third with Sandra Panem, who began research in biochemistry a few years after that, in the latter 1960s. The interviews deal with various aspects of their scientific and private lives in those differing times, places, and fields of interest. These interviews cannot, of course, convey anything like the full diversity of contemporary women's lives in science. But they do suggest the complexities of being a woman of science and how these have been changing in the last generation or so.

Part of the great intellectual migration from Nazi Germany, Salome Waelsch arrived in the United States just before World War II without a job and with something less than a command of English. It was a time when American women scientists had few opportunities for employment in science. They could choose between a minor research post (not always salaried) in a male scientist's laboratory or a teaching post at a women's college, where limited facilities would inhibit significant research. (In her native Germany, the alternatives were even more limited.) Waelsch describes the steps she took to develop a research program of her own and to establish an independent career. Hers is a tale in microcosm of the obstacle courses traveled by women scientists.

Waelsch is persuaded that, while careers of men and women scientists tend to differ significantly, their cognitive orientations are much the same. In her view, women do not think differently about science. Women conduct research, formulate their analyses, and respond to research contributions in much the same way as men. (As will be seen, these views are at odds in certain respects with those set out by Evelyn Fox Keller in Chapter 10.) Waelsch also emphasizes that these are hard times for both young men and young women in science; both suffer from the growing uncertainties and vicissitudes of obtaining scarce jobs and adequate support for research. This interview calls attention to the need for systematic comparison of the circumstances of women and men scientists rather than continuing with the more familiar practice of limiting studies to women or men alone.

Chapter 4 also recounts a life in science. Andrea Dupree is an astronomer born three decades after Waelsch. Like Waelsch, she became a research associate after her doctoral degree, and she, too, then had to find a way to establish herself as an independent and productive investigator. Now head of a major research group at the Harvard Smithsonian Observatory, one of the nation's top astronomical laboratories, Dupree began as a somewhat reluctant graduate student. Her account exemplifies Cynthia Epstein's observation (in Chapter 11) that career aspirations are, in large part, a function of the opportunities individuals encounter. Dupree's growing contributions and advancing career were no doubt made somewhat easier by improving opportunities for women, but as she makes clear, she still found it necessary to insist on being treated fairly. This does not happen on its own. She observes that, at least in astronomy, men and

women differ markedly in cognitive preferences, with men choosing theory more often than women. She goes on to observe that male astronomers
are generally more aggressive advocates of their ideas than women and
that they are less apt to hear what women say than women are to hear men.
Such strong impressions of differences in gender preferences in directions
of scientific inquiry and in styles of scientific discourse in the field of
astronomy deserve more general systematic study. Should these patterns
prove to be widespread, they would plainly distinguish women's contributions to science from those of men and would also testify to the differential reception of women's ideas.

In Chapter 5, Sandra Panem describes her training as a biochemist and
virologist at the University of Chicago. Like Waelsch and Dupree, Panem
began her career as a research associate. Unlike them, however, she was
appointed to an assistant professorship and headed up her own laboratory.
The recipient of a continuing series of grants to support her research, she
worked on the effects of interferon (then a hot but controversial subject)
and also served on the university's Tenure Policy Committee. She reports
her belief that she understood how the tenuring process worked and that
she could assess her chances for promotion. Her interview describes her
encounter with tenure review, making clear that her perspective may have
differed considerably from that of the other participants. With a distanced
objectivity usually associated with observers rather than participants, she
proceeds to dissect the event. She tells of the effects of competition for
space in the university, the competition between departments for positions,
the history of tenure decisions in her department, the controversy surrounding the research she was doing, the advent of affirmative action, and its
effects on the outcome of the decision in her case. She also tells of her
errors of perception and strategy, of the aftermath of the episode, and of
its effects on the later stages of her career. Panem's experience provides a
personal counterpoint to the accumulating evidence on gender differences
in rates of promotion to tenure.

Are Women Less Productive Scientists Than Men?

The essays in Part II take up the comparative research performance of
men and women scientists. Jonathan Cole and Harriet Zuckerman do so by
studying the effects of marriage and motherhood on the research work and
careers of women scientists. The other essays in this section ask whether
the observed gender differences in rates of publication arise from the
differential treatment of men and women in the organizations in which
they work, as William Bielby and Mary Frank Fox propose in Chapters 7

and 8, from differential socialization and aspirations, as Stephen Cole and Robert Fiorentine suggest in Chapter 9, or if they inhere in prevailing cultural conceptions of scientific knowledge and of gender, as Evelyn Fox Keller contends.

In Chapter 6, Jonathan Cole and Zuckerman, both sociologists, draw upon detailed interviews with women and men scientists of similar ages and research accomplishments and upon systematic examination of the scientists' publication histories to conclude that marriage and motherhood cannot adequately account for gender differences in research productivity. Married women, it turns out, publish more than single women, and mothers publish more than childless women. Furthermore, women's rates of publication, on average, do not decline following childbirth and during the years when they are caring for young children. Counterintuitive as these findings may at first appear, they are nevertheless consistent with the women scientists' reports of their experiences. They report that the obligations of marriage and motherhood are considerable but that they do not take their toll on women's research. Cole and Zuckerman suggest that women scientists engage in assiduous "status-set management" in order to maintain their research: they reduce their other obligations and activities to the bare minimum and concentrate almost entirely on work and family. These findings suggest that the research practices of scientists and certain attributes of the environments in which they work—for example, preferred research styles, access to collaborators and funds, and positions within universities and laboratories—are more powerful influences on rates of scientific publication than are extra-scientific influences such as marital and family obligations.

In Chapter 7, William Bielby, also a sociologist, asks whether women's careers in science really represent a "special case" or are simply another case in point. Are women scientists largely judged and rewarded according to universal standards, as some research has found? Or do they, like their counterparts in other occupations, encounter distinctive structural barriers and cultural stereotypes which curb their advancement? Bielby observes that in many work organizations, structural barriers—such as discriminatory employment and promotion practices and cultural stereotypes about women's work abilities—account at least as much as women's limited skills and work experience for their holding jobs with lower pay, prestige, and responsibility than men. He suggests that the emphasis in much (though far from all) sociological research on careers in science has focused on the extent to which men and women scientists are rewarded in accord with their contributions, that is, on the extent to which science is governed by meritocratic norms. This emphasis, he holds, has led to close examination of the connections between individual role performance (what he calls "supply-side factors") and rewards. But researchers have neglected to study organizational histories, personnel policies, and the constraints imposed on organizations by their environments. Thus, the effects on women's careers of the "blatantly exclusionary policies" of particular universities have hardly been studied. Nor has there been systematic study of the impact of forces external to universities and research organizations, such as the policies of funding agencies which encourage equitable treatment of men and women or of uncertain economic conditions which invite employment of part-time and temporary workers, who are more often women than men.

Bielby suggests that the few studies of scientists' careers which have examined the effects of organizational attributes find that they do exert an independent influence on outcomes. He also maintains that the studies which focus mainly on differential rewards give short shrift to the role of gender stereotypes in shaping day-to-day interaction between men and women and their careers. Such stereotypes—for example, the notion that women are biologically less capable than men of doing important scientific work—affect not only the perceptions and expectations of men and women scientists but also the social relations between them. In light of these limitations in past research, Bielby proposes that new comparative research on the careers of men and women scientists take these directions, rather than exploring further the extent to which science is a "special case."

Mary Frank Fox, sociologist, proposes a thesis in Chapter 8 that complements Bielby's. In an analysis of the considerable empirical literature on academic and scientific careers, she contends men and women who work in the same environment (laboratory, department, or university) do not necessarily have the same chances to advance their research. Nor are their contributions necessarily evaluated according to the same criteria. She focuses particularly on the different opportunities men and women have for collegial interaction and research collaboration in certain universities; she notes the effects of collaboration, especially, on rates of publication, and these, in turn, on rates of promotion. She points out that when standards used in allocating research resources are "flexible" (as they are in universities that make such allocations negotiable), the male members of the faculty stand a better chance to acquire the resources they need to get on with their research. Fox emphasizes that detailed knowledge of organizational customs and practice is required to understand differentials in published productivity of men and women scientists as well as their different ranks, resources, and salaries.

Stephen Cole and Robert Fiorentine, also sociologists, take a different tack in the next chapter. Focusing on the difficulties of observing and measuring gender discrimination, they maintain that differences between men and women in "outcomes" (for example, in admission rates to medical school, which they examine here) tell nothing about the extent of gender discrimination. The same outcomes might result from quite different sources: from biased selection practices or from self-selection, from men and women applying for posts or for admission at different rates or from a combination of these and other processes.

Cole and Fiorentine draw upon their research on the persistence of men and women in premedical programs and on rates of admission to medical school to conclude that self-selection and differential qualifications of men and women, as well as discrimination, operate in varying degrees at different points in the process of preparing for and then applying to medical school. In combination, these processes account for the phenomenon of fewer women than men in medical school. Research which centers on differential outcomes and neglects the processes which bring them about is apt therefore to produce erroneous conclusions about the extent of discrimination in a given case. Comparable evidence for the careers of scientists, Cole and Fiorentine note, could be assembled. They also suggest that women may persist at lower rates than men in training for scientific careers for the same reasons as they have observed among premedical students. They explain such differences in rates with a "theory of normative alternatives," which holds that the culture gives women more leeway in choosing careers than men and less often subjects them to pressure for career achievement. Cole and Fiorentine call for systematic analysis of the effects of differences in "gender culture" on the achievements of men and women in various societies.

In Chapter 10, Evelyn Fox Keller, a theoretical biologist and now also a philosopher of science, agrees that much thinking and research about women scientists fails to recognize the overwhelming influence of culture. However, she emphasizes the impact of culture on our conceptions of science and gender—both of these being, in her view, cultural constructions. Prevailing norms of masculinity, she argues, have been covertly absorbed into science. This holds for scientists' ideas about nature and for the standards they apply in gauging scientific achievement. She sees science as being prevalently "a monolithic venture—defined by a single goal and a single standard of success." This conception of science, Keller maintains, has made it impossible for women scientists to succeed in their struggle for equal treatment.

She observes that feminists long believed that if only women scientists could demonstrate that they did not differ in any important respects from men, equity could be achieved. In effect, this meant that all signs of femininity had to be "eradicated," with even the use of feminine first names disappearing to be replaced by the gender-neutral use of initials. This strategy, Keller observes, did not achieve its advocates' objectives. Nor has the opposite strategy of earmarking differences between men and women succeeded since differences are readily translated into inequalities. Keller calls for a reconstruction of the cultural definitions of both science and gender. She argues that since science is a cultural construct, it can be reconstructed to be gender-neutral, being biased neither in favor of men or women. To her mind, current developments in the history and philosophy of science have led precisely to such reevaluations. Scientific knowledge is no longer viewed as a strict representation of nature but as a fluid construct, shaped by social forces; it can have multiple goals and multiple standards. In much the same way, recent feminist scholarship has led to the view that gender norms are socially constructed. There is enormous variability, it seems, in individual compliance with those norms—such variability that the differences among men and among women may turn out to be more salient than the differences between them.

Keller urges that more precise and nuanced conceptions of science and gender be adopted. She also urges a recognition that *de facto* discrimination against women is often justified with the argument that feminine characteristics are antithetical to "good" science. Their introduction into the enterprise of science, some men claim, must undermine its integrity. Such feminine characteristics are, Keller argues, either irrelevant to the advancement of science or can make their own contribution to it.

Careers of Professional Women: An Obstacle Course

The papers in Part III extend the discussion beyond the boundaries of science. Both focus on the extent to which the obstacles women scientists face resemble those encountered by women in other occupations and by other disadvantaged groups, more generally.

In Chapter 11, Cynthia Fuchs Epstein, a sociologist, examines the ways in which the achievements of professional women have been constrained by society and culture. The comparatively low rates of women's participation in the learned professions and creative arts, she argues, have resulted neither from women's unconstrained choices nor from their distinctive abilities. Rather they are the outcome of priorities imposed by the society at large and by the professions themselves. She goes on to observe that appraisals of women's professional and cultural contributions are linked with social definitions of the character of a contribution. Thus the judgment that some contributions to science are dull or routine may, in some measure, result from their having been the work of women. Certain assessments of Rosalind Franklin's basic contributions to the discovery of DNA are presented as a case in point.

Such cultural definitions are reinforced, Epstein holds, by various modes of social control—in the form of rules, customs, conventions, and laws—which have severely limited women's access to education and training for the professions, particularly in the leading institutions. Thus, Harvard Law School did not admit women until 1950. In such institutions, customary practice continues to restrict the access of women to key apprenticeships with professionals working at the forefronts of their fields.

Historically, the learned professions have helped to maintain the image of fragile, dependent woman; the law, for example, was long undecided on whether ''women could be considered 'persons.' '' So, too, certain medical specialties have taken women as models of vulnerability, while the sciences, social and biological, have looked for evidence not merely of sex and gender differences but of women's deficiencies, thus providing seeming justification for excluding them from advanced education.

Such images have tended to give way as more and more women entered the professions. Nevertheless, women are still concentrated in a limited range of professional and scientific specialties, often those with least prestige. And they still encounter resistance from men who find their work presence discomfiting. In Epstein's words, professional women remain ''outsiders within.''

The study of women professionals, Epstein concludes, reveals much about the enormous changes that have taken place in women's roles and also gives the lie to images of women promulgated by the very fields they sought to enter.

Chapter 12, by the legal scholar Owen Fiss, analyzes two dominant theories of discrimination—''process theory'' and ''substantive theory''—along with a third, ''hybrid theory,'' which links the two. Each, he shows, has posed dilemmas when the courts have attempted to apply them in civil rights cases involving blacks and women.

The process theory of discrimination holds that the application of ''invidious'' criteria (for example, race, sex, or religion) in decisions involving social selection (hiring, or school admission), constitutes a violation of the law. It is discriminatory because the use of such criteria ''corrupts'' the process of selection. The substantive theory, in contrast, focuses on the consequences of using such criteria. Excluding blacks or women from employment or schooling deprives them of equal opportunities for employment or schooling and is therefore discriminatory. Use of process theory leads to a focus on the criteria of selection procedures, while use of substance theory leads to a focus on the outcomes of selection procedures. Process theorists are concerned mainly with maintaining fair selection procedures, while substance theorists are concerned with achieving equal outcomes.

In practice, Fiss observes, choice of one theory or another matters little since both usually produce the same legal judgment. However, the choice of theory matters a great deal in instances involving the use of what are called ''facially innocent'' criteria in selection, that is, criteria which appear to serve legitimate interests. In judging applicants for graduate work in science, for example, the use of test scores on mathematical or spatial abilities would appear to be connected with successful performance and thus serve a legitimate interest. Once it was determined that the tests serve legitimate interests and have been fairly applied, process theorists would conclude that their use was not discriminatory. Substance theorists, in contrast, would conclude that such tests are discriminatory, if it can be shown that a disproportionate number of blacks or women have been excluded even though the criteria of selection are defined as legitimate and the process fair. Substance theorists hold that individual and group well-being are closely connected and that the law must eliminate practices which permit the subordination of particular groups. Where substantive theory prevails, disproportionate impact on particular groups rather than on a given individual must be proved. Discrimination is said to apply primarily to groups and only derivatively to individuals.

The third or ''hybrid theory'' of discrimination focuses on practices that perpetuate past discrimination—for example, the use of literacy tests in voting where certain groups had been deprived of access to schooling. The use of the hybrid theory allows the courts to forbid the use of what appear to be permissible criteria if they have contributed to past discrimination. It also allows the courts to decide that antidiscrimination laws can be applied and remedial action taken in favor of individuals other than those originally subject to discrimination, such as their descendants.

Fiss argues that all empirical inquiries into the incidence, evolution, and intensity of discrimination presuppose a definition of discrimination linked to one or another of these theories. The definition adopted by researchers affects the questions they ask, the evidence they use, and ultimately the conclusions they draw.

The final chapter in the volume, by Jonathan Cole, a sociologist, and Burton Singer, a statistician, sets out a ''fine-grained'' theory to account for gender differences in published productivity and its dramatic increase over the course of careers. The "theory of limited differences" hypothesizes that scientists' rates of publication are determined by sequences of specifiable external influences on individuals—"kicks," as the authors call them, after terminology used in kinetics—and by the reactions of individuals to these kicks, all this occurring in a highly competitive community having limited resources and rewards. Kicks can be positive or negative. Having a grant application refused would be an instance of a negative kick, receiving a fellowship an instance of a positive kick. Such kicks evoke reactions in the scientists who experience them, and these, in turn, affect their rates of publication in varying degrees. Some kicks are more often experienced by women, some by men. Even if only a small proportion of women experience a particular negative kick, leaving the great majority unaffected, and even if the effects of particular "kick-reaction pairs" are insignificant, over the course of careers, these "limited differences" between men and women can have a significant impact on their rates of publication owing to cumulation and to continuing effects over the long term (kick-reaction pairs, as Cole and Singer put it, have "memory"). As the mathematical model set out in the Cole–Singer chapter demonstrates, large gender differences in published productivity develop over the long term, even when very small proportions of women experience negative kicks more often than men and respond to them accordingly.

The theory takes into account influences of gender differences in socialization, attitudes and preferences, discrimination (socially structured differences in opportunities), as well as individual differences in personality in estimating scientists' chances of experiencing certain kicks and assessing their reactions to them. The theory is consistent with the results of quantitative and statistical studies of social stratification in science and with qualitative data in which few women scientists report that they have been discriminated against even though they know others who have. Applying and testing the theory requires a program of data collection geared to gathering information on kick-reaction pairs over the course of careers. The Cole–Singer paper, like others in the volume, calls for more detailed, longitudinal data on individuals and the development of their careers, data that goes well beyond those on which prior research has been based.

Twenty years of research on scientists' careers has revealed much about the structure of the scientific community and the extent and character of its stratification. But we have yet to understand the sources of gender differences in productivity, career attainments, and peer recognition. As the essays in this volume attest, more nuanced explanations are required than those explored so far. Such explanations plainly hold both scientific and practical interest, in particular as they account for the evolution of gender differences in a domain marked by intense competition and a strong commitment to achievement. They will further illuminate the predicaments posed for *all* professional women by their positions in the outer circle. The papers that follow sketch out a rich agenda for those future inquiries.

I

THE
OUTER
CIRCLE

Women's Position in the Scientific Community

HARRIET ZUCKERMAN

1. The Careers of Men and Women Scientists: A Review of Current Research

MORE THAN SIXTY YEARS after its founding, the National Academy of Sciences elected its first women member: Florence Sabin, an anatomist and embryologist. But then Sabin's entire career was a succession of "firsts." In 1902 she was the first woman appointed to the faculty of the Johns Hopkins Medical School, then probably the most distinguished medical school in the United States. She was also the first woman full professor there (1917), the first woman elected president of the American Association of Anatomists (1924), and, upon leaving the Hopkins in 1925, the first woman to be made a full member of the Rockefeller Institute of Medical Research (now the Rockefeller University) (Breiger 1980; Rossiter 1982).

This array of posts makes it clear that Sabin became an insider and member of the scientific establishment. She was also an outsider and social pioneer, being not just the first, but often the only woman to be included in the circles in which she moved.[1] Sabin was atypical, if not unique, among women scientists of the time in the extent to which her work was recognized and rewarded by the scientific community. By contrast, the historical record shows that many accomplished women were either ignored or actively discouraged. Those honors which came to them at all came very late (Rossiter 1982).

Reprinted from Harriet Zuckerman, "The Careers of Men and Women Scientists: A Review of Current Research," in *Women: Their Underrepresentation and Career Differentials in Science and Engineering*, ed. Linda S. Dix (Washington, D.C.: National Academy of Sciences, National Academy Press, 1987), with permission from the National Academy of Sciences, National Academy Press.

How much has the lot of women scientists changed since the Sabin era? This review of research on the careers of American men and women scientists and engineers will show that there is no simple answer. Rather, as I shall indicate, the pertinent data, drawn from current studies, show three separate but interconnected patterns: First, there are *persisting differences* between men and women scientists on average, in role performance and career attainments when viewed cross-sectionally. These differences are almost always in the direction of comparative disadvantage for women and are usually combined with marked intra-gender differences. Second, there are signs of *growing convergence* between men and women in access to resources, research performance, and rewards. That is, there is evidence for increasing gender similarity over the last decade and a half and especially between younger men and women. Third, there is also evidence for *growing divergence* between men and women of the same professional age in published productivity and in some, but not all, aspects of career attainment; that is, for growing intra-cohort differences as members move through their careers.

This report divides into four parts: the first briefly describes what is known about the population of American scientists and engineers. The second inventories the main findings of research on comparative career attainments of men and women, earmarking the three patterns of similarity and difference just noted. The third reviews the principal explanations or "theories" that have been proposed to account for gender differences in careers and assesses how well the data square with these explanations. The last identifies some directions for further inquiry, based on "specified ignorance" (Merton 1987) or what we now know we need to know and why we want to know it.

Selected Demographic Characteristics of American Men and Women, Scientists and Engineers

This review of data on the size, composition, and growth rates of the population of American scientists and engineers is not intended to be comprehensive. Rather it has two purposes: to set in context the research on careers in the second part of this paper, and to caution against the use of simple bivariate distributions of gender and salary, or gender and rates of unemployment, or gender and organizational rank in describing the relative status of men and women scientists and engineers. Such bivariate distributions mislead as much as they inform since they mask marked differences between men and women scientists in professional age, education, and scientific field, these being independently related to salary, unemployment, and rank. These differing distributions of men and women must be taken into account in gauging the extent of gender difference in the career attainments. So much then for intent.

Where do things stand with respect to sheer numbers? In 1984 about 4 million Americans were working as scientists and engineers, and of these, 513,000, or 13 percent were women (National Science Foundation 1986: 61; hereafter NSF). This is not a large number, but it is 2.8 times as many as there were a decade earlier, when just 185,000 women (6 percent of the total) were working in these occupations (National Science Board 1977: 152; hereafter NSB).[2] Indeed, the number of women entering the scientific and technical population has grown even more rapidly than this population as a whole, which has almost doubled in the same period (NSB 1977: 152). Since the number of scientists and engineers typically grows by increments of young people beginning their careers (few new entrants are older people moving from other jobs), women scientists and engineers as a group are apt to be younger on average than men. This is consistent with the report that women are twice as likely as men (60 percent vs. 27 percent) to have fewer than ten years of professional experience (NSF 1986: 4).[3]

Women are also less apt than men to hold advanced degrees, and this is so in every field of science and engineering. In all, 19 percent of men scientists and engineers hold doctorates as against 11 percent of women, but such differences vary considerably from field to field (NSF 1986:1). Gender differences in educational attainments of men and women suggest that women on average probably hold high-ranking positions less often than men when the doctorate is a required credential for high rank.

Even so, the proportion of doctorates being awarded to women has been growing since 1960, particularly in the last fifteen years. By 1985, 24 percent of doctoral degrees in science and engineering went to women as against just 7 percent in 1970 (Vetter and Babco 1986: 20–21; see also Vetter 1981). Such marked increases have occurred against the background of decreases, not just in the relative but also in the absolute numbers of men receiving doctorates (see Figures 1.1 and 1.2). It is not clear now whether the prestige of scientific occupations will decline as they become more attractive to women and less attractive to men. What is clear is that high growth rates in the numbers of doctorates earned by women means that women doctorates are considerably younger professionally on average than men.

Men and women differ also in the fields they choose; and this is so both for holders of doctorates and for those with lower level degrees.[4] Women

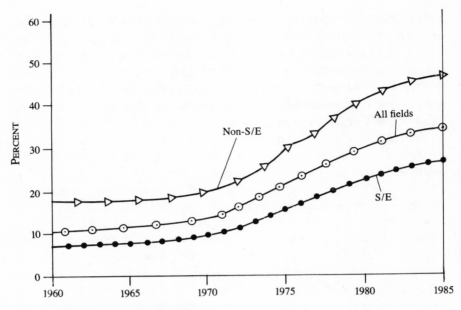

1.1   Shares of doctorates earned by women.

Source: Survey of Earned Doctorates, National Research Council.

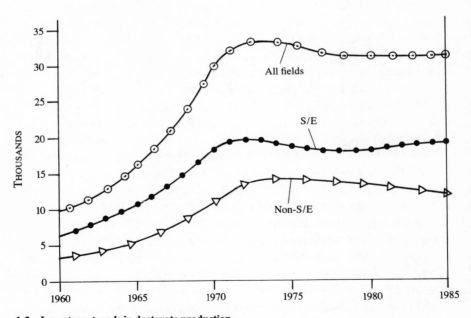

1.2   Long-term trends in doctorate production.

Source: Survey of Earned Doctorates, National Research Council.

are concentrated in the social and life sciences with comparatively small numbers working in the physical sciences and engineering (NSF 1986: 61–62, 71–72), as Figure 1.3 shows, and this has been so for decades. They comprise just 7 percent of all doctorates in the physical sciences and 2 percent in engineering in contrast to 17 percent of doctorates in the life sciences and 22 percent in the social sciences, these composite fields

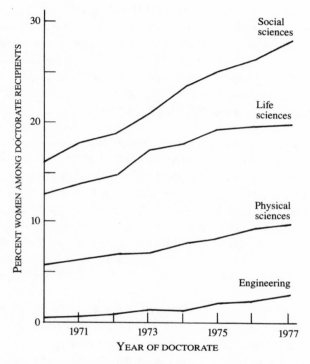

1.3 Doctoral degrees in science and engineering awarded to women, 1970–1977.

Source: Survey of Earned Doctorates, National Research Council.

having once been described as the "dispassionate" and the "compassionate" sciences. It is not clear whether gender differences also exist in specialty choice. There are marked differences in the representation of women among specialties in psychology (Russo and O'Connell 1980: 31) but no significant differences in specialty choice of men and women chemists (CEEWISE 1979; Syverson 1980, p. 24). In any event, the marked differences in the distribution of men and women among fields must be taken into account in examining salary differentials and differentials in rates of unemployment since these differ greatly field by field. Recent data show that rates of unemployment are higher and salaries lower in the fields in which women are more numerous, but this has not always been so. Thus, field differences in salary and unemployment do not arise exclusively from their gender composition.

Men and women scientists and engineers also work in different sectors of the economy. Relatively more men than women work in industry (69 percent vs. 53 percent), and relatively more women than men work in educational institutions (23 percent vs. 10 percent) (NSB 1985, table 3.7). A similar pattern of underrepresentation in industry and overrepresentation in education holds for women doctorates, with some variation by field, but gender differences in sector of employment are not as great among doctorates as among all scientific and technical workers (NSB 1985, table 3.8).

While rates of unemployment for scientists and engineers are relatively low compared to those for all American workers, they are consistently higher for women than for men (e.g., 3.4 percent in 1984 vs. 1.3 percent) (NSF 1986: 5). For doctorates, rates of unemployment are even lower: 2.5 percent for women and 1.2 percent for men (NSF 1986: 7). While such percentages are comparatively small, they must loom large for those unable to find work. Women are also more apt than men (8 percent vs. 2 percent) to be working part time or outside science and engineering, a difference that results largely from the small numbers of women in fields like engineering, which have comparatively good full-time employment prospects for both sexes (NSF 1986: 6). Rates of underemployment for Ph.D.s are lower for men and women—1.2 and 2.5 percent, respectively (NSF 1986: 7).

All this means that women's work histories are briefer, on average, than those of men of the same age and more often marked by part-time work. These patterns hold in every field. They cannot be attributed solely to women's domestic and parental responsibilities. Labor force participation rates (that is, the proportion working or seeking work) of men and women scientists and engineers are now high and almost identical (96 percent for men and 94 percent for women). Similar patterns hold for doctorates (NSF 1986: 4; 131ff). Yet larger proportions of women than men doctorates interrupt their careers for a year or more (17 percent vs. 5 percent), and, as Centra (1974: 32) shows, these spells of unemployment last longer for women than men. On average, however, women remain out of work for a brief time (Astin 1969: 58). Yet career interruptions do not explain much of the difference in salary paid to men and women scientists (Lewis 1986). Contrary to widely held belief, women's work patterns are weakly related to their family obligations; women scientists with young children under six are more apt to be working or seeking work than those with older children (NSF 1986: 5). However, women's family obligations are generally assumed to affect their work histories, and these assumptions affect women's employment opportunities.

From this brief report, it should be evident that the number of women in science and engineering has risen rapidly in the last 15 years. However, women scientists and engineers are concentrated in different fields and in different sectors of employment than men, are younger, and have less professional experience and less education. These differences cannot be ignored when comparing the career attainments of men and women scientists and engineers.

Research on Career Attainments of Men and Women Scientists and Engineers

Studies of career attainments in science and engineering not surprisingly reflect researchers' distinctive disciplinary interests, styles, and methods of inquiry.[5] These studies are sharply focused:

- On scientists rather than engineers, with the notable exception of the research program of LeBold and his group (LeBold et al. 1983). This limitation is significant, since engineers actually outnumber scientists and comprise 56 percent of those working in science and engineering posts.
- On academics rather than industrial or government scientists, this also being a significant limitation. Just 12 percent of all scientists and engineers worked in educational institutions in 1983, although a larger share of doctorates (53 percent) did so (NSB 1985, tables 3.7 and 3.8).
- On holders of the doctorate rather than those with lower level degrees, these being just 11 percent of the population of scientists and engineers (ibid.).
- On the two most recent decades.
- On individual career histories, not the effects of institutionalized processes of evaluation and allocation of resources and rewards on the careers of men and women scientists or their subjective experiences and attitudes.

Thus the findings of research reviewed here are partial, limited to particular subgroups and a particular time (see Zuckerman and Cole 1975 for an earlier review). They are also complicated. To make gender comparisons clear, I take up phases of the career, one by one, rather than emphasizing their connections in a model. Such models have been proposed (Cole 1979; Reskin 1978; Helmreich and Spence 1982; Long 1987), but longitudinal data on careers of men and women are largely limited to the first 15 years, making models for the entire career more schematic and speculative than empirically grounded.

INITIAL QUALIFICATIONS

The qualifications of men and women beginning careers in science are now similar in three important respects. First, among doctorates, the only group for which data are available, the intellectual caliber of men and women insofar as it is measured by standardized tests and academic performance is much the same. Women do as well or better on such tests as men (Cole 1979: 62; CEEWISE 1979: 23–25). However, measured ability seems unrelated to research performance in science (Bayer and Folger 1966). Second, the proportions of men and the proportions of women getting degrees from top-ranking research university departments do not differ overall. This is so when departmental rankings are measured by receipt of research funds or by ratings of the quality of doctoral programs (CEEWISE 1979: 37; 1983: 2.7). Such differences, however, appear in certain fields. Significantly fewer women than men get degrees from top-ranked departments in mathematics and physics but significantly fewer men than women get degrees from such departments in microbiology and psychology. Given the long-term effects of scientists' doctoral origins on their later careers (Long 1978; Long, Allison and McGinnis 1979), the overall similarity is worth underscoring. Third, men and women are now much the same age when they get doctoral degrees. In the doctoral cohort of 1981, the median ages of men and women were 30.3 years and 31 years, respectively, with variation, of course, among fields (CEEWISE 1983: 2.3). In the mid-1960s, however, new women Ph.D.s were markedly older; the comparable figures were 30.9 years of age for men and 32.5 for women (Harmon 1978: 54). Women also used to take longer to get their degrees, and a larger share than men were forty or more on receipt of the degree (Harmon 1978: 54–55). By 1981, however, women began careers substantially later than men only in the medical sciences. In all other fields, their ages were approximately the same—that is, within a year of one another (CEEWISE 1983: 2.8).

Thus recent women doctorates are quite similar to men: in where they got degrees and when and in those attributes that are measured by standardized tests of ability. Men and women are also equally apt (30 percent and 28 percent, respectively, in 1981) to take post-doctoral appointments in the sciences and engineering (Coggeshall, 1981: 148), and about the same proportions were accepted for post-doctoral fellowships by top-rated institutions. Such fellowships help to transform recent graduates into independent scientists and provide the opportunity to establish a program of research before teaching begins. The gender similarity here suggests that men and women start their careers as equals in these respects as well.

However, finer-grained data indicate that the gender similarities in post-doctoral appointments are not as great as they seem at first. The reasons that men and women give for taking post-doctoral fellowships differ. Married women particularly emphasize the geographic location of the fellowship more than do single women and men, married or single (Coggeshall 1981: 151). Moreover, Reskin's (1976) studies of chemists suggest that women may receive post-doctoral fellowships as often as men, but those they hold have less prestige. This, Reskin conjectures, is indirect evidence that more women than men accept these fellowships because real jobs are unavailable. Further analyses of these same data show that holding a prestigious fellowship helped men more than women to get tenure-track positions after completing their fellowships (Reskin and Hargens 1979: 118). Other studies show that men who had held post-doctoral appointments earned more than women in every field seven years after the fellowship, after taking sector of employment into account (Coggeshall 1981: 156). Such differences in earnings doubtless reflect gender differences in academic and organizational rank and possibly gender differences in rates of publication, which begin to appear at this stage of the career.

JOB HISTORIES

What happens to men and women scientists and engineers when they search for their first jobs? In engineering, men and women fresh out of college encounter quite similar job opportunities, at least as gauged by salary and by proportions given supervisory and technical responsibility in the year following graduation (LeBold et al. 1983: 22–23). Circumstances are more complex for new college graduates seeking jobs in the sciences. Women now do about as well as men (again using relative salaries as an indicator) in the aggregate, in every field but the biological sciences (Vetter 1981: 1318). There is evidence, though, that grade levels and salaries differ greatly for men and women scientists starting work in industry and government (CEEWISE 1980: 4–5). Among new Ph.D.s, gender differences in job opportunities are less marked. The once common pattern of women doctorates having to choose between teaching in a women's college or serving as a research associate in a university laboratory no longer holds. Although women continue to work in academia more often than men, an increasing share of new women doctorates find jobs outside the academy; and similar proportions of men and women doctorates now plan to work in industry (CEEWISE 1983: 2.13).

Confining our attention to academics and to doctorates, current data show, contrary to expectation, that men and women find jobs in much the same kinds of educational institutions. About the same proportions of men and women work in top-rated research universities, whether reputational measures or rankings based on receipt of funds for research are used. In 1977, for example, about 13 percent of women doctorates in the sciences and engineering employed in academe were affiliated with the top 25 institutions in terms of funding as against 14 percent of men (calculated from CEEWISE 1979, table 4.3).[6] These data are consistent with those reported by Cole (1979: 70) and by Ahern and Scott (1981: 47) for matched samples of men and women doctorates and are significant given the important effects organizational context has on scientists' research performance—though these effects have been studied only for men (Long and McGinnis 1981).

However, being on the faculty at MIT or Berkeley is quite a different matter from being employed at these same institutions as a research associate or in another off-ladder position. Too few studies combine data on institutional affiliation with data on organizational rank. Those that do show that women are slightly overrepresented among assistant professors, given their numbers among Ph.D.s generally, with some variation by class of institution. In 1977, 19 percent of assistant professors in the sciences at the top 25 universities were women, 15 percent in the second tier, and 18 percent in other institutions (CEEWISE 1979, calculated from table 4.3), this at a time when women earned 12–13 percent of all doctorates in the sciences and engineering. However, a larger proportion of women than men also held instructorships, lectureships, and other off-ladder posts, the outcome, in part, of fewer women proportionately holding the doctorate (Bayer and Astin 1975: 797).

The same data look altogether different when examined from the perspective of the distributions of men and women Ph.D.s *among* academic ranks. These cross-sectional data show, of course, that women are heavily concentrated in the lower ranks as compared to men. For example, 37 percent of women scientists and engineers held assistant professorships in 1977 as against 22 percent of men, while the situation is reversed at the top of the academic ladder (CEEWISE 1979, calculated from table 4.3), where 39 percent of men hold full professorships vs. 15 percent of women (CEEWISE 1979: 61).[7] Such concentrations are often taken to mean that women are not promoted and are kept in lower ranks. In fact, women do not get promoted to high rank at the same rate as men, but such cross-sectional data cannot show this, since they do not take into account the differing age distributions of the pools from which men and women professors are drawn.

So far then, comparisons confined to new Ph.D.s and to newly hired assistant professors show that women become assistant professors at about the rate that would be expected given their representation among new degree holders.[8] For this limited group and for the current period, it would appear that gender parity has been achieved, and this is especially so at the top-rated institutions. At the same time, it is clear that earlier cohorts of men and women scientists encountered quite different job prospects when they began their careers. Gender parity in hiring new Ph.D.s in academia or industry has not, the evidence suggests, been around for long.

LATER JOBS

What has happened to earlier age cohorts—to men and women who began their careers a decade or more ago? Cross-sectional data on the ranks they have attained in educational institutions, industry, and government are less precise than one would like, but they are relentlessly consistent; such women, on average, started out in lower ranks than men, and the disparity in their ranks continues.[9] Available longitudinal data on men and women of the same professional ages do not contradict the cross-sectional evidence. For example, Ahern and Scott (1981, calculated from table 2.5) report that among matched men and women scientists who received Ph.D.s in the 1940s and 1950s, 86 percent of men had become full professors by 1979 as against 64 percent of women (a ratio of 1.3: 1). In the cohort that got Ph.D.s in the 1960s, a smaller proportion of both men (52 percent) and women (30 percent) had become full professors by 1979, but men proportionately outnumbered women by a ratio of 1.7:1 (ibid., calculated from tables 2.5 and 3.5, p. 18, 25). And finally, among those who had gotten degrees between 1970 and 1974, just 6 percent of men and 3 percent of women had become full professors. However, 41 percent of the men and 29 percent of the women in this young group, a ratio of 1.4:1, had been promoted to associate professorships (ibid., calculated from table 4.4, p. 33). In short, the evidence suggests that men from these older age groups are ranked consistently higher than their age peers among women.

Gender differences in advanced ranks turn up in all classes of academic institutions but are most accentuated in the major research institutions (just the opposite of what is observed among assistant professorships). On average, the higher the prestige of the institution, the lower the proportion of women in full professorships and the smaller the proportion of women assistant professors who are promoted to associate professorships (CEEWISE 1979: 61; CEEWISE 1983: 4.7). These differences between institutions need to be treated with caution, however, since women employed by the higher-ranked institutions may be younger on average than those employed in other institutions.

Among those who earned degrees in the 1940s and 1950s, women are disproportionately bunched in "off-ladder appointments," that is, in instructorships and nonfaculty posts, including research associateships and other miscellaneous research jobs (Ahern and Scott 1981: 18). Estimates vary on the proportion of all women in these posts, ranging from 13–20 percent, or about two to three times the proportion of men (CEEWISE 1983: 4.15). Such positions are insecure, especially in times when research money is scarce. In many universities, holders of off-ladder appointments may not apply for research funds as principal investigators; they are dependent on others and cannot set their own research programs. Perhaps the only unmitigated virtue of holding such appointments is that they require less frequent attendance at committee meetings than do regular faculty posts.

Finally, trend data on the distribution of women among academic ranks indicate that important changes have occurred. Since the 1970s there have been increases in the proportions of women scientists in every academic rank in all classes of educational institutions, with the rates of increase being largest in the major research institutions—this being partly the result of so few women being on the faculties of these institutions before the 1970s (CEEWISE 1983: 4.6–4.13). At the same time, it is worth emphasizing that the top research institutions grew very slowly, if at all, in the 1970s and early 1980s. Increasing numbers of women faculty did not result simply from increasing overall faculty size. These trend data, like those examined earlier, show growing similarities in the career attainments, particularly in academic rank, of successive cohorts of men and women scientists. The gap between them is narrowing, but it has not been eliminated, especially at the highest ranks.

TENURE

As with rank, so with tenure (the two being strongly intertwined): academic women scientists and engineers are less apt than men to be tenured and to hold tenure-track jobs (CEEWISE 1983: 4.15). Among those in tenure-track jobs in 1983, two thirds of men as compared with 40 percent of women were already tenured (NSF 1986: 4). Such marked cross-sectional differences shrink, of course, when professional age is taken into account, but they do not disappear. Among the matched men and women scientists studied by Ahern and Scott (1981), women in each age cohort were somewhat less apt than men to be tenured. Among those who took degrees in the 1940s and 1950s, 98 percent of men and 88 percent of women were tenured by 1979; more of course than among those who took degrees later in the 1960s. In this age group, 89 percent of men and 78 percent of women had received tenure (calculated from tables 2.6 and 3.7;

p. 19, 27). Gender differences in rates of tenure are significant in some sciences but not in others. Among younger men and women, the tenure picture is more complex. Among those who got degrees between 1970 and 1974 and were associate professors, women were just as apt as men to be tenured—in all fields, not just the sciences (ibid., p. 31). But women scientists, it turns out, were less likely than men to have become associate professors (29 percent vs. 41 percent) and thus less likely, in the aggregate, to be tenured (ibid., calculated from table 4.4, p. 33). The differences in the tenure status of men and women professors appear to be narrowing (CEEWISE, 1983: 4.14), suggesting movement toward gender parity in this important aspect of academic careers.

TIMING OF PROMOTION AND TENURE

For academics, it is not just holding high rank and tenure that matters but how long it takes to achieve them. Women are promoted more slowly than men, and those promoted are slower to receive tenure. According to Ahern and Scott's careful study, 24 percent of women compared to 14 percent of men who had earned degrees in the 1960s and were tenured by 1979 waited nine years or more to be promoted to tenure. (These data are for all fields.) (ibid., p. 27). Correlatively, younger women doctorates also wait longer for promotion than men. A smaller share of women (44 percent) than men (62 percent) who got Ph.D.s between 1970 and 1974 and were assistant professors in 1977 had been promoted to associate professorships three years later. However, these data do not tell the whole story. Among those young academics who had already been made associate professors, equal proportions of men and women were promoted to full professorships in this same period (ibid., p. 35). The dominant pattern in all cohorts, however, is that a larger share of men are promoted to high rank than women of the same age group, and they are also promoted more quickly (CEEWISE 1983: 4.13). Taken together, the greater incidence and more rapid promotion of men enlarge gender differences in attainments among age peers over time. Other data suggest that gender differences in "time to tenure" are more pronounced in top-rated universities than in others—that is, the more prestigious the institution, the longer women wait to be promoted (ibid., p. 40).

In the absence of comparative data on research performance of men and women, it is difficult to say what accounts for gender differences in rank and tenure.

In the case of rank and tenure, the general pattern of growing disparities with age holds up to a point and then *narrows* toward the end of the career. Large majorities of men and women who remain in academia long enough

do eventually get tenure and most do eventually get promoted. Yet disparities in rank and tenure persist among older men and women scientists.
Among those who have had their doctorates for 30 years or more, as we
have already seen, a considerably larger share of men than women hold
full professorships, although the vast majority of both are tenured (ibid.,
p. 18).

SALARY DIFFERENCES

In light of gender differences in rank, it comes as no surprise that
women scientists and engineers earn less than men. On average, in 1984,
women's median salaries were 71 percent as large as those of men (NSF
1986: 7).[10] The disparity varies among fields, is smaller in engineering
than in the sciences, and also somewhat smaller among Ph.D.s. In 1983,
for example, women doctorates earned 78 percent as much as men. But
since salary is closely tied to age, experience, and rank, and since women
scientists and engineers on average are younger, have fewer years of experience, and hold lower ranks, some differences in salary are to be expected. However, among doctorates in the sciences "about one-half of the
differential in female-male salaries remains unexplained after standardiz-
ing for field, race, sector of employment and years of professional experience" (NSF 1986: 8).[11]

In academia, this pattern holds for men and women associate and full
professors. Women earn less than men in the same ranks who got degrees
in the same years. The most recent data available (1981) show that salary
differences between men and women are larger for full than for associate
professors but absent for men and women assistant professors (CEEWISE
1983: 4.21).

Among full professors, however, women's salaries are not lower than
men's across the board; they are as large in some fields—computer sciences, earth sciences, and engineering—but far smaller in others. The
largest difference—a median of $6,200 per year—appears in the medical
sciences, and this is followed closely by a difference of $5,800 in eco-
nomics (14 percent less than men's) (CEEWISE 1983: 4.22). Among
associate professors, similar patterns obtain.

Although it is useful to consider rank in comparing the salaries of men
and women scientists, doing so makes for systematic underestimates of
gender differences in salary. Women are less apt than men to hold high
rank, and those who do are promoted later in their careers. Such differences
are erased when comparisons are confined to men and women in the same
ranks. Comparisons of the salaries of men and women in any given year
also systematically understate long-term gender differences since salary
differences accumulate year by year. Even modest median differences in

annual salary quickly add up to large differences when considered over the course of scientists' careers.[12]

As with rank, so then with salary, the three patterns noted earlier of persisting gender difference, growing gender parity—particularly among the young—and increasing difference over the course of men's and women's careers hold fairly well. It is still not clear whether the salary gap between men and women narrows toward the end of their careers, as women catch up to men in rank. However, some hint that this does not occur can be found in Ahern and Scott's data (1981: 77). They show that deficits in women's earnings, as compared to those of matched men, increased among older Ph.D.s and are greatest for those in the sample who have had the doctoral degree longest (16–21 years).

Convergences in rank, tenure, and (to some extent) salary among men and women over the past decade and a half coincide with and may be the outcome of efforts to comply with affirmative action legislation. Such convergences may be consequential for women's later careers. There is good reason to suppose that high rank provides greater resources and opportunities for research, including less teaching, more access to graduate students and post-doctoral fellows, and more space, that *sine qua non* of research in the modern university. Thus, recent improvements in rank and the opportunity structure for research for women, associated with affirmative action, may enhance women's future research performance while improving their current rewards.

ROLE PERFORMANCE

Gender differences in rank, tenure, and salary need to be examined in light of possible gender differences in the quality of role performance, that is, how well men and women do their jobs. Do men and women who do their jobs equally well receive the same rank, tenure, and salary? This question is central because of the twin claims that women are not rewarded at the same level as men because they do not do their jobs as well or because women have to be twice as good as men to get the same rewards. This question is central analytically, but it cannot be answered in quite the way it is put. There are no satisfactory measures of the performance of scientists and engineers in their multiple roles as teachers, administrators, researchers, managers, and citizens of the scientific community. Thus, while we know that this question needs answering, we also know we cannot answer it now.

RESEARCH PERFORMANCE

Some data are available on one significant aspect of scientists' and engineers' role performance: the extent to which they contribute to the advancement of knowledge through publication. The number of papers individuals publish and the number of times they are cited—that is, the number of times their papers are used in the research literature—are crude measures of research performance. However, many studies show that they are correlated with other, more direct measures of extent of contribution, such as peer judgments and honorific awards (Cole and Cole 1973; Gaston 1978; Garfield 1979, 1982; Zuckerman 1977). This body of materials suggests, in the aggregate, if not in any individual cases, that the extent of publication and citation do register differences in evaluation of contributions to scientific knowledge.

The use of publication and citation counts is by no means uncontroversial (Edge 1979). And it is important to note that some who find them acceptable measures for men scientists and engineers object to their use in the special case of assessing the relative research performance of men and women. These critics claim that few women hold high rank in the top research universities and thus few women have as much access as men to the resources needed to do research that leads to high rates of publication and citation (CEEWISE 1979: xiv). There is merit in this criticism insofar as it emphasizes gender-related inequalities in opportunities to do research and to publish.

At the same time, this criticism is not entirely valid. First, the numbers of women in high-ranking positions is not too small for comparative analysis, and, of course, the numbers of women in lower-level positions is substantial enough to allow for comparisons with men who are similarly situated. Second, research performance is almost universally taken into account in science in decisions on allocating resources and rewards. Like it or not, research performance is important to the careers of men and women scientists, and it is crucial to know how men and women compare in this respect. Third, it is not only feasible to compare men's and women's performance (using these admittedly crude indicators), but it is necessary to do so if the sources and consequences of gender differences in research performance are to be identified. And last, should we find that men and women whose research performance is much the same are differentially rewarded, this would provide justification for concluding that gender, rather than functionally relevant criteria, really does affect the allocation of resources and rewards—better justification than would be available if such assessments were not made because they were considered unacceptable and even taboo.

With that lengthy preamble, what is known first about the comparative access of men and women scientists to resources for research and second about their research performance? Do men and women have equal access to research support? The evidence here is exceedingly sparse. Limited data on funding at the National Institutes of Health during the early 1970s (National Institutes of Health 1981:23–25) shows that a small proportion of women apply for research funding, smaller than their numbers in the pool of life scientists. But among those that do apply, women's applications are as apt to be successful as men's.[13] There are small gender differences in the size of awards the NIH made, but these appear to result from women applying more often than men to programs with small budgets. Access to research funding is crucial to research performance and not nearly enough is known about patterns of application for funds by men and women, how they fare in the budget-cutting process, and how much money they are ultimately awarded.

When it comes to rates of publication, more than 50 studies of scientists in a variety of scientific disciplines, types of institutions, and different countries show that women publish fewer papers than men of the same ages, on average, 50–60 percent as many (see Cole and Zuckerman 1984 for a review of the studies since 1975). The weight of the data is persuasive. Moreover, we find that gender differences in publication are smaller earlier in the career than later, as Figure 1.4 indicates. Data on matched pairs of men and women scientists who received Ph.D.s in the same departments in the same years in the same fields show that gender differentials in publication start early in the career and grow as scientists get older, and this has been so for some time (Cole and Zuckerman 1984). Such differentials are of course reduced considerably when rank and type of institution are held constant, but they are not eliminated.

Detailed analysis of men's and women's publication patterns indicates that the aggregate gender differences are mainly the outcome of differences in the proportions of men and women who publish at a very high rate. As Figure 1.5 shows, a smaller share of women than men turn up among those who publish large numbers of papers.

However these data, when juxtaposed with those for earlier age cohorts, suggest that this pattern may be changing. Just 8 percent of women scientists who got their degrees in 1957–1958 published as many as 20 papers in the first dozen years of their careers, as against 26 percent of women who got their degrees in 1970–1971 (Cole and Zuckerman 1984: 229). Should these increases continue among still younger cohorts of women scientists, aggregate gender differences in publication will begin to narrow.

Figure 1.5 also shows considerable intra-gender variation in publication. A great many men publish few papers and a small fraction publish many. The same is true for women. The degree of inequality in rates of

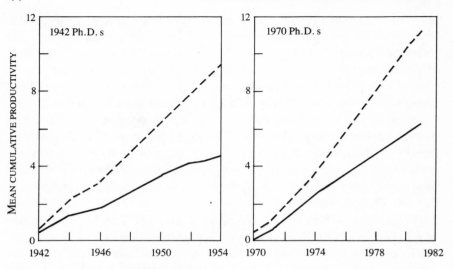

1.4   Mean cumulative productivity of men and women scientists who earned Ph.D.s in 1942 and 1970.

Source: J. R. Cole and H. Zuckerman, "The Productivity Puzzle: Persistence and Change in Patterns of Publication on Men and Women Scientists," in P. Maehr and M. W. Steinkamp, eds., *Advances in Motivation and Achievement,* pp. 217–256. Greenwich, Conn.: JAI Press, 1984.

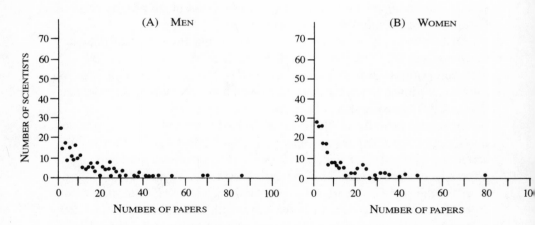

1.5   Distribution of total publications of men and women scientists who earned Ph.D.s from the same departments and in the same fields, 1968–1979.

Source: J. R. Cole and H. Zuckerman, "The Productivity Puzzle: Persistence and Change in Patterns of Publication on Men and Women Scientists," in P. Maehr and M. W. Steinkamp, eds., *Advances in Motivation and Achievement,* pp. 217–256. Greenwich, Conn.: JAI Press, 1984.

publication within each gender is much the same. Gender is therefore a poor predicter of published research performance. Knowing whether scientist-authors are men or women will tell little about their rate of publication.

The reasons for gender differences in publication are complex and not well understood. Two comparatively simple explanations—that women coauthor papers less often and therefore publish less than men, and that women have a harder time than men getting papers accepted for publication—do not square with the evidence in hand. On the first, women are as apt as men to publish coauthored papers (Cole and Zuckerman 1984). On the second, the evidence usually cited for this claim is a report showing that "blinded refereeing" (that is, removal of author's names) of papers increased the acceptance of women's papers for presentation at the meetings of the Modern Language Association (Lefkowitz 1979: 56). It is not clear whether data of this sort from the humanities are quite apropos. More recent and pertinent evidence on the disposition of manuscript submitted to one journal in sociology between 1977 and 1981 shows no independent effects of gender on publication decisions (Bakanic, McPhail and Simon 1987: 638).

Increasing disparities in rates of publication by men and women as they grow older may indicate that women's access to resources relative to men's diminishes with time and possibly that their commitment to research and publication wanes as they receive fewer rewards and incentives to continue. These are reasonable conjectures that require further inquiry before they can be properly assessed.[14]

IMPACT OR INFLUENCE OF RESEARCH

The extent to which scientists' research is cited in the literature is often used as an indicator of the impact or influence of that research. Just as analysts differ on the usefulness of publication counts as in assessing role performance, they also differ on the usefulness of citation counts (see Cole and Cole 1973; Edge 1979; Garfield 1979; Zuckerman 1987). Recent work by Ferber (1986) suggests that citation counts are biased against women. Her studies show a tendency for men to cite men and women to cite women, at least in economics and psychology. Given the small number of women in the pool of citing authors, women's papers receive lower rates of citation. This interesting finding needs to be followed up in other fields. However, the high frequency of mixed gender author sets in many sciences should reduce the effects of this bias, should it occur outside of economics. Other data suggest that women's lower rates of citation have quite a different explanation. Analysis of citations to the work of matched men and women in six sciences indicates that gender differences in citation are a function of their differences in rates of publication (Cole and Zuckerman 1984; Contra Helmreich and Spence 1982). Papers by women authors are cited, on average, just as often as papers by men authors; therefore, aggregate differences between men and women in numbers of publications account for their differences in citation. [15]

GENDER, RESEARCH PERFORMANCE, AND RANK

Are differences between men and women in research performance sufficient to explain women's lower ranks and slower rates of promotion? The answer is that they are not.

Evidence from carefully analyzed studies by Bayer and Astin (1975) for academics generally, by Reskin (1976; 1977) for chemists, and by Cole (1979) for scientists in five fields show, other things being equal, that men and women with equivalent records of research performance do not hold the same ranks. Men are apt to hold higher ranks than women, and this, Cole (1979: 57–58) reports, is especially so in the prestigious research universities. When it comes to rank and affiliation then, comparable men and women do not get equal rewards. And, as I noted earlier, high rank and appointments in universities which facilitate research, are in one sense rewards, but, in another sense, they are resources for further work. To the extent that this is so, these gender differences in rank and affiliation may help account for women's lower aggregate rates of publication later in the career.

HONOR AND REPUTE

So much for role performance. How do women fare relative to men in the allocation of recognition for contributions? Again, the evidence on hand is fragmentary. Cole (1979, chap. 4) has examined the "reputations" of men and women scientists; that is, how visible men and women scientists are, how their work is assessed, and how often they are named as major contributors to their fields. Women, he finds, are, on average, less visible than men; their work is perceived to be of lower quality and they are rarely mentioned as being major contributors. Cole also finds that while department, rank, awards, age, and doctoral origins contribute to scientists' reputations, research performance is far and away its strongest correlate. Once research performance (here the extent of publication and citation) is taken into account, being a women does not detract from or add to a scientist's reputation (1979: 119). However, he also finds that women scientists of the very first rank, Nobel laureates and others of Nobel class, are less well known than men having the same social statuses (ibid., p. 120). In the absence of further evidence, it is difficult to say why this is so. Moreover, Cole reports the process by which men and women achieve renown appears to differ (ibid., p. 142), again a finding that is not readily explained with the data in hand but plainly is important in understanding gender differences in career attainments.

Scientists are formally honored by an array of prizes and awards. These may seem trivial, but they are not—for recipients, the groups that give them, or the scientific community at large. They reassure scientists that the work they have done matters, they provide incentives for future work and, indirectly, help in the intense competition for resources (Zuckerman 1977).

There are no great differences in the sheer number of awards conferred on men and women scientists; most have few or none (Cole 1979: 59). But among those who have, it is difficult to estimate whether women are underrepresented and if so, by how much, because little is known about the demographic composition of the pools from which award winners of various sorts are drawn. Women now comprise 2–3 percent of the major academies and 2 percent of Nobel laureates, the top-most levels of the reward system, as Table 1.1 shows. We know that academicians and laureates are typically selected from among older, full professors or senior

TABLE 1.1 *Numbers and Proportions of Women Who Are Members of National Academies of Science and Winners of Nobel Prizes in the Sciences Compared with Proportions of Women Among Holders of Doctoral Degrees in the Sciences in England, France, Germany, and the United States*

	NUMBER OF WOMEN	ALL MEMBERS	PERCENT OF WOMEN	WOMEN AMONG DOCTORATES[a]
Académie des Sciences (1982)	3	130	2.3	19.0 (1970)
Deutsche Akademie der Naturforscher Leopoldina (1982)	21	1000[b]	2.1	4.8 (1973)
National Academy of Sciences (1986)	50	1477[c]	3.4	9.8 (1960–1969)
Royal Society of London (1982)	29	909	3.2	9.3 (1968)
Nobel laureates (1901–1989)	9[d]	407	2.2	[e]

[a] Data on proportions of women among holders of doctoral degrees (or in the case of England, holders of "higher diplomas") are given for the years indicated in parentheses. Those for France, Germany, and England are from OECD 1984, 26–31, and for the United States from Harmon 1978, 117–119.
[b] Estimated figure reported by the Akademie.
[c] Excludes foreign associates. Total membership may be slightly smaller owing to delayed reporting of deaths of members.
[d] Nobel laureates in physics, chemistry, physiology. Madame Curie, laureate in chemistry for 1911 and in physics for 1903, is counted twice. Eight different women have been named Nobel laureates.
[e] There are no reliable estimates of the proportion of women in the pool from which Nobel laureates are drawn.

researchers in major research universities and institutes and that the number of women in these posts is small. Thus the extent of underrepresentation of women may be less marked than the last column in Table 1.1 implies since the populations of those holding doctoral degrees in the sciences, country by country, include a larger share of young people and particularly of women than the population of potential laureates and academicians. It is not possible to say, then, whether women are now underrepresented in the major national academies, and if so, by how much. In the United States, we do know that the proportion of women elected to the National Academy recently parallels their representation on senior faculties (CEEWISE 1979: 102), suggesting that women are now being elected in proportion to their numbers in the pool. However, this conclusion holds only if it is assumed that women are on senior faculties in the appropriate proportions. Some would question whether this is so.

In recent years, the number of women elected to the National Academy of Sciences has risen sharply, by 5 times since 1972, while the membership as a whole has grown 1.6 times. However, research done some years ago on the ages of members at the time they were elected to the Academy (1865–1969) showed that women were elected nine years later than men in the same fields, yet there was no reason to believe that the research that led to their election was done any later (Zuckerman and Cole 1975: 98). Comparable data for the period since shows that the average age of women at the time of election has been dropping, though it has not quite reached the averages for men.

On the basis of this limited evidence, it appears that two of the general patterns noted at the outset hold here: cross-sectional differences persist between men and women in reputation with the evidence on awards being scanty but suggesting some indication of growing gender parity. The absence of longitudinal data on awards makes it impossible to say whether gender disparities grow as scientists age, but the overall tendency, observed among men, of conferring awards on those who already have them (Zuckerman 1977) would seem to suggest this might be so.

Taken together, what does the evidence tell us about career differences of men and women scientists and engineers? What is known suggests that women occupy lower ranks, advance more slowly, are paid less, and have less managerial responsibility than men (CEEWISE 1980); but apart from differences in rates of publication, little is known about their relative role performance. Among academics, a consistent picture emerges: women scientists' career attainments, for the most part, do not match those of men. In some measure, these differences are attributable to their being younger, having different work histories, and being in different fields from men. Older cohorts of women scientists' differ from men in rank, salary, research performance, and reputation, with these differences being almost always in the direction of comparative disadvantage for women. However, there are also signs of growing parity in career attainments, particularly among younger scientists. It is not yet clear whether this trend will ultimately reduce disparities in men's and women's attainments as they grow older.

Some Explanations Proposed for Gender Differences in Career Attainments

There is little disagreement that women scientists' career attainments, on average, do not equal those of men. There is, however, much disagreement about the explanations for gender inequalities. In general, these fall into four classes and emphasize:

- Gender differences in scientific ability
- Gender differences arising from *social selection*, based on
  a) gender discrimination
  b) gender differences in role performance and the allocation of resources and rewards
- Gender differences arising from *self-selection*, including
  a) marriage and motherhood and their consequences
  b) gender differences in career commitment
- Outcomes of accumulation of advantage and disadvantage[16]

How well does the available evidence square with each of these explanations? Not well. The evidence on all is ambiguous, not because the theories are unclear but because the data are complex, often vexingly incoherent, and frequently partial.

GENDER DIFFERENCES IN SCIENTIFIC ABILITY

There is no support, as I noted earlier, for the view that the different career attainments of men and women scientists result from gender differences in ability or competence. To the extent that these can be measured by intelligence tests or academic performance, abilities of women scientists equal or surpass those of men. However, there is evidence that girls do less well than boys in mathematics (see Kahle and Matyas 1987) and also that girls turn up in disproportionately small numbers among youngsters with high scores on tests of mathematical ability. Benbow and Stanley (1980; 1983) conclude that superior male mathematical ability, "an expression of both endogenous and exogenous variables," accounts for this finding. However, there is also enough evidence for marked gender differences in socialization and exposure to mathematics to raise serious questions about this and to warrant further examination of these effects on variability in achievement scores of boys and girls. For our purposes, these conjectures are not entirely pertinent. Men and women who do science are a highly selected sample of all adult men and women. Data on youngsters, even highly selected ones, are not helpful in understanding field and specialty choice much less differences in career attainments. Yet women do, more often than men, select fields and specialties of science which are comparatively less demanding mathematically, but there is no systematic evidence indicating why this is so.

PROCESSES OF SOCIAL SELECTION

Explanations of gender differences in career attainments that emphasize gender discrimination, on the one hand, and women's poorer research performance, on the other, both rest on notions of social selection. Social selection processes involve decision making about individuals (here about their careers) over which they have no control. They contrast with processes of self-selection, in which decisions are controlled by individuals and are not, except perhaps indirectly, attributable to socially structured arrangements for selection.

Gender discrimination occurs when the unequal treatment of men and women is based on the functionally irrelevant criterion of gender. Discrimination can affect men's and women's career attainments when their opportunities for role performance are unequal, when their role performance is judged according to different standards, and when they are differently rewarded for the same quality of role performance. Gender discrimination, as with social discrimination generally, treats some social status "as relevant when intrinsically it is functionally irrelevant" (Merton 1972: 20). Proponents of the view that gender discrimination best explains the unequal career attainments of men and women point to instances of women having poorer facilities and resources for research, to their being judged by harsher standards, and to their being promoted and paid less than comparable men.

It is no easy matter, however, to assess the extent of gender discrimination and how it affects scientists' careers. Discrimination is often subtle and therefore difficult to identify, much less measure. It can be entangled with other forms of prejudice (based on age, for example), and because it appears throughout the career, a full accounting of its effects is hard to make. As a consequence, researchers have come to rely on indirect rather than direct measures of discrimination. They have assumed that differences in the career attainments of men and women which remain after taking *all* functionally relevant criteria into account *must* be the outcome of gender discrimination. Measuring discrimination by the use of residual differences has its problems, not least that it requires that appropriate evidence be available on all functionally relevant criteria which could account for gender differences in career attainment. (On "residualism" in its various guises in the law and in social science research, see Cole 1979: 36ff.)

When this mode of analysis is used, the evidence shows that gender discrimination affects promotions, tenure, and salary allocation among academic men and women with similar records of research performance. It also shows that discrimination is receding, especially for younger women. However, available data are limited only to position and to salary and do not register gender discrimination in informal social interaction or its subjective effects on women (see, for example, Briscoe 1984; Keller 1977).

There is also evidence suggesting gender discrimination may operate in different ways for different groups of women. Women who make important contributions to science may fare less well relative to comparable men than do the journeywomen of science relative to their performance peers. Cole's (1979: 120) studies show that women who have done important scientific work are less apt to be considered major contributors to their fields than are comparable men. Conversely, it has also been suggested that it is the journeywomen of science who fare poorly compared to men in the absence of clear-cut evidence of their role performance (Zuckerman and Cole 1974). Thus, the incidence and dynamics of gender discrimination in science have neither been satisfactorily described nor fully explained.

Discrimination can affect the career attainments of men and women, and it can also produce gender differences in the processes by which such attainments are reached. Reskin (1978) and Reskin and Hargens (1979) suggest that the connections between role performance and rewards for women are less consistent than those which apply in men's careers; women are more often rewarded for poor performance than men and less often for good performance. And, as noted, Cole (1979) observes that the processes of reputation building differ for men and women. If so, then women's incentives for high-level role performance are less clear than men's, and they may, as a result, perform less well.

Social selection, as I have indicated, also includes differential evaluation of men and women's role performance on functionally relevant criteria and differential treatment on this basis. Some believe that women are judged fairly and that gender differences in career attainments result from women performing less well than men. As I have repeatedly noted, we do not know how men and women do in their various roles as teachers, administrators, managers, and citizens of the scientific community. We do know that, on average, women publish less and are cited less than men. To the extent that these actually gauge research performance, then women's poorer performance is related to their lower career attainments, especially in those institutions which put a premium on publication. However, as we have seen, women's research performance does not explain all such differences; indeed, some remain after research performance and other relevant variables are taken into account. Moreover, the conclusion that women perform less well than men does not take into account unequal opportunities to do research. The sources of differential role performance may well reside in structured inequalities of opportunity. How much, we do not know.

GENDER DIFFERENCES AS OUTCOMES OF SELF-SELECTION

Career attainments are of course shaped by decisions individuals make for themselves, by self-selection as well as by social selection. Women's decisions to marry and to have children, to take on their distinctive domestic and parental roles, are said to interfere with their scientific work and to lead to career decisions that benefit their families but damage their careers. The evidence here is mixed:

- Women scientists and engineers are more often employed part time, less often looking for work, and out of work longer than men. Women attribute these work patterns to their family obligations more often than men. However, actual family obligations (having young children) are a poor predictor of unemployment among women scientists and engineers.[17]
- Married and single women academics are less mobile geographically than men (Marwell, Rosenfeld and Spilerman 1979), and married women say that decisions to move are affected by their family obligations (Coggeshall 1981). Since promotion and pay increases are often tied to a change in employment, women's limited geographic mobility may, in part, account for gender differences in career attainments.
- Marriage and motherhood are widely believed to account for women scientists' lower rates of publication (Lester 1974: 42). However, on balance, the evidence suggests that this is not so. Married women Ph.D.'s in the sciences publish as much as single women, and having successive children is not associated with reduced rates of publication (see Astin and Bayer 1979; Ferber and Huber 1979; Helmreich, Spence

et al. 1980; Wanner, Lewis and Gregorio 1981; Simon, Clark and Tifft 1966; Centra 1974; Cole 1979; Cole and Zuckerman 1987; Luukkonen-Gronow & Stolte-Heiskanen 1983; Toren 1989; Kyvik 1990; Joas 1990 but also see Hargens, McCann and Reskin 1978 for contradictory evidence).

- Moreover, marriage and parenthood are not uniformly associated with lower rank and salary among women. In some fields and classes of institutions, the correlations are positive and in others negative. Overall, however, they are not large (Ahern and Scott 1981, chap. 6). This is so in spite of the widespread belief that married women have poorer career opportunities than single women.

In short, women's domestic obligations are not *the* simple explanation of gender differences in career attainments since, in many respects, married women and women with children fare as well or better than single and childless women.

GENDER DIFFERENCES IN CAREER COMMITMENT

There are little or no systematic data on the career commitment of men and women scientists and engineers, especially those holding doctoral degrees. That is, there are no data bearing on whether women care less, more, or the same as men about rank, salary, responsibility, and recognition. There is indirect evidence that women are more apt than men to prefer teaching over research (or at least they did two decades ago [Bayer 1973]) and for living in urban as against less populous areas (Marwell, Rosenfeld and Spilerman 1979). But the connections between such career-related preferences and career commitment are not established nor are the ways these preferences are shaped by opportunities, perceived and real. This hypothesis lacks any support, pro or con.

ACCUMULATION OF ADVANTAGE AND DISADVANTAGE

The notion of the accumulation of advantage and disadvantage has been repeatedly used in studies of stratification in science (see Zuckerman 1989). It is plainly pertinent to the disparity observed in the attainments of men and women scientists. Accumulation of advantage occurs in science when certain groups receive greater opportunities to enlarge their contributions to knowledge and then are rewarded in accord with those contributions. Recipients are thereby enriched at an accelerating rate, and, conversely, nonrecipients become relatively impoverished (see Merton [1942] 1973; Zuckerman 1977: 59–60, 1989). The accumulation of advantage helps to account for the observed cross-sectional differences between men and women scientists in research performance, rewards, and recognition. It can also account for observed intra-gender variation (not all women are equally disadvantaged, nor are all men equally advantaged), and (not least) for the growing divergence in performance and attainments of men and women scientists as they grow older. It is also consistent with, or more precisely does not exclude, the third pattern we observed of growing parity in attainments of men and women, especially among the young.

To the extent that processes of accumulation of advantage and disadvantage are supplemented by self-selection, by women making decisions which they believe benefit their families but have the effect of damaging their careers, disparities between their attainments and those of men will be amplified and accentuated. The ideas of accumulation of advantage and disadvantage have been further elaborated and examined empirically not just in science but in a variety of other occupations (Zuckerman 1989: 169). But before we can conclude that accumulation of advantage and disadvantage and related processes of self-selection really do explain why the career attainments of men and women differ, more detailed data will be needed on how advantage and disadvantage actually accumulate.

A Limited Research Agenda: Domains of Specified Ignorance[18]

Here is what we need to know, and why:

1) On the research performance of men and women scientists: We need to know why women publish less than men and the extent to which this results from discrimination, differential access to the means of scientific production, and women's preferences or choices. More specifically, we need to know the relative access men and women have to such important research resources as funds, space, time, appropriate co-workers, and instrumentation. How do women's organizational ranks and institutional affiliations, in combination, affect their resources for research and their research performance over the course of their careers? We also need to learn whether there are significant differences in the research strategies and practices of men and women scientists. Are there greater differences between men and women in these respects than among them? If so, do they affect how much research is done and its significance?

2) On disparities in career attainments of men and women: Why do these disparities grow as men and women get older? Does the accumulation of advantage explain this pattern fully, or is there evidence also for other explanations such as women's growing discouragement and reduced aspirations? (See Zuckerman and Cole [1975] on the "Triple Penalty" against women, which links discrimination to reduced aspirations.) Longitudinal

studies on multiple age cohorts are needed to answer these questions.[19] Why do women fare better in certain sciences and less well in others? Is it the case, as Rossiter (1978) suggests, that women do better in new and growing fields? Are the "cultures" of some fields less consistent with feminine values than others, as Keller (1985) and Traweek (1984) imply?

3) On gender discrimination: its incidence, forms, and consequences: To what extent is discrimination, conditional, that is, targeted, or is it practiced against all women, regardless of their status characteristics? How does discrimination in its less blatant forms affect women's informal relations with their colleagues and their networks of associations? Rose (1985) observes that the networks of young men and women scientists differ not only in composition but also in how useful they are believed to be. More research attention needs to be paid to the consequences of informal associations for scientists' careers. And finally, how does the experience of gender discrimination affect women's motivation and career commitment? Do men have equivalent experiences not associated with gender that have similar effects?

4) On changing labor markets: In what measure are the career attainments of men and women scientists and engineers determined by changing labor market conditions? Is the move toward gender parity likely to continue, or will it wane if jobs become scarce?

5) On the career consequences of marriage and parenthood: To what extent is women's limited geographic mobility (perceived and real) related to their poorer attainments? How do the complex arrangements dual career couples make affect their attainments? This is of no small moment given the fact that a majority of women scientists who are married are married to men scientists.

6) On commitment to careers: So little is known about the career commitment of men and women scientists (and other professionals) that this is a thoroughly uncharted area. It would indeed be useful to know whether men and women differ or are the same with respect to career aspirations, concern with promotion, income, and fame. Believing that they do, some attribute gender differences in career attainments to these attitudinal differences. Others contend that such differences are negligible and that structural barriers faced by women account for differences in career attainments. In either event, this domain of specified ignorance about attitudes requires further examination and needs to be linked to evidence on the behavior of men and women scientists.

So much for the future research agenda. Based on what is now known, how much have women scientists' careers changed since Florence Sabin's time? They have changed considerably, but change has been slow and uneven. Sabin's career, then a succession of "firsts," is now less atypical. It would have been inconceivable in 1925 that women scientists would be hired as faculty members by universities at about the rate they were getting Ph.D.s, and it would have been thought highly unlikely that women would be promoted into senior posts in all classes of universities at the rate they have. Still, their career attainments continue, on average, to be more modest than those of men in all sectors—in academia, industry, and government—and the gap in attainments grows as men and women age. Moreover, while some distinguished women scientists and engineers have become insiders and members of the scientific establishment, those who have often feel themselves to be outsiders and on the margin. It is not clear at this juncture whether parity will be achieved in the careers of men and women scientists and engineers and, if so, when.

HELEN S. ASTIN

2. Citation Classics: Women's and Men's Perceptions of Their Contributions to Science

RESEARCH AND PUBLICATION are the essentials in the production of knowledge. The extent to which research productivity affects status attainment continues to attract the attention of scholars of the sociology of knowledge. Whether research productivity affects directly status attainment of scientists has been examined by a number of scholars in the past (Crane 1965; Hagstrom 1971; Cole and Cole 1973; Gaston 1973; Reskin 1977; Long 1978; Astin and Bayer 1979; Allison 1980). However, the direction of the relationship between productivity and status attainment continues to remain somewhat unclear. For example, Crane (1965) and Long (1978) maintain that institutional location is more important to productivity than is early productivity during the graduate school years. On the other hand, Cole and Cole (1973) and Astin and Bayer (1979) argue that early productivity determines institutional placement, which in turn affects long-term productivity.

Another area of interest is the relation of gender to both research productivity and status attainment (Reskin 1977; Cole 1979; Astin and Bayer 1979; Cole and Zuckerman 1984).

The present study will explore how environmental and contextual variables and experiences contribute to gender differences in research productivity. More specifically, this study will explore the ways in which often-cited women and men scientists plan, execute, and promote their significant published work.

Background of the Research

Women in the aggregate publish less than men do; and this fact has generated numerous questions and hypotheses. Do women produce less because they have fewer research resources at their disposal or because they allocate more time to other job tasks and requirements? Are they isolated from important collegial networks? Does their professed lower level of interest in research and publication erode their productivity? Or is it the tug of family obligations? Or gender discrimination, and other institutional barriers? Thus far, research has been able to provide only partial answers to some of these questions:

- In general, women are reported to have fewer institutional resources such as research funds and graduate students who serve as research assistants.

    In a study that examined facilitators and inhibitors to research productivity among highly productive scholars, men were much more likely to identify resources and graduate students as important facilitators to their productivity than were the women (Astin and Davis 1985). Moreover, data from various national surveys have demonstrated differences among women and men with respect to financial support for their research activities and availability and use of research assistants. For example, Bayer (1973) in his normative report indicates that while 30 percent of men report financial support for their research activities, only 14 percent of women do so. More recent data (1984) from the Carnegie Survey of Faculty indicate that while 26 percent of men report having research assistants, only 11 percent of the faculty women say that they have such assistance (unpublished tabulations).

- Marriage does not appear to be a barrier to research and publication. As a matter of fact, in previous studies we have documented the positive effects of marriage on research productivity (Astin and Bayer 1979; Astin and Davis 1985). Cole and Zuckerman in a recent study reported in Chapter 6 of this volume have concluded that marriage and family obligations do not account for differences observed with respect to women's lower publication rates.

- Women are reported to be somewhat more isolated from important collegial networks. Reskin, in her thoughtful analysis of "Sex Differentiation and Social Organization of Science," suggests that "since the [collegial] role applies to relationships between scientific researchers of approximately equal status [and] because women's lower gender status is inconsistent with the implicit status equivalence of colleagues, sex stratification itself can block normal collegiality between male and female researchers" (1978: 9). Furthermore, Helmreich and Spence and their colleagues in interpreting sex differences with respect to research productivity speculate that "women may be more isolated from the national 'old boy' network and thus out of touch with the *'invisible college'* through which much exchange of scientific information takes place" (1980: 907). However, these statements are speculations that call for further research on the issue.

While we are still exploring the hypothesized external or structural barriers to women's research productivity, we also realize that women's published research may not receive the same degree of recognition as similar research produced by men. A number of investigators have documented the fact that women's and men's work are not equally recognized by such usual academic rewards as salary, promotion, and professional recognition (Astin and Bayer 1979; Ferber, Loeb and Lowry 1978; Cole 1979). Furthermore, in an article entitled "The Productivity Puzzle," Cole and Zuckerman hypothesized that *differential reinforcement,* that is, more limited recognition and use of their research may account in part for women's lower productivity. Their research leads them to conclude that "women scientists in the 1970 cohort are *slightly* more responsive than men to the lack of reinforcement and *considerably less* responsive than men to positive reinforcement" (1984: 243). In order to explore further how reinforcement may affect research productivity, we conducted the pilot study summarized in this chapter.

There is no question that perceived reputation—one's standing in the discipline or profession—influences a scientist's career progress, and that citations provide collegial recognition for one's scholarly contribution to the field. By looking at how highly cited published research by women and men scientists is conceived and by examining the reasons given by these scientists for its recognition, I hope to contribute to a greater understanding of gender differences in productivity. This analysis should also shed some light on how differential reinforcement affects women and men scientists.

While citations represent an indicator of impact, they also provide reinforcement for a scientist's work. For example, in summer 1982, a subsample of 543 highly productive academic scholars[1] were mailed a semi-structured survey questionnaire with questions about barriers and facilitators to research productivity (Astin and Davis 1985). The survey also assessed what these scholars considered their most important piece of published work and why they considered it important. The two reasons given most often by both women and men were: that the piece broke new ground and explored a new area; that the piece was widely cited. In other words, they considered their own work important because colleagues often chose to cite it in their own publications (Astin 1983).

Since the number of citations is considered a dimension of one's status in the field by both those who cite the piece as well as the scholar's own perceptions of its importance, I designed the present study to further explore citations as an important element in research productivity. The underlying assumption is that citations act as rewards that in turn provide incentives to further productivity in research. The study was designed to determine what types of scholarly work tend to be highly cited and in how such work was originally conceived. I also wanted to know whether there are gender and field differences with respect to how scholars conceive their best work and how they carry it out. The study examines reasons why a piece might be highly cited. I chose to look at these attributions because they represent a person's causal explanations for a success-oriented event such as published research that is highly cited. It was hypothesized that attributions have a direct effect on expectancies about future performance and subsequent achievement behavior. Thus, any identified gender differences in attributions of success (i.e., reasons for high citation counts of their work) might give us some clue as to how women and men react to this form of reinforcement.

Citation "classics" and essays prepared by the authors of the classics were the primary sources of data for this study. A citation classic is a weekly feature in *Current Contents*.[2] Publications labeled as *citation classics* are identified through the *Science Citation Index* (SCI) and *The Social Sciences Citation Index* (SSCI) data bases.

A citation classic is described by the publishers of *Current Contents* as a publication that is "highly cited": "a large number of citations to a particular publication usually indicates that the cited work has made a significant contribution to the development of scientific knowledge in its field." The publishers of citation classics also believe that such publications have "a lasting effect on the whole of science."

How does a publication become a citation classic? The publishers of *Current Contents* scan the citations in the various publications and select pieces that are highly cited. More specifically, certain areas of research are singled out by the publisher each year, and a search is made to identify the papers and books with the highest number of citations within these fields. The number of citations necessary to make a piece a citation classic depends on the size of the field, or the number of papers published in it. The authors of these highly cited publications are invited to prepare an essay about their publication in which they discuss (a) *what prompted the research;* (b) *any obstacles they encountered in research and publication;* and (c) *why they think the publication has been so highly cited.*

The present study is designed to examine gender and field differences with respect to the above three questions.

SAMPLE

A sample of 56 essays was used in this study. It was drawn from 589 essays which appeared in *Current Contents* published between March 1984 and July 1986. The areas covered among the sample of 56 essays included life sciences, physical, chemical, and earth sciences, agriculture, biology, and environmental sciences, and social and behavioral sciences.

The 56 essays selected for the present study included all 28 essays authored by women during the two-year span and a random sample of 28 essays authored by men and matched by field with the women's essays.

Twenty-two of the essay contributors were from foreign universities, including Canada.[3]

PROCEDURES

Descriptive information about the authors of each of the essays, as well as any other authors of the 56 citation classic publications,[4] included academic field and sex. For the classic we also determined if it had single or multiple authorship, the number of authors, the type of publication (book, article, chapter, etc.), the year of publication, and the number of times cited. These were all treated as categorical variables in the analysis.

The response categories for coding the essays were developed by first reading the responses offered by the authors to the three questions: How was the research conceived? What obstacles, if any, did you encounter? Why do you think your work is so highly cited?

Two analyses were undertaken. The first compared responses of men and women to the three questions asked of the authors. The second analysis compared responses by field of study.

Results and Discussion
SAMPLE PROFILE

Of the 56 scientists who wrote an essay about their citation classic over the two and one-half-year span (March 1984–July 1986), 56 percent were from the natural sciences (life sciences, physical, chemical, and earth sciences, agriculture, biology, and environmental sciences) and 44 percent from the social and behavioral sciences. The median year of publication of the citation classic was 1967 for men, with a range of 1946–1976; for women the median year was 1970 with a range of 1943–1979. Women's median citation count was 202 with a range of 40–2,125; men's median citation was 215 with a range of 65–1,850.[5] The citation count for each participant represents the number of publications in which it has been cited since the piece's publication date. Looking at the statistics about year of publication suggests that women in the sample tended to produce their highly cited articles during the more recent years. This is probably because the women's movement and affirmative action have increased women's overall productivity and thus their greater visibility. It may also be because more and more women are entering scientific fields. In an earlier study comparing the research performance of academics in 1972 and 1980, we found an increase in women's research activity (Astin and Snyder 1982). Also, in a later study (Davis and Astin 1987) we failed to find any significant sex differences on four different citation indices among a recent (1982) cohort of academic scholars.[6] These recent findings are consistent with earlier findings by Cole (1979) and Cole and Zuckerman (1984), that when women and men are equal with respect to the quantity of research production, differences in citations cease to be significant.

The issue of multiple versus single authorship was also examined. Previous research on the matter of collaboration and its effect on productivity (Cole and Zuckerman 1984) found no evidence that women are less likely to collaborate than men. Thus, their lower productivity could not be explained on that basis. The present study confirms their findings, that women indeed collaborate as often as men do: 44 percent of the citation classic publications were single-authored by either sex; the remaining 56 percent that were coauthored tended to vary somewhat in the number of coauthors depending on whether the author of the essay was a man or a woman. Twelve percent of the women had five or more coauthors compared to none of the men. Also, more men tended to coauthor only with other men (77 percent); while only 10 percent of the women had coauthored with women only. This finding is not surprising since women, being a minority, have many more male than female colleagues, therefore more opportunities for coauthorship with men.

In research of gender and research productivity, journal articles have been studied more than other types of publications. However, it is important to examine other forms of publication as well. Thus, Davis and Astin (1987) examined the extent to which various forms of publication play a role in the scholar's professional standing, and chapters in books were found to be strong and consistent predictors of reputational standing. Likewise, in the present study we also examined the type of publication that had become a citation classic. Among citation classics we observe the following distribution of type of publication:

	WOMEN	MEN
Book	12%	0
Review article	16	22
Article	72	78

Indeed, we do observe some sex differences with respect to type of publication that represents a citation classic. However, before we conclude that women's books are more highly cited than men's books, it would be necessary to ascertain whether the women in this sample are more likely to produce books, and whether the men are more likely to produce review articles (such data were not available to us in this study).

It is important to note that both chapters and review articles are important vehicles for creating visibility. (Later in this chapter we address the issue of review articles as it relates to the explanations given by these scholars for why their publications were so highly cited.) It will also be important in the future to explore further the gender differences with respect to books as a vehicle to collegial recognition. Questions to be addressed should include the following: Do women's books have a greater influence in the field? Are the topics women choose to write books about of greater general interest to other colleagues?

ESSAY ANALYSIS

The first question we addressed in the content analysis of the essays was *How was the research conceived?* Five types of responses were coded.

1) *Personal interest or experience.* For example, "I was visiting my grandmother in a nursing home. I was struck by how little control she and other residents were permitted." Personal interest or experience was mentioned as an impetus by 12 women and 12 men.

2) *To solve a problem* was also a popular response to the question (14 men and 10 women listed this as an impetus to the research). "The work was done to answer a difficult clinical question [that] I didn't know . . . embarrassed by my ignorance [I] decided to try to search for an answer." "Wigglesworth's findings prompted me to look for a function of the NC (neuro-secretory cells) of the brain of the adult female blow-fly."

3) *Dissertation research* was cited by five scholars (all women). For example, "This review paper was an outgrowth of my doctoral dissertation research," or "Our dissertation investigations . . . led to an interest in cytotoxonomy and cytogeobotany."

4) *Assisting graduate students* was cited by just one man and one woman. The male participant reflected on the question and responded as follows: "The paper began as a handout to students in a graduate course in causal models." The woman scholar tells us that "when I completed work on the Family Problems Scale . . . I did not intend to construct another test. However, a graduate student . . . decided to validate the Family Problems Scale's interpretation in terms of Ego Development, using a sentence completion test. . . . To complete her project, we needed a tentative scoring manual."

5) *Invitations* to prepare reviews on a topic were also perceived as the impetus to the preparation and publication of the citation classic. "When invited by the 'Annual Review' to write on this subject in 1978, I leaped at the chance." (One man and three women indicated invitations as a reason in undertaking the work.)

When we look at these responses, we find that women tend to cite dissertation research and invitations somewhat more often, while more men indicate the need to solve a problem as the impetus to undertake the research that led to the citation classic. While only one of these sex differences—dissertation—is statistically significant ($p < .05$ using a one-tail binomial test), it is interesting to speculate on the larger pattern of sex differences. Are women's responses more "passive" and men's more "active"? Do men see themselves as being more actively engaged in problem solving? Do women see their work more as an outcome of circumstances (the need to complete a dissertation or a response to an external invitation)? Such an interpretation would be consistent with some of the research on attribution theory, which indicates a tendency for men to make internal attributions (internal locus of control) and women more likely to exhibit an external locus of control (Simon and Feather 1973).

The essay contributors were also asked to report any obstacles or barriers they had encountered. Only women mentioned a lack of collegial support for their project or topic. Such an obstacle is represented in the comment: "Most of the work for this paper was done at home on nights and weekends in order to minimize certain criticism for failing to be fully engaged in the 'right' kind of research, meaning physiology." Other comments include: "I discussed my idea with a number of senior colleagues. One or two found it intriguing, but in general it was discounted." "The unpopularity of the subject was such that after I completed my thesis . . . an eminent Oxford professor strongly advised me to change my field of study." It is significant that an earlier study of facilitators and barriers to research productivity also found that women were much more likely than were men to report colleague behavior as a key factor in their scholarly work (Astin and Davis 1985). Again, we have evidence here that women are more likely to make more external attributions, that is, focusing on the endorsement or lack of it by colleagues.

Four men and three women mentioned that their publication was initially turned down by publishers. For example, one author wrote that "we experienced considerable difficulty in getting it accepted for publication". Another one said, "Getting it published was not easy." Out of the seven participants who mentioned difficulties with publishing their work, five were from the social and behavioral sciences. This is not surprising given the high rates of rejection in the social sciences (Zuckerman and Merton 1971). Other obstacles included problems with sample maintenance over time and problems with equipment and facilities. However, the majority of both men and women did not respond to the inquiry about obstacles, and 7 of the 56 participants mentioned only facilitators.

The last question—perhaps the most interesting one—deals with the participants' reasonings about the importance of their work. Participants were asked to indicate why they believed their piece was so frequently cited. Based on their responses, we developed nine categories that could absorb the various explanations that reflected their attributions for their work's high citation count. Table 2.1 shows the frequency with which

TABLE 2.1 *Attributions for High Citation Counts (in percentages)*

REASONS	WOMEN	MEN
First of its kind	48	15
Scope and applicability	32	15
Integrates knowledge and provides direction for further research	28	7
Raises fundamental questions	16	7
Easy to read; simplicity	12	19
Timing was good for the topic	12	22
Provides theoretical framework	8	15
Journal it appeared in is highly respected and widely read	4	7
Other researchers disagree with the findings and conclusions	4	7

each category was mentioned by the participants. The majority of participants believe that the primary reason for the high citation count is that the publication is the first of its kind. "I believe that my paper has been cited so frequently because it is the first demonstration of neuroendocrine function in an adult insect." "This was probably the first scoring manual for any projective test to attempt both logical and empirical justification of ratings." "It is likely that this paper has been cited frequently because it was the first to demonstrate definitively the localization of these important secreted platelet proteins." This attribution "first of its kind" is not unlike the response given by a totally different sample of respondents to a survey in 1982. When a group of highly productive academics was asked to indicate why they chose a certain one of their publications as their most important one, they said that they chose it because they considered it a "ground-breaking" piece of work. In their words the "newness" of research is considered a critical part of its impact on the field (Astin 1983).

Applicability and scope were also cited often by the participants. One essay author believes that the publication is cited often because "it provides a simple, highly reproducible technique for the determination of microgram amount of phosphate." Another suggests that many experimenters "can probably find a fact or statement within this publication that supports or appeals to them."

The third frequently cited reason for the publication's high citation is that the publication integrates knowledge and provides directions for further research. Usually this attribution is made by persons whose highly cited contribution was a review article. Two comments exemplify the nature of this type of attribution: "This frequently quoted review article probably provided [readers] with a convenient summary of the state of the art at a time of rapid expansion." Another one indicates, "Each chapter was an evaluative review of all existing research on that topic . . . an integration of findings."

When we examine gender differences, we find that women and men differ somewhat on their attributions. Men are somewhat more likely to believe that the high citation count was the result of timing of publication and its readability. For example, "Our paper was published at a time when the popularity of causal modeling via path analysis was at its peak in sociology." "The success of my review can be explained on the basis of being a thorough piece of work, much needed at the time." "It communicated the problem and the solution to researchers in a nontechnical, here's why-and-how-you-do-it style" (41 percent of the men compared to 24 percent of the women make such attributions). These are attributions of circumstance rather than attributions about the significance or importance of the publication. Women are more likely to indicate that the piece was the first of its kind and that it had wide applicability (see earlier quotes). In part the attribution "first of its kind," often given by women, could be the result of recent changes in the scholarly enterprise, such as the emergence of the "new" scholarship (research on women). Indeed, women's contributions to this emerging area could rightfully be viewed by them as the "first of its kind."

Men have a propensity to attribute theoretical significance to their work; and they often tend to provide circumstantial rather than substantive reasons for the piece's high citation count. Women, on the other hand, interpret the high citations as the result of their having produced work that facilitates others' efforts, by integrating knowledge, and providing further direction, by being widely applicable to the work of many other scholars.

What have we learned from these causal attributions by women and men scholars? When we examine their perceptions about what led them to undertake the highly cited research in the first place, women appear to be responding to others rather than being driven by their own quest. That is, they are less likely to undertake the work because they are interested in solving a problem, but rather that the work was the outcome of the dissertation or they were invited to prepare the piece. Furthermore, when it comes to explaining why their work is so frequently cited, the women appear to be more interested in how their work can be useful to others (their research can help and the findings can be applied by others). They also make more positive attributions about the importance of their work than do the men. They see their research as integrating knowledge and providing direction for further work: "a useful procedure for calculating the affinity of the drugs for the receptor"; "the hope that this approach might lead to a new type of cancer immunotherapy."

It is interesting and somewhat surprising that women in this study make more positive attributions about the importance of their work than do the men—a finding that runs counter to scientists' characteristic "norm of humility," their insistence that their work builds on the work of others. How are we to interpret this observation? Do women—because they receive less recognition and validation—congratulate themselves more for their achievements? Is it that they are not as thoroughly socialized into the norms of science? One should not forget however that many of their remarks also point out that their research has wide applicability, integrates knowledge, and provides direction for further research. In other words, it can be of assistance and help to other scholars and researchers.

The findings summarized thus far underscore the importance of external reinforcement in women's participation in research and publication, as suggested earlier by Cole and Zuckerman (1984). The women in this study do not appear to take the frequent citation of their work for granted. On the contrary, they are more sensitive than men are to the environmental cues that suggest colleagues' acceptance and validation of their achievements. We might speculate here that women's past experiences of gender discrimination and differential treatment in education and the work place may sensitize them to external validation, which in turn can affect their motivation to engage in further research and publication efforts. However, the extent to which external validation or lack of it affects directly women's research productivity ought to be examined further by the use of longitudinal samples.

GENDER AND FIELD DIFFERENCES

As indicated earlier, 56 percent of the essays written by the participants in this study's sample were in the natural sciences (physical and biological), and 44 percent were in the behavioral and social sciences.

Is the scholarly field which a woman chooses as important as her gender when it comes to determining impetus for research, perceived obstacles, and attributions for the high citation count of their publication?

Social scientists, independent of sex, are much more likely than natural scientists to indicate that the research was driven by a personal interest (15 vs. 9 indicate so). Both women and men natural scientists, on the other hand, indicate the need to solve a problem as a prime motivation (21 vs. 3 say so). This could reflect something intrinsic to the fields (perhaps "problems" in the natural sciences are more easily and clearly defined), but it could also reflect differences in a person's initial motivation for pursuing "hard" vs. "soft" sciences as a career. While the analysis by field appears to mask the gender differences with respect to obstacles encountered during conception of research, two of the women natural scientists and one social scientist report lack of collegial support compared to none of the men. Problems with publishers tend to be mentioned more often by social scientists, women and men, than by natural scientists of both sexes (five social scientists and two natural scientists mentioned problems with publishers). Again this finding is not surprising considering the fact that there is more room for debate about appropriate methods and about theoretical and interpretive statements in the social sciences than in the physical and biological sciences. Another interpretation is that social scientists are more likely to produce books that are harder to publish than are articles produced by natural scientists. Or, as mentioned earlier, rejection rates are higher in social journals than those in the natural sciences.

The results reported thus far suggest that overall, field may be more of a factor than gender in the experiences reported by scientists who produce highly cited research. It is possible that the norms of science within fields are a stronger determinant of the actual experiences scholars have with respect to the conduct of research and publication process than is gender. However, when we examine the attributions offered by these scientists about the high citation count of their research, gender differences persist. In other words, women more than men believe that their work is significant because it is the "first of its kind" and because it has wide applicability. Likewise males continue to indicate the "timeliness" and "readability" of their publications as crucial factors.

Concluding Comment

If indeed women in the aggregate are less productive than the men are, what accounts for it? Early research on gender differences in achievement indicated that such differences were the result of *differences in needs* for achievement and gratification (McClelland, Clark and Lowell 1953; Maslow 1954; Horner 1972). But recent theories on women's and men's approach to work argue that women and men are motivated by the same work needs (survival, pleasure, and contribution) and that differences in socialization and the structure of opportunity produce different expectations (Astin 1984). Even if both genders are motivated by the same needs, there is evidence that certain needs might be stronger in women than in men. In one study, college men and women were asked to give reasons for their future work interests (Astin 1979). Women were more likely to answer that their future career would enable them to contribute to society, to work with people or ideas, to help others, and to have an opportunity for self-expression. Men, on the other hand, were more interested in occupations that offered high pay, prestige, and rapid advancement. More recently, Cunningham and Cunningham (1986) developed scales to measure the three needs hypothesized by Astin (1984). In their study they were able to demonstrate that while both men and women express these needs in similar ways, women's contribution needs are slightly stronger than men's. Thus evidence from the study reported here showing that women feel their work is cited because it has wide applicability and is useful to others may reflect their greater contribution needs.

How are we to interpret another finding concerning why men and women pursued their research in the first place? While men look at their research efforts as a consciously motivated effort, women view theirs as happenstance: "it just happened," it was simply their dissertation, or they "were asked to do it."

Studies on locus of control and attributions for scholarly success indicate that the male has more of a sense of internal locus of control—a belief that he is responsible for his actions. This tendency has been attributed to differences in opportunities to control the results of one's behavior. If men receive more reinforcement for high quality work than women do, then women may indeed begin to believe that they have less control over the consequences of their behavior. Good work will not necessarily be rewarded.

Women's typical educational and career experiences may serve to reinforce their perception that they have less control over the outcomes. If this is so, it becomes easier to understand why women will be less likely to connect their behavior to its consequences. Our data indicate that, even when they are as successful in their research as men (authoring "classics"), women exhibit less "ownership," that is, they are less likely to connect their behavior with the outcomes. Their work was conceived and undertaken because it was either suggested by the dissertation mentor or because someone invited them to do it. These findings suggest that differences in early socialization and in the structure of opportunity (as reflected in the differential reward structure) have produced differences in expectations about work. These different expectations, in turn, help explain why women and men scientists have different approaches to producing and publishing research.

Even so, more recent evidence suggests that the gap in research production is narrowing. After all, affirmative action and the women's movement have created more opportunities for women to enter scientific careers and to apply for and receive more research funding. While differentials in the reward structure still persist, they have somewhat diminished. Such changes in the structure of opportunity will continue to narrow the gap between women's and men's research productivity.

3. Interview with Salome Waelsch

NOW PROFESSOR OF GENETICS AT the Albert Einstein College of Medicine, Salome Waelsch has worked in the field of developmental genetics for more than a half-century. Her research has focused on the role of genes in mechanisms of development and differentiation and has been widely recognized, especially in recent years. Born in 1907 in Germany, Waelsch received her Ph.D. at the age of 25 from the University of Freiburg, where she worked in the laboratory of Professor Hans Spemann, a distinguished experimental embryologist and soon to be Nobel laureate. After taking her degree, she moved to the University of Berlin, where she was a Research Assistant in Cell Biology for a year. That same year, she married Rudolf Schoenheimer, one of Germany's most promising young biochemists. In 1933, as the Nazis rose to power, she and her husband moved to the United States. He would take an appointment at the College of Physicians and Surgeons at Columbia University; she would have no post for three years. In 1936, she began work as a Research Associate in L. C. Dunn's laboratory, a position she would hold for 17 years. Widowed in 1941, she married the Columbia University biochemist Heinrich Waelsch in 1942 and later had two children. Between 1953 and 1955, she conducted research in the Department of Obstetrics at P&S again as Research Associate.

In 1955, with the founding of the Albert Einstein College of Medicine, she was appointed Associate Professor of Anatomy. Three years later, she was promoted to Full Professor. She went on to found and chair Einstein's Department of Genetics in 1963.

Waelsch was elected to the National Academy of Sciences in 1979 and the American Academy of Arts and Sciences a year later. She continues to do research and to run a laboratory. In 1988, the National Institutes of Health approved continuation of her research grant for another five years.

ZUCKERMAN: In a conversation we had a few weeks ago, you told us that someone ought to respond to a *New York Times* article which suggested that women did science differently from men.

WAELSCH: It's very interesting. I have been asking people, both men and women, about this issue. I can't say that the people to whom I happened to be able to talk would be a representative sample, but by and large, the women said that women do approach science differently, whereas the men said that they didn't.

ZUCKERMAN: Did you get a sense of the way in which people thought about it?

WAELSCH: I gave a lecture yesterday at Hunter College, and by chance the Department of Biology there has a lot of women, and we talked. And so I just asked them. They felt that the approach to science on the part of women was better, more intuitive, more imaginative, more idealistic. Whereas the men whom I asked felt there was no difference.

COLE: Does there seem to be any difference in the ages of the men and women that you asked?

WAELSCH: I don't think it's age—it's other things. The women whom I asked yesterday were very strong feminists, and the men were very competitive and aggressive. They were the ones who didn't think there was any gender difference in the approach to science, or maybe they did not want to say what they were thinking. One of the big problems is that when you look at women scientists, you look at a group that has been selected from childhood. Whether a boy becomes a scientist or not is determined, I think, by different criteria than with girls.

ZUCKERMAN: It may be that the selective factors are very different; I suspect they are. But how would that influence the way women, for example, choose problems to work on?

WAELSCH: I don't think it does. I started out studying with Spemann, and Spemann was a strong male chauvinist; you were told what to work on for your Ph.D. thesis. He gave me a problem that was very boring; in retrospect it was an insult to have been given such a project for my dissertation. Whereas a young man who was my colleague was given a very exciting problem, namely to find out whether a particular germlayer in the vertebrate embryo was responsible for the formation of the pattern of the extremities.

The work was to be done with two species of salamanders, with different patterns of limb development. I was asked to sit down and describe the pattern of development of each species, whereas the young man was asked to transplant tissues between them, a really exciting experimental problem. I was asked to provide the groundwork, which was the most boring thing that you could imagine.

I think one's doctoral project plays a role in one's training, in one's development. That certainly held me back very much.

ZUCKERMAN: How did doing a descriptive thesis hold you back?

WAELSCH: It held me back because I never had the chance of learning the very exciting techniques of transplantation on embryos that Spemann had developed. Because I was very busy with the description, and because I was only a student. I didn't have the chance or the intelligence to go and learn techniques of transplantation myself.

After my degree, when Spemann accepted me as a post-doc, I would have learned them. But then Hitler interfered.

COLE: Do women, today, experience the same kind of restricted options?

WAELSCH: No; certainly not openly, not in my field. I do not think that any senior investigator would give a woman graduate student an inferior project. But there are other problems today.

In my time, there was absolutely no question that women were given less challenging Ph.D. projects. Another woman whom I knew was also given a boring project. She did what I did, which was to play around at night with experiments outside her thesis project. She came up with a very exciting finding which really laid the groundwork for a lot of later research on mechanisms of embryonic development, embryonic induction, and so on. At first, she was considered crazy when she came up with these data. Then somebody repeated her experiment and confirmed her results. The paper that eventually came out reporting the work did not carry her name. It was totally omitted.

COLE: You were talking about selection of women from childhood. If that's the case, do women who go into science exhibit many of the same characteristics? And if so, how then would you account for hypothesized differences in intuitiveness or other attributes that women are sometimes claimed to exhibit?

WAELSCH: I don't think women exhibit them preferentially. Take Barbara McClintock, she is unique. There is no woman and there is no man with whom you can compare her. What she has done is in no way an expression of femaleness or whatever you want to call it. She's just really an exceptional genius, irrespective of sex.

I knew Gerty Cori only superficially, but I knew much about her because my husband knew the Coris in Prague. Now of those two very successful scientists Carl Cori seems to have been the intuitive and imaginative one, who also liked to enjoy life, while she was the hard and persistent laboratory worker. I know another couple, also experimental scientists, where the husband had a truly artistic approach to science—intuitive, and imaginative. She was much more down to earth, and she succeeded in making both of them work hard to prove their creative ideas. So in these two cases the male partners were the more intuitive ones.

ZUCKERMAN: You're saying that there's too much variation within each sex to draw blanket conclusions.

WAELSCH: Exactly. The women you look at, even today, have different developmental patterns in their history than the men, because they're women but these are due to their different histories not to differences in fundamental traits such as intuitiveness.

COLE: Some observers of science have argued that science, since the seventeenth century, has developed gender biases; that is, there is a tremendous stress on traits that tend to be associated with males. For example, a greater premium placed on objectivity as opposed to subjectivity or intuitiveness.

ZUCKERMAN: They also argue that the extreme emphasis on causality, on one thing causing something else to happen, is masculine.

COLE: Do you have any reactions to that?

WAELSCH: I can't accept that. I don't know the evidence for that. Where is there evidence that causality or attention to causality is more masculine? Maybe it is but I just don't know.

COLE: I don't think there's very much evidence for it. These authors suggest it almost as a proposition: that if traits which are frequently associated with women (for the moment let's not say they're proven or unproven) were more incorporated into the scientific enterprise, then science would look different and it would reward different kinds of work. In other words, women would be more comfortable with a science that had fewer of what are claimed to be masculine characteristics. It's more a philosophical than empirical argument.

WAELSCH: Women need an awful lot of organizational talent in order to manage. Even if they don't do science; just to manage children and husband and cleaning and cooking and what not. Some people have said to me that women organize their laboratories better than men do.

ZUCKERMAN: Women seem more often to have smaller labs, even in the same fields as men. Now, this could be the outcome of a great many things.

WAELSCH: I also thought this was the case, but in discussions with male colleagues of mine several names of women scientists were pointed out to me who have huge labs and expect to get even bigger ones. That's what they want.

COLE: Again we return to your theme that there is greater variability in work styles within sex than between the sexes.

WAELSCH: Precisely.

COLE: Science is terribly competitive. It puts a tremendous premium on being competitive. Do you see any differences in the level of competitiveness of men and women scientists? Are women less attracted to highly competitive environments than men?

WAELSCH: Look at Einstein Medical School, where I've been for 32 years. The women there are as competitive as the men. Almost without exception. And so maybe it's not men and women. Maybe it's something else in the environment.

Many women and men at Einstein come from similar backgrounds: second or third generation Eastern European Jews. This is a very competitive group of people. There's no difference between the men and women at Einstein.

ZUCKERMAN: Do you think it's related, by and large, to their being pretty good scientists?

WAELSCH: They are, of course, very good scientists.

COLE: One might respond by saying: well, of course, these are people who have succeeded. They had to be competitive in order to succeed. That's the real requirement in the game. But what about the men and women who don't succeed; why do fewer women succeed? Maybe they don't succeed because those women are less competitive.

ZUCKERMAN: Good scientists will tell you that being a good scientist requires a very competitive spirit in this day and age. It isn't really clear what the causal relationship is. Maybe you have to be competitive in order to succeed, but maybe succeeding also helps you be competitive. Some people have remarked to us that the young women post-docs aren't as driven to succeed as men.

WAELSCH: But that is also said about the men. I hear this again and again about today's post-docs, that they're not driven. That goes for both men and women.

Nonetheless, in my own experience, of those who did not succeed, there were probably more women than men, and they were noncompetitive.

ZUCKERMAN: Are you competitive?

WAELSCH: I think I am. I like to succeed. I get a lot of pleasure out of the lectures that I give that are successful. I never did anything in order to get elected to the National Academy, but I have to admit that it pleased me very much when I was elected.

ZUCKERMAN: Were you ever in a situation where someone tried to deprive you of credit for your work?

WAELSCH: No; but Columbia for years deprived me of any chance of a career. It was Heini, my husband, who said, what are you doing there? Why don't you get out and do something on your own? I was totally repressed there.

COLE: But you seem to have reacted to negative experiences, when you became aware of them, in a very positive way.

WAELSCH: But again, that was the environment in which I grew up. I come from a Jewish family who were immigrants in Germany, in East Prussia. My father died in the flu epidemic of 1918. My mother lost everything in the inflation. So I wouldn't be here if I hadn't been competitive and hadn't reacted to adversity in a positive way, that is, by fighting.

ZUCKERMAN: The fact that you're clenching your fist now isn't going to get on the tape.

WAELSCH: I had to fight on the street against the children who ran after me and sang dirty anti-Semitic songs. I grew up fighting.

ZUCKERMAN: Do you think that women who get to be scientists, in the first place, may more often be survivors than the men are?

WAELSCH: Yes; but again, and I'm talking as a geneticist, for reasons of the environment and not for reasons that are encoded in their genes.

COLE: How do you react to the following proposition: Women tend to be less competitive in their careers in science, because the culture says that there's another option for women. If they don't succeed in science, they have the possibility of succeeding as wives and mothers, whereas for men, the culture doesn't provide that as a viable option.

WAELSCH: That cannot be denied. There's no question that this alternative has been and still is being taken by women. We know that female graduate students who may not be quite up to the top level often get married, have families, and disappear, whereas men persist. That's in the culture.

ZUCKERMAN: That means that there should be a larger share of men than women who are sort of mediocre.

WAELSCH: There are not very many mediocre women, or not as many as there are mediocre men. There's no question about that.

COLE: If someone gave you a representative sample of papers and concealed the names of the author from you, would you be able to say which of these papers were authored by women or by men?

WAELSCH: Absolutely not.

COLE: If you were to classify that sample of papers as theoretical or experimental, as opposed to descriptive, do you think that you would have
different proportions of men and women authors in the two groups
(above and beyond their representation in the field)?

WAELSCH: I don't think so, not in my area. But in absolute terms there are
particularly more women in my field. But if you extend that to mathematics, not to mention theoretical physics, there is no question that
you'd find a predominance of men.

I think that the way an article or paper is written depends on how
you were taught in school. I don't think there's any gender difference
here.

What people have said to me is that if you look today at mamma-
lian, primarily mouse, developmental geneticists, most of them it
seems are women. I don't know why that is but it seems to be a fact.

There are Anne Mclaren, Elizabeth Russell, Janet Rossant, Eva
Eicher, Mary Lyon, Dorothea Bennett, myself, and others, all women.
I do not understand their predilection for mouse developmental ge-
netics.

COLE: Would you say the sex discrimination is the same in all age groups?

WAELSCH: I'm talking principally about scientists above the age of 50, but
there are also younger ones.

ZUCKERMAN: It may be that women attract women.

WAELSCH: Yes, that may be one reason.

COLE: What about the amount of mathematics one needs for these different
specialties?

WAELSCH: There's no question that this may be a factor. The ability to
handle quantitative aspects of problems is definitely needed more in
the specialties where women are underrepresented.

COLE: This is a pattern in all fields. The higher the level of mathematics
required, the greater the proportion of men.

WAELSCH: That's very interesting, and I cannot judge whether that is
genetic or environmental.

COLE: Consider a field such as astrophysics, which is very mathematical.
There are many women who are very good in mathematics in astrophysics. The theoretical work is nonetheless dominated by men. Why?
One prominent woman astrophysicist suggested that the social organization of theoretical astrophysics had a lot to do with this sex ratio.
The male theorists would go at each other with a kind of vehemence
that was absolutely extraordinary; women didn't want to participate

in that. Women tend to shy away from those kinds of interactions. She suggested that this was the image which theory had in the field, and that it was one of the reasons why women were less apt to go into it.

WAELSCH: I haven't seen anything like that in genetics. But I do not want to go to the extreme and say that men and women are completely alike. In physiological developmental genetics, such a difference doesn't apply. But I do not know any woman in population genetics. It is primarily a mathematical, statistical area, and I cannot think of any women in it.

For example, recently a paper was sent to me for review, "Population Genetics of Alcohol Resistance and Susceptibility in Mice." I was hardly able to read it. When I returned the paper with suggestions of names of references to whom it could be sent, there wasn't a single woman among them. I didn't try to think of a woman, but a woman didn't occur to me. I listed names of men only. So there are the mathematics again. The women seem to be in the more biological areas of genetics.

COLE: Do you think that there are any sensibilities that women have as a result of their nonscientific experiences, that are somewhat different from men's, that they might bring to the laboratory?

WAELSCH: I can't think of anything. As far as the power of observation, for example, is concerned, I've known as many male students as female students who make good observations and subsequently suggestions for experiments. There intuition becomes important, and I do not think that that is sex limited. I know as many men as women with a really high degree of intuition.

I often say to my younger collaborators that I have chosen areas of research where I have a certain monopoly. I have established a system, and it would take too much trouble for other people to get into the system.

I doubt this is something that a man would not do.

ZUCKERMAN: Is this area somewhat off to the side of the main line of development?

WAELSCH: It may appear so, but eventually I at least think that it has implications that are relevant to the central problems of development and others feel similarly.

COLE: Do you tend to avoid or take on what might be described as risky problems, that is, problems in which there tends to be a high payoff but a low probability of success?

WAELSCH: I have not changed problems very much, so I can't say I'm jumping around; I'm following the same line, namely that of trying to analyze mechanisms of development and differentiation, and I take different approaches. It's not an area where there are high risks because it's not a question of getting results. It's the interpretation of the results that's important. The biological material is there. It performs the experiment, so to say.

COLE: There is a logic of progression to the problem.

WAELSCH: Right; and a lot depends on the interpretation of the biological results that present themselves. And the results generate the next set of problems.

ZUCKERMAN: Because you've been working on a system for so long, have you developed a kind of second sense about what the interpretation ought to be?

WAELSCH: I wouldn't say that, because what I am reading now and making use of for my interpretations is really very recent stuff. What I'm trying to do is to integrate the most recent developments in molecular genetics and molecular biology into the developmental biology and genetics that I have been consumed with.

COLE: When you think about the way in which the day-to-day activities of your lab are conducted, are they comparable to the way other labs at Einstein in biology are conducted, or is there something different about your lab?

WAELSCH: I think it's different, but it's hard for me to say. It's probably a combination of my personality and the material. I run the lab with an iron hand.

COLE: More so than other people do?

WAELSCH: Yes.

ZUCKERMAN: What does that involve?

WAELSCH: I try to get to the lab before anybody else does and to leave after everybody else has gone. So I'm trying not to be one of those who travels all over the country and the world and lets the people in the lab do the work. I work just as much as they do.

ZUCKERMAN: And the purpose of getting in early and leaving late is to set an example or have control?

WAELSCH: Fundamentally, I don't trust anybody else.

COLE: When papers coming out of your lab are written, are they always initially drafted by you or by your post-docs or by your students?

WAELSCH: I write them. They give me a first draft and the data. I am the one who actually writes the papers. Perhaps because I'm not very tolerant.

But I also collaborate with people in other parts of the country and around the world. For example, I am collaborating very closely now with a lab in Heidelberg. That's a totally different story. Those people are top-notch molecular biologists, and they're doing the molecular analysis of these systems. We write the papers together. They write their part, I write my part. We put them together. It's completely equal. There are drafts which go back and forth.

COLE: To what extent are you capable of checking their work or they yours?

WAELSCH: That is a very important question. I have total confidence in them, as I had when I collaborated with Carl Cori, using experiments and procedures that I could never check. But at the moment, I have a situation where I am myself unable to repeat the experiments made by someone who was in my lab and has now moved abroad. And so I am sending this paper around, and I ask various colleagues to review it as rigorously as if it were sent to them anonymously by a journal.

COLE: If you think of the drafts of papers that have been produced in collaboration with the people in Heidelberg, would you say that the final product, compared to the original draft, has been toned down in terms of its claims?

WAELSCH: No. It is more a question of style, clarity, organization, and high standards for writing a paper. We usually settle the content before we decide that the paper is ready to be published. That is, we decide ahead of time how many times the experiments have to be repeated and how sure we are of the reliability of the results. Then we say, don't you think it's time now?

COLE: Do the results of these experiments yield only one possible interpretation in the end?

WAELSCH: No. You can never categorically state that a certain interpretation is the only possible one. The data could mean this or that, and so you then have to go on and produce the next set of data.

ZUCKERMAN: And the paper specifies that it could mean this or that?

WAELSCH: Right. It makes clear what you would like to think, but you always have to say, on the other hand, it could possibly be so and so.

ZUCKERMAN: I am curious about one comment you made, that you were never interested in building an empire. You said that the lab was never more than five, and is now three. Why didn't you have an empire?

WAELSCH: The type of work that I'm doing doesn't really require it. Perhaps more than that, it may not even be compatible with an empire. Furthermore, I wouldn't like it.

COLE: Could it also have to do with what you were saying earlier, that it's very difficult for you to delegate and to trust other people's work?

WAELSCH: It could well be. I do not easily delegate.

COLE: Do you still do as much benchwork as you did?

WAELSCH: No, it has changed. I do the genetic and developmental work with the mice, but I can't do the molecular and biochemical work, because I don't know how. That is done by my collaborators. I have to rely on them, but I also check on them and I have a lot of colleagues who help me with that. I send these young people to them, so that I'm sure that what they're doing is acceptable.

I have never had any trouble with funding. Developmental genetics doesn't require or perhaps even tolerate empires. I have grants of close to $300,000 a year, but that includes everybody's salary.

COLE: You raise your own salary?

WAELSCH: I had to raise my own salary before, and particularly now, of course, as Professor Emerita. The research is expensive because mice are expensive. We pay room and board for them. I get approximately four fifths of my budget from NIH and one fifth from the American Cancer Society. I am now in year 40 of funding by the Cancer Society, and in year 33 by the NIH.

ZUCKERMAN: Going back to the period of 1933 plus a couple of years, when you came to the United States; your husband was immediately offered a job here right after the Nazis edict. Did you expect to have a job?

WAELSCH: No. I didn't have a job in Germany either, because universities had a rule against nepotism. The most I could do was to continue working with Spemann on a joint project, in which my husband would also have been involved. But it was without pay. So I didn't have a job.

ZUCKERMAN: It was acceptable to work and be a wife, but not to get paid for it?

WAELSCH: Right. Here, at Columbia, I worked. But don't forget what time that was. It was the Depression and there was no money. So I was very glad that there was someone who let me work in his lab. And when I came to Dunn I worked in his lab without salary for a year, and then he asked me how I would like to have a salary. But there was no NIH, there was no Cancer Society, there were no outside agencies and money didn't exist.

ZUCKERMAN: Where did the money for your first paycheck come from?

WAELSCH: Columbia must have had some kind of funds. I didn't ask. It must have been an interdepartmental project.

But then later, the National Research Council came into existence. There was a so-called Committee on Growth, and a representative came to Columbia and offered Dunn some money. I got paid from that fund. That was the Committee on Growth which later became part of the American Cancer Society. So my first grants were from the Committee on Growth and then the Cancer Society.

ZUCKERMAN: From what we gather, it wasn't just Columbia, but at no major research institution were there regular faculty posts for women in the biological sciences or any other field.

WAELSCH: That is true. Berta Scharrer, for example, was in the same situation. She didn't have a faculty appointment until she came to Einstein. Jane Oppenheimer did have faculty appointments at Bryn Mawr, but not at major research institutions. My friend Sarah Ratner was like myself, a research associate at P&S [Columbia's College of Physicians and Surgeons] at the time. I don't know what the situation was for Barbara McClintock.

ZUCKERMAN: Was it even conceivable at the time to think that there was something better?

WAELSCH: No, it wasn't. That is a good word, it was not conceivable. I can tell you that it never occurred to me. That I could perhaps have a better job wasn't anything that occupied my mind. I was perfectly happy working there in Schermerhorn Hall at Columbia, for $1,500 a year, and I really enjoyed the work thoroughly. It was only much later that Heini Waelsch called my attention to the possibility that I could have a faculty appointment.

It is not quite true that this hadn't occurred to me, because I did go to Dunn and tell him that there had been young men here during my time who had climbed the academic ladder and I asked him why I still remained a research associate. He told me there was no chance for advancement for me. I went to Francis Ryan, and he also told me the same. There was absolutely no chance to advance.

So then it was actually Heini who discussed the problem with Jane Oppenheimer, who was a very good friend of ours. Ernst Scharrer was taking over Anatomy at Einstein at that time, and Jane was a very good friend of his. She called his attention to my existence. Scharrer left the University of Colorado at Denver and went to Einstein because he saw a chance there to do away with academic prejudices, e.g., that against women on faculties of universities and medical schools. In the late 1950s, he appointed another woman, Helen Wendler Deane, as Associate Professor in his department. Deanie had come to the Pathology department at Einstein after her contract at Harvard as Assistant Professor was not renewed as the result of the effects of the McCarthyite spirit at that time. In a way, I believe that Harvard was glad to find this reason for getting rid of her since it solved their problem with the academic career of a woman who was really out-standing in her research in histochemistry. So, Ernst Scharrer appointed three women, Berta Scharrer, his wife, Deanie and myself to senior positions in the Anatomy Department.

COLE: When Dunn and Ryan said to you that there was no chance of a regular position, were they reflecting their own views, the university's position, or both?

WAELSCH: I would think that it was the university's. In Dunn's case I don't think he theoretically approved of that position, but he certainly wasn't willing to fight it.

COLE: Tell us a little bit about being a research associate in those days.

WAELSCH: There were positive and negative aspects. Dunn would point out to me that, after all, I really enjoyed the research while he provided the money to support it—which should have been a pleasant situation for me.

I thought that was very condescending. Of course I enjoyed doing research, there was no question about it. But you also want the responsibility of having your own laboratory.

While Dunn certainly was an exceptionally brilliant man, science wasn't necessarily his primary interest; he actually told me once that he was writing poetry in which he was at the time more interested than in science.

Once I reminded him that I was not only doing research but also teaching (because he didn't really care much for his developmental genetics course and he let me teach it, which I enjoyed very much). But, I said, I do everything that is expected of a young faculty member. Why can't I get an appointment? So he sent me—I'll never forget it—a check for $150, for my teaching. I was very angry; I did not accept the check.

This shocked me. I found no real understanding for my situation. Ryan certainly was not interested in promoting women. Harriett Taylor, a very brilliant graduate student in the department, had a very hard time with him.

COLE: What were you able to do in terms of publication as a research associate at Columbia?

WAELSCH: I had no problems with publications.

COLE: Was Dunn's name on most of those papers?

WAELSCH: I think on all of them. But I wouldn't say that he didn't contribute. He certainly contributed. There wasn't anything that we didn't discuss; he was a very brilliant person and certainly he deserved credit.

ZUCKERMAN: When people read those papers, did they see them as collaborative papers by you and Dunn?

WAELSCH: I don't know. I didn't think or care about that. I really did not feel repressed or anything like that. I just simply didn't see why I couldn't get an assistant professorship. I was doing everything that an assistant professor did.

COLE: Were you becoming reasonably well known, despite the fact that you didn't have a regular faculty position in your field of genetics?

WAELSCH: Yes, because I was invited to give papers at prestigious conferences, papers which were given just by me.

COLE: They would invite women to give papers at these conferences?

WAELSCH: Yes, maybe because developmental biology had more women than other fields. I was on very good terms with a lot of biologists, some of whom had come from Columbia as graduate students.

COLE: In those days, were there two worlds: a world of the production of scientific knowledge and a world of careers? Were these worlds separated for women? Could one develop a reputation in the world of science and do very good research and be admired for that research, and still be left in the gutter with regard to one's career?

WAELSCH: Absolutely, and in my case what probably had a lot to do with it was the fact that I was not a single woman. I didn't worry about the future.

ZUCKERMAN: Had you been single, might you have left Columbia earlier?

WAELSCH: I might have left and gone to some women's college, because I would have had to build up my career. But this way, I wasn't worried. It was only when my consciousness was aroused to the fact that I should have had a position that I began to wonder about it.

COLE: Had you been single and moved to a women's college, would it have hindered your development as a researcher?

WAELSCH: Yes, because the research opportunities there were certainly not what I needed for stimulation. I learned an awful lot from the people around me, and not only in my department but in other departments as well.

   If I look at my colleagues at women's colleges, they haven't really had a chance to develop. It's much harder, and I can't think of people who went that route who have really been creative researchers.

ZUCKERMAN: So, in effect, you were damned if you went to a women's college and damned if you stayed a research associate.

WAELSCH: Yes, but I also didn't have that choice because I had children and had to stay in New York. I wanted to stay in New York. But Columbia was out. NYU I don't know about. And so when Einstein came up, that was a big opportunity.

ZUCKERMAN: I was struck by your description of getting out of Dunn's lab and going to P&S to work alone—might you have done it five years earlier?

WAELSCH: I was not ready yet. This is a totally separate story. The psychology of L. C. Dunn is a totally different problem. He was in his way a very domineering person. And he held me there. I didn't even realize it, and was not interested in my career or development. It was very good for him, because he completely relied on me, and things went very well. And in a sense my own welfare didn't interest him.

COLE: Did the people at Columbia try to impede your career once you left?

WAELSCH: Nobody did, no.

ZUCKERMAN: If you had not had the chance to go to Einstein, and had been offered a job at Hunter, would you have gone? Would you have left the rather peculiar position you had at P&S for that?

WAELSCH: Whatever answer I give you now would not necessarily be what I would have done. I don't know. A lot of very good people went to Hunter. At P&S, I was totally isolated, and it was only by chance that I got to do my own thing, something which I enjoyed doing. I was in Obstetrics, and a very nice man named Howard Taylor was Chairman of Obstetrics and wanted to have some genetics in the department. He asked me to teach and talk to the residents, which I did. But that would have petered out. Otherwise I was isolated. So I think that if Hunter had offered me a regular job, I would probably have gone.

I just wanted to, first of all, establish my independence at P&S, and then try to find something. But there was very little in New York. So I was just very lucky.

COLE: If we might turn to the issues of your research productivity and reproductivity, would you tell us a little bit about how your children affected your scientific work?

WAELSCH: Why don't you ask me about how I affected my children? It's much more to the point.

My children did not really affect, except positively, my research activities, because we maintained a very strict routine and organization. Children require that you stick to a routine. What it taught me was that I could do even more in the time available to me. The first year that I lived across the street from the lab, I came home for lunch, but except for that, I really worked hard, from eight to five, every minute, so that I could devote myself to the children after work.

ZUCKERMAN: Do you think you worked with any greater concentration than people who didn't have children?

WAELSCH: Now that I have nothing except my work, I'm much less concentrated than I was at that time, but then again, I was younger. And when I moved to Einstein, it became even more necessary to stick to a schedule. Heini, who was the kind of person who in Prague had lived his life from noon to 2 a.m., also had to get used to the change. So if the children affected me at all, it was in a positive way.

COLE: Was Heini very helpful?

WAELSCH: Yes, in every way. Not only by continuously encouraging me, by never raising any question about whether this was the right thing to do, as many people did at that time, but also by helping with the children. He wasn't a particularly good cook, but he was wonderful with the children.

COLE: What would have happened had you had children with your first husband, in terms of its effect on your productivity?

WAELSCH: My first husband wouldn't even have encouraged me to go on, because he was a very troubled person, so consumed by his own problems that he would not have been able to offer me the necessary help and support. We wouldn't have had children.

COLE: So the correct pairing makes a big difference.

WAELSCH: Absolutely.

COLE: And you also had help with the children, I take it.

WAELSCH: Yes.

COLE: You made a very interesting observation that because of changes in the economy and various other aspects of society, it is more difficult for young women today to handle children than it was when you were bringing them up. Why do you believe that's so?

WAELSCH: In retrospect, the degree of devotion that our baby sitter, Ellen, brought to her job and to the children is hard to find nowadays. The type of woman who brings up other people's children doesn't exist anymore. The difference now is the unavailability of people as highly qualified as she was.

COLE: You imply that women today might be able to spend as much time in the laboratory, but the effects on children might be greater.

WAELSCH: I am not sure of that. But actually, the effect is on the mother. I was able to work without worrying about the children. And actually, when it comes to the effect on the children, those growing up today with different types of childcare may perhaps become more independent. My children were not encouraged to be independent, because the baby sitter was very overprotective. So in retrospect, it was wonderful for me, but possibly not for the children.

COLE: Of course, now there are many more good childcare places for upper middle class people, although not for society as a whole.

WAELSCH: I can't really judge it. I have young friends who take their little boy to such a childcare place, which is very good. But it costs a fortune and it's not the same. You see, our baby sitter cleaned, she cooked, she washed. I didn't have to do anything. She took care of the children, and she took them to nursery school.

ZUCKERMAN: Did that leave you free to do anything other than your work and look after your family? For example, when you wanted to give a lecture somewhere, did you feel that you could be away from the family?

WAELSCH: Yes. I could even go abroad. When I was invited to Edinburgh for several months, I was able to go because the baby sitter took care of the children. Our pediatrician made house calls just to see how things were going. I had no worry. Heini stayed in New York for the first two months of my absence and came over for the third month, so that then the baby sitter was alone with them.

COLE: How old were your children at the time?

WAELSCH: Very young; Peter was one year, and Naomi was four.

COLE: When your children were very young and you were working across the street, would you say that your family life and your laboratory life accounted for, let's say, 90 percent of your time?

WAELSCH: A hundred percent.

ZUCKERMAN: Did you do anything else?

WAELSCH: No. Heini always said that people who wanted to be with us had to come to our house. We didn't go out.

COLE: You said once that one of the reasons why you were scientifically productive while you were bringing up your children was that you were not encumbered or distracted by unhappiness. Was it because of the pleasures you got from family life, from having the children?

WAELSCH: Yes, and also because of the very good relationship with my husband. We were very happy together and never bored for a second. We just didn't have enough time to spend with each other. There was no stress. We were very happy.

ZUCKERMAN: Our general findings are that having children is unrelated to the extent to which women are productive scientists. People have told us that's because we haven't found out anything about the price husbands have paid. Do you think your husband's work suffered because he pitched in and helped you with your obligations?

WAELSCH: No, I don't think so. The question then comes up, whether a scientist becomes more productive or less productive depending on his happiness or unhappiness. I don't know. Maybe unhappiness makes a person more productive.

But Heini was satisfied and was happy and enjoyed his life and his work. I don't think that helping me distracted from his work in any way. We were also lucky. The children were healthy. It is totally different if there are children with problems, congenital or psychological. We were just lucky.

COLE: We also have been confronted with the following question, which we can't really answer. Although we find no adverse effects of family on women's or men's scientific productivity, what about adverse effects on children?

WAELSCH: That is a totally different story. That is a whole subject in itself. And since you can never test this in a controlled experiment, you can't really know. If I had to do it over I don't know what I would do. I would not do it in the same way, but I don't know in which way I would do it differently. The choice of schools, maybe, but I really don't know.

COLE: What are the things that you have doubts about?

WAELSCH: For example, I have doubts about whether our baby sitter, wonderful as she was, provided enough stimulation for the children. Maybe if I had stayed home it would have been different. When I suggested that to Heini, he said to me, "Well, you would drive them crazy in five minutes if you stayed home."

But I think that they might not have been sufficiently stimulated. I'm not a child psychologist. Maybe children should be, or maybe they should not be, stimulated. I honestly don't know.

ZUCKERMAN: Do the children of men scientists who are married to women who are not full-time scientists differ at all in their psychological well-being from the children of women scientists?

WAELSCH: I can't say. For instance, my children wouldn't have dreamed of going into science. When I asked them about that—after they got old enough to discuss it—they answered that they really didn't want to live the kind of life that we lived, totally obsessed with science.

ZUCKERMAN: Are they obsessed with what they do?

WAELSCH: Well, interestingly enough, they both are very devoted teachers. Peter teaches drop-outs from high school and other adults at an educational institution in Cambridge. My daughter has a regular teaching job in a school in Westchester, and loves it. But science is something they never considered. I don't know whether one can generalize, but I know of families where the wife stayed home and the husband was a scientist, and the children did become scientists.

COLE: You have told us about the extent to which you were treated fairly in terms of authorship and credit for your contributions, and you told us about the extent to which you were denied opportunities for your career. Is there anything more you would say about the discrimination you experienced because of being a woman?

WAELSCH: Don't forget, I'm not only a woman. I'm also a Jew. It's very difficult to separate these two targets of discrimination. When you asked me earlier about the problem of getting a job in Germany, I may have told you that in 1932 I went to see a famous geneticist to inquire about a job and got the answer, "You—a woman and a Jew—forget it."

In terms of what happened at Columbia, I would say that the career of Sally Hughes-Schrader, who had a position similar to mine and was not Jewish, is proof that it wasn't my Jewishness, but my femaleness that was responsible there. At Einstein, obviously I am not discriminated against as a Jew. In addition, Einstein is also probably one of the least discriminating places when it comes to women.

COLE: What about in terms of things like salary?

WAELSCH: There have been claims at Einstein that women's salaries may be lower than those of men. There's no question that salary differences exist not necessarily between men and women but between people. There are men who are underpaid and there are women who are underpaid. When I became chairman at Einstein, my salary was raised. But I could not tell you, today, whether at that time I got the same or less salary than the average male chairman. I know some men there who are more underpaid than women.

COLE: Were you promoted, do you think, at the same rate as men?

WAELSCH: Yes. I went straight from research associate at Columbia to associate professor at Einstein. I remained associate professor for two years, and then I was promoted to full professor. We all had so-called tenure, which didn't mean much because we brought in our own salaries.

ZUCKERMAN: Some people might say, well, sure, someone who is as good a scientist as Salome Waelsch doesn't have these problems. But it's the average women who are discriminated against; they are treated worse than the average man.

WAELSCH: I was average. I really was not some big shining light. I was an average scientist. Maybe I profited from the protective hand of Ernst Scharrer, who was so totally unprejudiced.

COLE: Tell us a little about your perceptions of the life that a young woman entering science today would face compared to what you faced when you came to America and began as a scientist.

WAELSCH: It's much easier today. People are very conscious of the fact that women are as acceptable as men in science. For instance, I know

young women scientists who are active researchers and whose professional difficulties have nothing whatsoever to do with their gender.
They may be in the most prestigious laboratories and supported by
the most prestigious NIH awards but still have serious problems.

COLE: What kinds of problems does she face that are not gender related?

WAELSCH: They may face personal problems, e.g., those of loneliness.
The demands of laboratory science make it difficult to develop a life
outside the laboratory. On the other hand, a satisfying social life
within the laboratory depends very much on chance. However, unless
these problems become too difficult such women may eventually rise
to prestigious academic positions.

ZUCKERMAN: Does that mean that you believe that the problem of women
in science is solved for this young generation?

WAELSCH: It's not solved, but I think it's easier. It's not solved because it
is not easy to be a female scientist and combine that with what is
called a normal family life. That goes for men as well.

Let's assume you are a male scientist and you marry a woman
who is a publisher. The situation is different if you are a woman
scientist and you marry a man who is a publisher. The life of a scientist
is unique in terms of the commitment and the time that's required,
and what man would want his wife to spend her nights in the laboratory? A woman scientist would have a very hard time if she wanted to
get married and have a family, unless she married a scientist.

COLE: What about the informal nature of relationships in the scientific
community: the casual conversations, the going out for drinks, the
conferences, the informal discourse which makes up a lot of scientific
life? Do you think that's as accessible to women today as it is to men?

WAELSCH: You'd have to ask people younger than me. But I have never
heard my younger colleagues complain about that. I'm not a great
meeting-goer. I go to small meetings. I don't go to the big meetings.
And at the small meetings, I've never seen any discrimination against
women, even at bars. Maybe other women have encountered more
discrimination and would disagree with me.

COLE: What about conversations between male and female scientists around
the laboratory or department, as opposed to between male scientists?
Do you think the content differs?

WAELSCH: Not where I am, no.

COLE: There's not a tendency for the men to talk shop, whereas men and
women talk about all kinds of other things, such as family, concerts,
or whatever it might be?

WAELSCH: It's nothing that I am conscious of. When we sit at lunch, say 10 or 12 people, there are both men and women; perhaps more men than women, but I don't know. And we talk about everything.

COLE: What represents today the single greatest obstacle to women's scientific careers?

WAELSCH: We have not even touched upon all the great difficulties that all young scientists face, men and women. First of all, the unbelievable insecurity which results from the lack of sufficient funding. The question is whether this may be even harder for women to face.

If I had to try for a new grant today and I didn't get it, then maybe I deserved not to get it at this point in my life. But if you're 40 or 50 years old and you've done good science, maybe not top-notch but still good, you don't know whether next year you will be able to go on. Things are terrible. Between writing grants and waiting to hear about grants, nobody has any sense of security anymore. Maybe that affects women more, I don't know.

COLE: From what we know, women are not turned down disproportionately.

WAELSCH: But even if they're turned down proportionately, maybe it's not worth it.

COLE: Do you think that they may react differently from men to the turndown?

WAELSCH: That's what I mean, they may react differently.

ZUCKERMAN: Do you think it interferes with people doing their work?

WAELSCH: Yes; I'm convinced it interferes not only with women's work but also men's work. It also interferes with a willingness to go out on a limb and start something original and add some new ideas. People can no longer afford to sit and think. You have to produce, to publish. They judge you by the number of publications; they do not read them, they just count them. It's a very bad time. Whether women are repelled by that more than men, I don't know.

COLE: Could we go back again to the things that you experienced very early on in Europe? Did they instill in you the kind of qualities of personality and character which carried over to your career and were helpful in producing a successful career?

WAELSCH: There is an awful lot of luck and chance. It isn't all that logical.

ZUCKERMAN: To what extent did the environments in which you found yourself—the fact that you worked in stimulating places—shape your development as a scientist? Would you have done very different kinds of things, do you think, if you'd worked elsewhere?

WAELSCH: I don't know whether it has anything to do with science. I don't easily give up and I'm very stubborn. I work hard and I've always worked hard, since I was a little child.

ZUCKERMAN: Is that stubbornness and persistence exhibited in your work?

WAELSCH: Yes, very much so. But I have also been very lucky in terms of the people with whom I have been associated, both professionally and in my personal life.

COLE: Do you think that the networks which were formed by virtue of your associations with both your husbands, as well as your mentors, were critical in your own scientific development?

WAELSCH: It was my husbands, definitely, more than whatever networks were established through my mentors. The time that I was at Columbia was what we called the days of the closed doors. We did not communicate very much outside the lab. Dunn was certainly a very brilliant, very stimulating man. But there were no networks.

ZUCKERMAN: Even amongst the younger people?

WAELSCH: Yes. There were students; post-docs didn't exist. And I became very friendly with them. But throughout the years, many of the people with whom I communicated were those whom I knew through my husbands.

COLE: Was there a group of women scientists who knew each other and stuck together in those days?

WAELSCH: Yes, there were the "girls"; there were Rita Levi-Montalcini, and Jane Oppenheimer, and Dorothea Rudnick, and myself. But it wasn't a network, and we were not particularly concerned with career opportunities.

COLE: Suppose the job at Einstein had never come along, and that you had been forced to choose between continuing as a research associate at Columbia or taking a job at a women's college close enough to New York. Do you think you would have chosen to stay a research associate?

WAELSCH: It's a serious question. But I didn't even have that choice anymore, once I cut myself off from Columbia and moved up to P&S. That was a temporary arrangement and I was taking a risk, and it was either going to work or not, and it did work. If it had not worked, I don't know what I would have done.

COLE: What if you had left science for a decent period of time, to tend to raising your children?

WAELSCH: I have never seen anyone, man or woman, come back from that successfully.

COLE: Do you think you would have been very unhappy in that situation?

WAELSCH: Well, as you have heard, predictions were that I would have driven my children crazy, and that would have made me unhappy.

ZUCKERMAN: Did you ever think of leaving science to do something else?

WAELSCH: Never. But don't forget, I never thought I would go into science. I just slipped into it. So it's not that science had obsessed me all my life. But once I was in it, I didn't see myself doing anything else.

COLE: How able were you to make friends with other scientists your own age?

WAELSCH: That was a very pleasant situation; scientists were among my close friends, and I made friends easily with various other colleagues as well. Pre-Hitler Germany was particularly wonderful that way. Socially there was a total acceptance of women and there was certainly in the students' social life no discrimination or restriction. The social life we had was absolutely unique.

COLE: This was true in terms of academic positions as well?

WAELSCH: No, only in the social life of the students. It was absolutely wonderful; there was never a dull moment.

ZUCKERMAN: So if you wanted to talk science with these young men, you were taken as an equal?

WAELSCH: A total equal. While Spemann discriminated as strongly as I told you, our own private social life was wonderful.

COLE: And when you came to the United States, was that matched here?

WAELSCH: No, not at all. In fact, we were quite shocked at how different it was here. The so-called sexual revolution which we had put behind us in the 1920s in Germany wasn't anywhere near to occurring here. The relationship of young men and women was quite different here at that time. German social life had taken care of that in the 1920s, which was for intellectuals a wonderful time.

4. Interview with Andrea Dupree

ANDREA DUPREE HAS BEEN Associate Director of the Harvard-Smithsonian Center for Astrophysics and led its Solar and Stellar Physics Division from 1980 to 1987. She is also an Astrophysicist at the Smithsonian Astrophysical Observatory, a post she assumed in 1979. Her research has focused mainly on stars and stellar systems and the interstellar medium; she has made significant contributions to both areas.

Dupree did her undergraduate work at Wellesley College and took her Ph.D. at Harvard University in 1968. She then became a Research Fellow in the Harvard College Observatory, was promoted to Research Associate there in 1974 and then to Senior Research Associate in 1975, holding this post until she joined the Smithsonian in 1979. From 1970 to 1983, she was also a Lecturer in Harvard's Department of Astronomy.

Dupree, the author of numerous scientific papers and editor of several symposium volumes, has been an active researcher and research administrator. She was awarded the Bart J. Bok Prize of the Harvard College Observatory in 1973, the Wellesley College Alumnae Achievement Award in 1982, and the Lifetime Achievement Award, in 1988, by the Women in Aerospace Association.

Dupree has served on and chaired national and international advisory committees in astronomy and astrophysics. She is also active in professional societies, including the International Astronomical Union, the American Association for the Advancement of Science, and recently served as Vice-President of the American Astronomical Society.

Andrea Dupree is a strong advocate for the interests of women in science at Harvard and in the astronomical community. She is married and has two children.

———

ZUCKERMAN: Why don't you tell us about your current position.
DUPREE: I am an astrophysicist and am, I think, officially titled Supervi-

sory Astrophysicist. That means that I have what's called a super grade, which is a level above the standard levels of physicists at the Smithsonian Institution. I have a full-time position, which is funded by the Smithsonian Institution, 12 months a year; this support generally includes full salary, some travel support, computer time, and things like that. I also serve as the Associate Director for the Harvard Smithsonian Center for Astrophysics.

The Smithsonian's Astrophysical Observatory has joined with the Harvard College Observatory and the Department of Astronomy at Harvard to make what is called the Harvard-Smithsonian Center for Astrophysics, which I believe is the largest complex for research in astronomy and astrophysics in the world. We have about 450 people there, and of these 450, about 150 are Ph.D. scientists.

I am both an Associate Director of the center and also a director of the Solar and Stellar Physics Division of the center. We work on problems in solar physics, stellar physics, interstellar physics, and galactic physics. It's really like a small department of astronomy. We have about 41 people, about 25 Ph.D.s. We conduct observational and theoretical research. Some of my scientists build instruments that fly in space, on the shuttle, or on various satellites or rocket payloads.

Our scientists are really more problem-oriented than technique-oriented. In other words, we don't just work in one energy region or one energy frame. Scientists might work at our large multiple-mirror telescope or the very large radio array in New Mexico, or with space instrumentation, or with big optical telescopes at the National Observatories.

My function is to lead this diverse group of people, encourage them, support them, and review a certain number each year who are under my direction. These are mainly scientists.

We have about a $3 million dollar budget, and of this, 60 percent is obtained from outside grants and contracts. One grant is from the National Science Foundation, but most are from the National Aeronautics and Space Administration, from the Air Force, from the Department of Defense, from the National Geographic Society, and from Smithsonian Institution funds.

ZUCKERMAN: How important is the division you head?

DUPREE: I think it's one of the three major divisions at the Center for Astrophysics. Others are more program-oriented.

I really like to think of my division as a collection of very good scientists who work independently or together, depending on the problem.

And I guess I have a very unusual view of science, and I think it influences how I direct the division.

Most people go into science because they, as individuals, like the challenge of trying to understand a part of the universe. I don't think most people go into science these days believing, I'm going to be one of the team of 50 people, and then together we'll understand the universe.

So I try to run a department or division of individuals and somehow make these individuals blossom and work well together and do good science. I try to facilitate their doing good science. I might suggest certain areas or topics to them, because of my contacts in Washington and on the national scene. I might see certain areas that are strong, other areas that are weak, or areas where there's money available, and I will talk to them about that. I try to facilitate their interaction and give them support when they need support, and try to make it a better and very productive place to work. I know what's going on among members of the division, and I could say, gosh, you ought to go talk to Jim or to Bob or to Sally.

And we have one of the highest percentages of women in our division, of which I'm very proud.

COLE: Is that the outcome of the Women's Movement?

DUPREE: I don't know. I'm not sure the Women's Movement has helped at all, except perhaps in government laboratories or in government grants or contracts, where in the past employers were required to consider equal opportunity and maintain appropriate representation, whether it's in admission to graduate school or to entry-level positions.

There were three things that happened to me in the past week which are relevant here, and looking back over them, I thought, well, things really haven't changed. They illustrate how the more we try to progress, the more somehow we're not making progress.

There was a letter that was written to support the application of a woman for a post-doctoral fellowship. This is a very competitive position at the Center for Astrophysics, where about 150 people apply and we finally select 3. Even in this day and age, letters are coming in for this woman scientist, in which they compare her to other women.

One professor wrote, I'm going to compare this woman to other women. And then, by the way, he did compare this woman to other men. But still, in his mind he was dividing the world up according to hormones and chromosomes.

Second, I had just told a reasonably prominent male astronomer that I couldn't participate in a certain activity, and he said, oh, that's all right, I'll get another blonde to take your place. That's exactly what he said. Then he laughed and said, Oh, that's really a male chauvinistic thing to say, isn't it? I said, you're darned right it is. I said, that's absolutely absurd, you shouldn't be thinking like that. But it was a very light situation and the moment passed quickly and it wasn't the time, really, to pound in on him. But it illustrates how we're valued in this world.

And third, I was on another committee and I was talking to a retired senior official from the State Department who was organizing the committee's activities. This particular committee's responsibilities involved traveling to China, to Japan, to the Soviet Union, to European countries, and I said that I'd be willing to go on some of those trips. He said, Oh, my goodness, would you be willing to travel alone?

That comment is just so out of touch in this day and age. My colleagues and I go to Europe at the drop of a hat. We have standard visas and passports and so forth. But that statement was coming from another generation where the reaction is, Oh, my goodness, here's this woman, will she be willing to travel alone? This is completely unorthodox.

So those three events, which just happened in the past week, struck me and made me wonder how much things have really changed.

COLE: Do these remarks come from men of varying ages, or, perhaps, are these ideas and values which will die out as a different generation comes into science?

DUPREE: These are men in their forties and older. It's not the younger scientists. I consider myself at the leading edge of the women's movement in this context, but there are many more to follow behind me, and I think that women who are now 30 or 35 work in a more cooperative way than women of my generation. I graduated from college in the early 1960s, when the women's movement was just really beginning. And when I went back to my college reunion, many women came up to me and said, you know, Andrea, we have nothing to do now. Our husbands have left us, and you are the only one who has a career. You have something to do, whereas, for example, one woman was selling in the five-and-ten, and this person has a B.A. from a prestigious women's college.

These women were just not thinking about careers, were not trained for careers, or were not out in the world of paid work. Many of them, right now, are floundering a little bit.

So I think I'm at the beginning of a different phase, and maybe men who are now in their late thirties or early forties, instead of in their mid-forties and upward, have more understanding or are more accepting.

COLE: You once talked about a well-known scientist who you happened to meet while you were vacationing at a Caribbean island. All of the conversation centered on how you found the island and rarely on science. Could you tell us a bit more about this episode?

DUPREE: This was a scientist, a member of the National Academy of Sciences, who I first met scientifically. Then it turned out that we had been on the same island at pretty much the same time and we've always gone back to the same particular island, so we had a shared experience there.

After discovering that, every time we would meet, he would bring up the island, although we had a lot of science to talk about. At first, I felt very flattered. I thought, Oh, this is wonderful, this world-famous scientist wants to talk to me. Then I began to look closer. Of course, I would respond in terms of, Oh, yes, the island is wonderful, and we talked about vacations, we talked about anything, other than science. I began to think, what was going on? I began to think of the image I'm giving to him; why is it that he never talks to me about science? He talks to everyone else about science. But my conversations with him seem to be reserved for these things other than science.

I'm more aware of the structures of various conversations now, and at least in my own division at the observatory we try to have a mix of various things, and I actually do engineer it to some extent. Every morning we have coffee at 10:30 in the morning, and I think it's very important that I arrange my schedule so that I'll be there, because people from all over the observatory come, as well as visitors from out of town. They all know about this, that if you'll show up at 10:30, there will be great conversation.

And it will definitely be about science. I really use it as a time to learn about science from the people who I work with, who I don't have a chance to see every day.

But it will also be about the America's Cup, it will be about the Red Sox, it will be about those New York Giants, or whatever. I'm conscious of it now, but I've gotten to the point where I don't care.

There are times that I like to talk about other things, and I'm not embarrassed about the fact that I'm not going to spend all my time simply talking about science. That's just part of me, to have other interests and to be able to communicate these other interests.

But it probably wasn't a very positive image that I was presenting to these more senior scientists. Many of these scientists have no social skills, or have them at an extremely low level. I used to worry about that. I used to worry about my "image," and for a while I went through a phase of really being very serious about everything; I didn't talk about good food and good music, or good clothing or traveling, or whatever. And then I decided, well, that's not me. Part of me is a variety of interests, and I will talk about those things if I want to talk about them. And I now feel secure enough in my own accomplishments that there are times when I want to talk about science, and that's fine. And there are times when I don't want to talk about science and that's fine too.

ZUCKERMAN: Do you think that this freedom is the outcome of your really having accomplished a great deal and that a young woman wouldn't be as free as that?

DUPREE: Definitely yes; no question at all. It takes a while to develop confidence. It's something that you really have to build from inside with experience.

I remember, as a young post-doc, going in to see an administrator at the National Aeronautics and Space Administration, and telling this woman that I had a hard time paying my own salary and being able to write enough proposals to pay for my own salary. This was very difficult because most of the proposals and funding levels are really structured for nine-month permanent appointments, and then you look for your summer salary at a standard rate. But I wasn't in that sort of position. I needed my whole salary.

And this woman said to me, "Well, if you have a problem with what I'm giving you, go write a letter to Congress and get more money in the NASA budget."

That was a totally insensitive and terrible reply to give to someone who was just starting out. At that point, all I could do was just smile and say, thank you very much, I'll think about that, and back out the door. If someone said that to me now, I would tell them that was absolutely ludicrous and not an effective answer whatsoever.

But as you get older and you have more accomplishments, things seem to build, and then it's not as hard to get grants and get funding as it was, obviously, when you first started.

COLE: We know from examining your curriculum vitae that you've been very productive scientifically over the years. Do you have much of an opportunity these days to actually do research, or is most of your work now administrative?

DUPREE: No, it's not all administrative. For example, I've got this marvelous paper going out tomorrow, and it's very exciting. We've discovered that Alpha Ori, which is a big super-giant, is pulsating regularly. This gives us an important clue as to the way all stars can lose material and evolve. No one ever knew this before, and we just found that out. So we're very excited.

Admittedly, a fair proportion of my time, probably 50 to 60 percent, goes into administration now. Moreover, I'm Vice President of the American Astronomical Society for the next two years. This involves going to a lot of meetings and doing a lot of organization on the national scene.

COLE: Have the types of papers you publish changed since you took on a great deal of administrative work?

DUPREE: Well, I was very active in the launching of a satellite in 1978, and that really did a great deal toward furthering my career. And then I became the Associate Director here in 1980. So there was a great burst of publication once the satellite went up, and we had a lot of the early results from it. Now, I seem to be going more to the review paper. I was horrified to look at the last three papers that I've submitted; they were all review papers, which is fine and that's prestigious, but I get bored doing review papers all the time.

ZUCKERMAN: You mentioned that at least some of the work you do now is what you call "charity." That is, you think it is part of your obligation to help younger people along, to do work that they would otherwise not have an opportunity to do, and to give them some reviews and papers to write.

DUPREE: That's true.

ZUCKERMAN: Does that mean those would have been papers you would have written, had they not been given away?

DUPREE: No question. You find out that the world of science at high levels is really very small. There may be 5,000 members in the American Astronomical Society, but it seems that the same few people are asked to do the same things over and over again, so that you find you just can't handle it. You physically couldn't handle it; mentally, you just don't want to handle it. And on top of that, I think it's very important to try to help younger people.

So I always try to make sure that, at any meeting I organize, there are some younger people, maybe even people from my own group, who have a chance to speak and tell of their results.

In that sense, it's giving away some things. If I had a list of everything I said no to, it might even be bigger than what I've already published.

ZUCKERMAN: You really didn't have a sponsor or a mentor of your own to "take care of you" the way you take care of younger people. How did the absence of a strong sponsor affect your early career?

DUPREE: I remember the spring term of the year when I finally turned in my thesis. I went to my thesis advisor and said, "You know, I think I'm going to get it done." He said, "Oh, my goodness, really? You're really going to finish it?"

Instead of saying, that's wonderful, that's terrific, what are you going to do about a job, there was just amazement. You mean you're really going to settle down and get it finished?

This thesis advisor was much older than I. Maybe because he was from a different generation he felt uncomfortable having a woman student. Maybe he couldn't look at me as more than a cute pair of legs to have around the lab or to have walking down the hall.

And it even affects salary. There were various pay inequities that became apparent in our group when salaries suddenly became public. When I went to investigate these inequities I was told with a laugh, well, we knew you had a husband who could support you, so we didn't see anything wrong with keeping your salary down at this low level.

And I said, look at what I've done, and look at what I've published, and look at what all these other people have done; how can you justify that?

So I was told, well, that's fine, and when your salary review comes up the next year, we'll think about it and we'll take all that into consideration, and I said, well, that's not satisfactory. So I went to see the director, to get an immediate raise.

COLE: And you were successful in that?

DUPREE: Yes.

ZUCKERMAN: How hard was it for you to buck what they were saying?

DUPREE: I was angry. It was so clear, it was so obvious. And the people who had supervised my salary admitted that. They said, oh, yes, we knew you had a husband and so we didn't raise your salary.

I had to make sure that I really had ten papers when everyone else had five, before I would even dare to suggest that maybe the papers were somewhat commensurate in quality or in number. I even checked the Citation Index to see that my papers were cited just as much as somebody else's, which you sometimes do in desperate situations. I was arguing for a proper salary. I was arguing for what I thought was commensurate, and it was.

COLE: You mentioned that younger women amaze you a little bit in their willingness to engage in aggressive behavior that you might not have thought of when you were younger. Do you think that their willingness to engage in such aggressive behavior is a function of greater numbers of women in the field, ideological or cultural changes which have occurred, a combination of both of those, or something else?

DUPREE: Certainly, the culture has changed. Also, in the handful of cases where I admire these younger women, I can see that they've generally worked with peers, and maybe it is there that I made my mistake. I was working with someone older, because at that particular time in that area, there was a job available, and I thought, isn't that wonderful? I loved astronomy and was so happy to have an opportunity to work in the field. I can get a part-time job and I'll work on my degree. This is a world-famous person who will support me. And if I do a degree with this world-famous person, then they will know that I will be world-famous, too.

But I do notice that many of these younger women are working with their colleagues, so that there may be more of an equal basis for collaboration, an equal give and take, as opposed to a more senior person setting the agenda.

That's one thing.

Also, many of these younger women that I'm thinking of are single. They are their own sole responsibility. Maybe that makes them more anxious to get ahead. I might have been just as aggressive if my life and my salary were totally depending upon it. Maybe a part of me thought, if it gets too tough, if the boss tells me he's not going to give me the raise that I really deserve, well, then I can always back out and do something else, or I can always stay home. I have other options available. Maybe that's part of it. I'm really speculating now.

COLE: To shift gears a bit, why are women astronomers less apt than men to do theory?

DUPREE: There are different levels of theory. Just thinking about it briefly, one can say that there's computer theory, where one might use computers very intensively and calculate, let's say, the structure of an atom or something like that. There's model-building theory, where one maybe uses computers, but also ties that in with observations.

And then there's the general kind of theory which is considered to be the "cutting edge," which is more what I would call conjectural theory. It involves synthesis and various assumptions and hypothesis or conjecture; you then try to convince your colleagues of the rightness of your version of the truth.

My observation is that women tend to be more prevalent in computer theory, and perhaps as much in the so-called model-building theory.

But there appears to be a dearth of women at the cutting edge, namely, in conjectural theory. I'm not an expert, and all I can give you are my perceptions, which are perhaps very naive ones.

First of all, to go into this conjectural theory, you need to have a solid foundation in mathematics and physics and astrophysics. Therefore, if you are coming from a purely descriptive astronomical background, you're going to have a hard time coping with this area and handling it.

That may be part of why we don't find a great number of women there.

ZUCKERMAN: So what you are saying is that women tend to cluster in the more observational and empirical side of astronomy, rather than in the specialties involving mathematics.

DUPREE: Not necessarily, because computer theory and model-building theory can be very mathematical.

I think you need an extra bit of chutzpah, or aggressiveness or assertiveness, if you want to go into the cutting edge of conjectural theory.

It's not necessarily the mathematics or the physics. We do see women there. Not only do you need the strong foundation in math and physics, which many women do have, but you also need another attribute to do this so-called heavy theory or conjectural theory. You need to have a strong ego, where you can't be easily bruised, and you need to be very articulate and aggressive.

The reason is that you're proposing your interpretation of the universe, and for that you need to have the recognition of your colleagues. You must assert that this is a good idea, the right interpretation, and that *you* thought of it, because all three of those things have to be accepted by your colleagues. It doesn't do your career any good to have the theory accepted, without anyone giving you the credit.

It's been my observation that women tend not to be as articulate and aggressive in promoting their own ideas as the men. You also need to be insistent and consistent about what you're talking about.

This relates back to the conversations between women and men scientists. If every time when you see a Nobel prize winner you try to convince him of a great idea and the Nobel prize winner asks you how you liked that snorkeling or scuba trip down to the Virgin Islands, you're not going to have a chance to promote your ideas.

So, to be a conjectural theorist requires a certain sense of inner strength, a certain sense of ego, and the ability to be verbal, to be articulate, and to be aggressive.

COLE: Is this difference between men and women determined by the general culture?

DUPREE: I think it is. All I can do is speak from my own experience. I know, for example, that it took a long time for me to go to meetings and speak up. I remember the first meeting I ever went to. It was a national meeting, and I was the only woman on a particular committee. I made some comment which, looking back, was a perfectly logical comment to make. And an Italian who was also on that committee said, oh, that's just stupid. He just cut me down.

It really wasn't stupid at all, but I thought that maybe he knew more than I did. So I was very quiet for the next two or three days in the meetings, for when I just opened my mouth to say something, I was getting cut down so rapidly and so finally. I sat on the committee a few more years and I saw that this fellow did that to a lot of people, in particular to a lot of women.

Then it just became clear, at least to me, what he was doing, and I could see him operate. And then I developed the inner reserves to say, oh, now, you're wrong, and give it back to him. But it takes a while to be able to do that.

I used to call this male assertiveness the rooster complex. That's a very large factor. But it is also different for younger women. I see some younger women who are very successful and very articulate; they won't let anyone get the better of them, and rather than retreat quietly, they will push an argument until the bitter end. It's funny, sometimes that makes me uncomfortable because I know that I couldn't do that, although people say to me, Oh, you're very aggressive, too. And one very prominent senior astronomer said, "You know, Andrea, you're almost as aggressive or you're as aggressive as a man."

ZUCKERMAN: Are there any women who work in speculative, heavy theory?

DUPREE: I can't think of any. There's a new Ph.D. who just graduated from Princeton this year, and maybe she'll be one. But there are none in that area.

COLE: Have women been absolutely excluded by the heavy theorists, who won't take them on as students, or is there a process of self-selection out of that specialty?

DUPREE: Both self-selection and selection by peers. Theorists love to rank all the other theorists in the world, for some reason. It's idiosyncratic of the letters of recommendation we get from theorists. They have no qualms about saying, there are six good theorists in the world. It's a simple declarative sentence.

COLE: Is there a good deal of consensus among them?

DUPREE: Not always.

ZUCKERMAN: But do they believe that they agree?

DUPREE: They do, they believe it. It's very interesting to just look at the tone of letters of recommendation. Theorists present the most grandiose letters of recommendation for other theorists. "This person will revolutionize our understanding of the universe as we now know it, and expand upon it." Yet when observational people write such letters, even men about men and men about women and women about men and all combinations, they're much more critical. There's almost a different personality type, though perhaps that's going a little bit too far.

ZUCKERMAN: Turning back to your comments on the small number of women in heavy theory, speculative or conjectural theory, the high prestige part of astrophysics.

DUPREE: Speculative theory is considered by many to be the high prestige part of astronomy or astrophysics. But the other fields of observational astronomy, where one is using large pieces of equipment, major telescopes, are also perceived as being at the cutting edge, and also are prestigious astronomy.

ZUCKERMAN: Do women participate in these kinds of enterprises in larger numbers?

DUPREE: You will see women in relatively larger numbers on the observational side of astronomy, rather than on the theoretical side. The statistics on access to the large telescopes are something else again, and I would have to go and check that. My impression is that women don't get access to the big telescopes in proportion to their applications. But that is something which the Astronomical Society is studying right now.

To do their observational work, astronomers need to have access to telescopes; optical and radio telescopes on the ground or telescopes in space. Certain telescopes are more competitive than others, and certain time, such as dark time, is more competitive than, say, bright time. So there are degrees of competition. You do see a higher percentage of women in observational astronomy. That may be because it requires mathematics and physics to participate actively and productively in theoretical astronomy. Generally, women's presence in physics is much less than it is in the astronomy part of astrophysics. I believe it's around 2–3 percent in physics.

Astronomy started out originally as a very descriptive science and didn't require much mathematics. So that if you look at the statistics back in the 1930s and right after the war, there were very high percentages of women; of those who received Ph.D. degrees, 15, maybe even 20 percent were women.

Now the character of the field has changed. We've been attracting more people from other fields, such as physics, where there is a smaller proportion of women.

COLE: If you were to think about the top 10 or the top 50 astronomers in the country, in terms of reputation, do you think women would be represented proportionally to their numbers in the field?

DUPREE: They would be underrepresented. There's no question. Women in astronomy and astrophysics publish on the order of 7–8 percent of the papers. I'm reconstructing the figures in my head. But ten years ago, when we looked at the faculty membership of the top five graduate schools in astronomy and astrophysics, we found that there were 120 positions but no women in those positions.

Out of perhaps 120 faculty positions at Caltech, Berkeley, Chicago, Harvard, and Princeton—those were the "top five"—there were no women. Currently, there is one woman at Caltech and there's one woman at Princeton. So there are 2 out of 120.

COLE: There are some who have been recruited but who haven't accepted these positions, isn't that true?

DUPREE: Yes, that's true. One had an offer from Harvard. Another had an offer from MIT, although Harvard refused to give her an offer. That's crazy, but true. And men, too, have had offers and haven't accepted them.

COLE: Does being successful at heavy theory depend upon the active persuasion of others about the veracity of your truth claims, on the basis of very limited data; whereas in lighter theory, there are more data to rely on?

DUPREE: I'm not going to fall into the trap of saying that women like the details and women like the numbers. But from my own experience and from talking to two female scientists in my group, we felt much more secure in working on the results, or working on a problem where one didn't have to speculate as much. One could calculate it and then one could compare it to the observations, and one could say, look, my model works or my model doesn't work and it's irrefutable because I've got the numbers right here.

This is related to problem choice. You could say that women work on problems that are small-scale, like the surface of the sun, and men work on the large-scale problems, like the structure of the universe.

But it's not as simple as that, because there are other factors that are very important. One of the factors is the kind of support you have, both emotional and financial support. In other words, do you have a tenured position where you have assured financial support for the next 10 or 15 years? That certainly gives you a better ability to face long-term problems than if you have to write a grant proposal every six months. If not, then you have to do a problem that you know is tractable. You have to do a problem where you can publish from that problem, where you can then go on to get more money.

In the early stages of your career, it's very difficult to take on these large-scale problems, even if you want to, because it takes five years for you to produce your first paper. That may be all right in some fields, to write a book at the end of five years, but in astronomy one really needs to have papers in scientific journals. You can't afford to pick a great big problem that won't show results for years and years because you won't be able to feed or support yourself.

ZUCKERMAN: Do you also think that women feel the price they pay for being wrong is very high?

DUPREE: I think the price that you pay for being wrong depends on the time that you've invested in it. Clearly, if you have a program that goes on for five years, and you're wrong at the end of five years, that's a serious problem. That's a serious impediment to your career progression.

On the other hand, if you're publishing many observational papers, one paper that's wrong won't sink your ship of professional expertise or your professional advancement. Unless of course it's an enormous blunder!

COLE: You've suggested that speculative work is apt to be 50 percent correct and 50 percent incorrect, as new technologies and everything else evolve, and that this is taken for granted in the community. If you turn out to be wrong because of new instrumentation which demonstrates new things, it is not considered to be a black mark on your record.

DUPREE: Certainly not. It's also considered to be an intellectual exercise of great achievement, if you work on these so-called cutting edge problems.

We had a colloquium on Thursday where a physicist came to tell us about strings, which are these great new theoretical constructions. He gave us the colloquium, and at the end of the hour he said, in spite of all this, I don't believe them—I'm working on it because it's fun. And everyone thought, isn't that wonderful?

Very few people can afford to do that. I think you need to be in a certain position to be able to do that.

One of the goals of a research scientist is to establish one's identity in one's area of expertise. It's a different kind of science if you work on a large problem, where you're one member of a 50-person team. And I don't think many astronomers like that. Astronomers are having a hard time right now coping with this change in the structure of our research, mainly because our facilities are getting larger and larger. One person can't build it, design it, operate it, analyze the measurements, and write all the papers. It's just impossible to do all that. So astronomy is getting to be somewhat more team-oriented, which many people are having a hard time coping with.

COLE: Do you find that the women in astronomy are as competitive about their ideas as men, or is that another component of this differential selection in subspecialties?

DUPREE: The women don't seem to be as competitive about their ideas. But I have seen women become very emotional about their ideas. I've seen women at meetings burst into tears, that's how emotional they are about adherence to their ideas, and I've seen them walk out of meetings, slam books on the table, and things like that. But the general reaction among the people who are left in the room is sort of one of laughter, that she's exhibiting an extreme form of behavior. And that's just really unacceptable in the groups.

COLE: What form of competitiveness might a male scientist exhibit that a woman scientist would not be apt to exhibit?

DUPREE: For example, verbal competitiveness. Or, I see a lot of sneaky dealing in references and citations, and there are many little subtle games that people play. Basically, everybody is looking for recognition from their peers, recognition for their observations. And identification of their name with that idea and that observation.

As a result, there are a lot of subtle games that people play in publication. They might cite or fail to cite an appropriate group or person. I tend to see that more in men's papers, especially in some of my main competitors, who at this point are not women. Of course, I'm very critical of it. I look and see that they didn't really cite things as completely as they should.

ZUCKERMAN: What do you do in return?

DUPREE: There's really not much you can do.

ZUCKERMAN: Do you cite them?

DUPREE: You might go a little light on their citations. But this is a personal thing. I certainly wouldn't exclude them completely. But that's my own particular attitude toward it.

COLE: Do you ever encounter what might be called intellectual blacklisting, where people's work is actually excluded from all references because others don't like their ideas and are trying to exclude them from being recognized or having other people even read those papers?

DUPREE: I've seen that in certain areas.

COLE: What would be an example?

DUPREE: First of all, you honestly don't want to cite a paper that in your opinion is completely worthless. There's no point to doing that. So you then rationalize, and you say, well, that paper is completely worthless. I am not going to cite it. Now, I might feel it's completely worthless, but you might not feel it's completely worthless. You might feel it has 10 percent worth saying, or maybe even 50 percent worth saying.

People play the citation game in various ways. You could heavily or lightly cite a group. You can also play in more subtle ways. You can cite someone else's review article instead of the original work. You can just list the reference by number, as opposed to saying, explicitly, Smith et al. did this.

Many research groups today rely heavily on their own works, on self-citations to their own work.

Some of it seems planned and premeditated. At other times, I think there are some groups who just aren't very smart. They simply don't read beyond their own circle of publications.

COLE: Do you notice these games when it comes to funding? You review their proposals, they review yours. Is there a game played there—a game of we know that we can hurt you, but we know you can hurt us. Therefore we'll go along with some stuff we don't really think is all that good because we've all got to get funded.

DUPREE: There's a certain amount of that. But now they're changing the review procedures for telescope time on some of the major space observatories. The peer review group sits around the table and discusses the proposal. If you have a personal vendetta against one particular researcher, it'd be pretty hard to get away with it. If you can show that the person's research is no good, because what he's proposing just doesn't make sense, that's totally acceptable. But it's a little hard to carry out something that's really irrational and erroneous in front of six of your peers. And you know that this information will get out into the community eventually. You know that eventually it will probably come back to you.

ZUCKERMAN: So there are constraints on efforts to squash people.

DUPREE: You can't really do it unless the science is wrong. You can, perhaps, make your particular input on an anonymous proposal review. On the other hand, the National Science Foundation asks for evaluations from roughly seven referees. If your review is out of line, then it's more likely that your report will be called into question or ignored.

I certainly have seen poor citation in proposals that I've read, and for some of them I just chided the author, saying, look, come on, you know you didn't discover that, it was discovered by somebody else and you should put that into your proposal and be honest about it.

COLE: When you think about men and women in the same specialty area of astronomy, do you have any sense that they practice science any differently?

DUPREE: I don't see that at all. Given a particular problem, men and women in astronomy approach it the same way. There may be a preselection of problems, as we discussed before, which depends on other circumstances. But in my opinion, men and women, given the same resources and given the same position and environments, will attack the problem in the same way.

COLE: Would you be able to identify men's papers or women's papers, with the names taken off?

DUPREE: No. I don't think you could tell, absolutely not.

COLE: Some people claim that the styles of work and the types of problems selected are different among men or women. For example, it's claimed that women tend to be more intuitive in their work than men. Do you find that a credible claim?

DUPREE: I have not found that at all. I think that the choice of problems—the size of the problem, the scale of the problem—are more clearly influenced by a scientist's position. If you're a full professor and you have a lot of resources available to you in terms of students and so forth, you choose and work on different kinds of problems. If you have had the experience to develop a broad perspective, your research could reflect that perspective. If you have the good taste to choose an important problem—that problem could be large or small.

ZUCKERMAN: Then gender enters into this question only insofar as women are less apt to be full professors. But if they were, then they would choose problems in the same way?

DUPREE: Definitely, no question. I couldn't tell the difference between a man's paper and a woman's paper. Given the same situation, where they're both post-docs, both assistant professors, or both research assistants or full professors, they would approach it in the same way.

Scientists in astronomy and astrophysics pride themselves on being very quantitative and on being very precise in applying known laws of physics to the universe. There's not much room for emotion. There is room for interpretation, certainly. But we like to think that we're quantitative and that we abide by physical laws.

ZUCKERMAN: But the issue of problem choice, of course, is one that isn't entirely governed by these kinds of quantitative matters. For example, women turn up in relatively large numbers in observational astronomy, which requires you to be away for long periods of time. Do you think there's any greater stress and strain on women trying to get that time to be away, than men would have?

DUPREE: From my personal experience, I would say so. Many of the male observational astronomers have a support system at home. They have a wife at home. Some of them have dual careers, admittedly, but that's how they do their job.

And maybe some of the women feel very uncomfortable traveling away for a long period of time. But I still think, given the same resources, given the same facilities, the same access to equipment, given the same time scale of their funding, that men and women will handle problems in the same way.

COLE: It seems that it's an anachronism to talk about male science or female science, because from what you and others have told us about the increased number of collaborations, it's rare to find a group that's all women or all men.

DUPREE: Certainly you can find all male groups. Or if you just look at publications in the *Astrophysical Journal,* there are lots of male-male authorships. And of course, you have many mixed groups. I'm on lots of papers with other men and other women and vice versa. But you hardly ever find all-female groups. I've never seen a group comprised only of women—except when I published with a female graduate student, so there was a paper by two women. There just aren't enough women. We're only 6 percent, and the chances of a couple of us working on the same thing at the same time is small.

COLE: Is that 6 percent growing or not? One doesn't get a sense of great growth in the physical sciences.

DUPREE: No. There may be a growth in the percentage of degrees awarded to women, but that's probably because the men are leaving to do other things where financial rewards and prestige are greater than in scientific research. Many men are going into biology or they're going into medicine, law, or investment banking, but they're not going into science or into astronomy and astrophysics.

COLE: Earlier, you were talking about the extent to which people put down women's ideas and don't give credence to their ideas. They would call it Jim's idea, rather than Andrea's idea. And how men in that situation would tend to say, no, this is my idea. Whereas you would be reluctant to do that kind of thing.

DUPREE: That's part of my socialization, I guess. Remember, it's often a lot more subtle. This is a fascinating thing. How do women learn what to do? I think one learns by observing, by practicing. There's a real art, in a large room full of men, to make your point and make it at the right time. You have to be able to jump in at the right time, with the right statement. It also helps to observe how many women differ from men in their activities and their behavior at meetings. For example, I find that women tend to speak quickly.

I'm a perfect example. I perhaps speak more quickly than anybody else in the universe, but women in general tend to speak more quickly. Women also don't give the large summation sentences. Many men sit back in their chair and say, "Well now, I think what we're really trying to say is this, or the thrust of this discussion seems to me to be B or C or D." Women very rarely put themselves in that particular position of summing things up, which must take a lot of courage. Maybe that's just part of the male way of operating.

I was struck by something else. I wonder whether scientists are so competitive with one another that they just can't turn this competition off. Let me give an example.

I went to a scientific advisory committee and sat there for two days, and we were talking about all kinds of ideas, and never once in that meeting did anyone refer to me by name, never once.

People would just interject remarks and ideas. They all knew who we were because we'd been meeting together for two years. But never once was my name used. Now, some of the men's names were used, but mine was never used.

The next day, I went to a meeting of totally different people. This was a committee in which there were men from the State Department, men from the White House, from Congress, and from industry, and immediately they started calling everyone by name. Four times during that one day, they said, "Well, now, as Andrea said"—only it wasn't just me, it was, "As Bill said" or "Jim said." A lot more credit was being given and a lot more names were being dropped. Afterward, I tried to think, why is that? That's because we weren't really competitive with one another in that particular position. We were giving advice to a third party. Whereas, even though the scientific committee was meant to give advice, the scientists couldn't give up their competitive outlooks.

But the atmospheres of these meetings were very different. For the second one, it was more that we're a team working together to try to solve this problem. And so we will interact and support one another, as we try to solve this problem together.

COLE: When you were at Wellesley were there any things that you learned that actually have proven to be helpful in your career as a scientist and as a woman?

DUPREE: Yes and no; maybe in the long run. I like to think of myself as being fairly well rounded, that I'm interested in a lot of things. I like to think that I can converse about the new art museum or architecture or a new style of cooking or the castles in the Loire Valley. I like to think that I know about those things; it's part of my image as a person and part of what I want to be.

I don't have a very strong identification with the "scientist's role." Whether it's just part of my upbringing or just part of my attitude toward life, as I've developed, I've learned to enjoy a lot of things other than doing science. I like to do science, too, and I do good science, I do excellent science. But there are many parts to us, and I certainly wouldn't be happy if I couldn't do the other things, as well.

ZUCKERMAN: Are those other interests denigrated by fellow scientists? I sometimes have the sense that they're not considered quite respectable.

DUPREE: They're considered light or frivolous or they're seen as distracting me from what I should be doing all the time, which is sitting home and solving equations.

COLE: What about the whole issue of sexuality among men and women scientists, both as a negative and also potentially, a positive force. Could it be at all positive that some men like to be around attractive women and think of women as women?

DUPREE: Certainly, men are very aware of the fact that they're around women. I think some of them don't know how to behave around women. They feel very uncomfortable and they feel very awkward, and maybe they like it, yet they don't like it because it makes them feel uncomfortable.

ZUCKERMAN: There is, of course, the stereotype that women scientists have ripped stockings and are unattractive. And it is said that women who are attractive are not taken very seriously. Do you think that apart from being a woman, being a pretty woman can be an additional problem?

DUPREE: That's happened to me, also. When I was young, I had more people react to me like that, especially since I collaborated with my older thesis advisor, who never really acted as a mentor. I was more like a butterfly on a pin or something. He didn't quite understand that I was really serious. Even recently, I've had people say to me, you know, Andrea, when we saw you going to those meetings and when we saw you giving papers, they were always jointly authored. And we thought, well, you're a cute little thing, but you're not really doing the work. I mean, it's really all your advisor or your older collaborator who is involved in this, who's doing the work.

That also comes back to my point, that maybe women who choose contemporaries to work with don't have this problem of being misunderstood.

COLE: So that's the negative side of it. Because you're a young, attractive woman working with an older sponsor, the allocation of credit is almost sure to go to that male older person.

DUPREE: No question.

COLE: This may be irrelevant to your experience, but one could also ask whether being an attractive woman might also be helpful in terms of one's career (by which I do not mean carrying on affairs).

DUPREE: Oh, no, *that* would be trouble. I think being a woman can be helpful because first of all, you're noticed. More people probably know who I am than I know who they are. There are many men who come up to me and say, remember, I met you at that meeting seven years ago. And honestly, I hate to say it, but to me, he had a suit on, he had a tie on, and I most probably couldn't distinguish him from all the others seven years ago.

There have also been instances where it's been helpful in the sense that, some man might invite you to go out for a drink, and you go and you learn something interesting. Or you'll develop a rapport with someone that you could then use for advantage later on.

Some of this is actually helpful. There are men at meetings and I would have dinner with one of them or drinks and it can be very useful. They are very powerful men and I learn a lot about what's going on.

It's never the powerful ones who want to seduce you. It's always the other ones who have nothing else on their minds.

So there are times when, I suppose, it's an advantage.

COLE: Does the issue of sexuality sometimes get in the way of doing science and getting on with scientific conversation, insofar as there are undoubtedly men who have nothing else on their mind?

DUPREE: Oh, there have been risqué remarks. But these were light and passed quickly. Even now, I'm sure a lot of men speak to me for that reason, because, all right, so I'm an attractive woman. But I don't have much interest or time. I can't spend time talking to people if there's nothing in it for me, in terms of professional accomplishments; that is, either professional accomplishments or activities that I find personally enjoyable. For instance, going out to a good dinner!

COLE: This is something which one doesn't talk about or even acknowledge. The question is whether or not, for younger women you have seen, their sex influences whether they get to collaborate with other people. Might it be that some men simply don't want to work with women because they're afraid that emotionally neutral relationships might become emotionally involved?

DUPREE: I really don't see that. I'm going to have to confess to being sexually neutral or emotionally neutral on this.

Some people have said, it's nice to have a pair of good-looking legs in the lab. I've seen that written on a letter of recommendation for an experimentalist, saying that she was pretty good and it was nice to see a good pair of legs in the lab.

But generally, nowadays, I don't perceive that. The older people have left. So I just don't see that, unless I'm oblivious to it and I may very well be. I have also been married during my career. It may be that more of this goes on among a younger generation. I'm told by two single women who talk with me off and on, about various people who have made sexual passes at them, either at meetings or at the observatory or where they work, things like that. But it doesn't seem to be a dominant thing or anything that influences their decision or others' decisions to collaborate.

ZUCKERMAN: Just to finish this off, you said that to get involved would really be trouble.

DUPREE: In a professional career, to be sleeping your way to the top is to be on shaky ground. It would also undermine your own sense of what you're achieving in the first place, and I would think you'd always be left with this feeling of, well, did I achieve what I did because I slept with this person or because I was really intrinsically good in my own right.

Astronomy is a very gossipy community and it's very small. There are various people in the community about whom there are always "stories" going around. Many of them are probably apocryphal, but you still hear them and I just don't like that. I always chastise people when I hear them running down someone. It's usually the same two or three people running down the same two or three individuals. There's no space for that or room for that. I don't think it's productive.

ZUCKERMAN: Turning to another matter, you were a research associate for some time. We've been told by certainly more than one woman we've interviewed that being a research associate has both benefits and costs. Among the benefits are being free of a whole lot of obligations. If you don't have problems raising your salary, it's not such a bad position. Is that consistent with your view of it or not?

DUPREE: At certain times in one's career, it might be satisfactory, because if one is associated with a sufficiently large grant or program, then one doesn't have to worry about the continuity of that program. It really depends on the person. I felt very uncomfortable being one unrecognized member of a large team.

I was a very junior member at Harvard, and they were particularly sensitive to who was a principal investigator and who was a coinvestigator; they had two coinvestigators and then the rest were just team members.

And it turned out that these team members, of which I was one, had no legal access to the experiment or to the data. It was only the coinvestigators who could have legal access to it. In other words, if I decided to pick up and leave for some reason, I had no legal rights to any of that material or anything that came from the grant.

And then I found out that because I was so junior, they wouldn't even share data with me. They were busy working on a large project but didn't even share part of the data.

There were also Harvard rules, which meant that I couldn't apply on my own for any particular grant. You had to be a professor or tenure-track assistant professor to apply for a grant on your own.

So I did something, I guess, rather illegal. It was about the time that I was being treated as one of these very junior members of this research team. I could see that there was another opportunity to work on a new instrument to actually gain observing time and to write proposals, and I knew that eventually money would flow through this effort.

For this new telescope the first proposal that was required was one just for observing time. One could be a very junior member and still apply for observing time.

So I sent off a number of proposals for observing time which were all successful, and then two years later, money started coming in. I didn't quite know what to do about this. But Harvard wouldn't turn down money. Money was considered to be a good thing.

So that was really the beginning of my separation from this large team project. But other people might find such team projects very satisfying. Teams are run in different ways, and the resulting science is very different. It's just that, for me, that wasn't a satisfactory situation.

COLE: What about the flexibility and the absence of teaching obligations in the research associate role?

DUPREE: That can be considered an advantage if you really want to do research. On the other hand, it can be a disadvantage if you're using the research associate role as a way to get a tenured faculty position or to move to an assistant professorship, because then they always ask, what teaching experience have you had?

Now, I'm sure the top-notch schools don't put too much weight on the teaching experience of their applicants or their assistant professors. On the other hand, some people want to get teaching experience. I personally enjoy doing teaching. I've had a chance to teach. And my feeling has always been that if I were made a faculty member, I would be happy to teach. But I'm not a faculty member, and so I will do research. And I've been delighted. I've really enjoyed that life.

COLE: Had you been a regular full-time faculty member, would you have been as scientifically productive as you have been?

DUPREE: Yes, because being a professor, one probably attracts even more students than those I've had. Our department is a relatively small one. We have 150 people who can supervise Ph.D. theses, and we take in 4–5 new students a year. So we have 18 or 20 students, something on that order. The chances of getting a student are few and far between.

I enjoy doing research, and I could teach if I wanted to teach. If I want to have students, I can have students, and I have had students. So I've been very happy as a researcher.

ZUCKERMAN: How hard is it to move out of a research associate post into a regular faculty slot?

DUPREE: It may depend on the school. At Harvard, it's extremely hard, just as it's extremely hard for the junior faculty at Harvard to get tenure. That's well documented. A Harvard assistant professor is looked upon as someone who will get a good background and get a professorship at another place.

If I had wanted to become a faculty member at another institution, I could have applied and undoubtedly would have been considered. I just can't predict what would have happened.

I was married—I'm still married—and I didn't have the flexibility of moving around. That's one of the best ways to achieve a permanent position and to increase one's standing; to have the lever or the threat of saying, well, I'm going to leave. And to mean it. You can't do it as an empty threat. You have to be ready to leave, and people are. I was never in that position, so I could never use that threat.

I received a permanent position at the Smithsonian in 1979, and I had received my Ph.D. degree in 1968. That's 11 years. That was long. Most of the other people in my cohort, who went right into assistant professorships, probably received tenure after 6 or 7 years. So 11 years is a long time.

But it's been getting longer, too. Nowadays, it's approaching 8 and 9 years. People are going into a long holding pattern in post-doctoral positions. They used to go straight into assistant professor-ships. Now, they go to one post-doc and then they go to another post-doc, so that people tend to wait until they're 8 or 10 years out.

COLE: Your husband is a physicist, and he was slightly ahead of you in terms of his career. That, I gather, led to a certain amount of geographic immobility.

DUPREE: That is correct.

COLE: Was that also a matter of your collective values? If you got an offer next year for a professorship at Princeton, would you seriously consider something like that, or would that still not be a reasonable option?

DUPREE: That still wouldn't really be a reasonable option. Unless of course it was a magnificent offer. We like Cambridge. We're pretty well settled.

Part of it is that my husband was ahead of me professionally. He was an assistant professor when we were married. I was still in grad-uate school. In fact, they were trying to get him to move to other universities, and he always said, Oh, I can't because my wife is in graduate school. He really didn't want to move and I was a convenient excuse. (Then they would say, Oh, no, we can get her into this graduate school, too.)

So it was really more of a collective decision. First of all, my husband was ahead of me. He was established. He had tenure by the time I graduated, at a good university, and he was happy with the research prospects there. And certainly, I had always had a position at the observatories in Cambridge. But I certainly didn't advance as fast as I could have advanced if I had moved around.

Not only would I have achieved a position, undoubtedly a tenured position, earlier and faster, but I also would have matured more rapidly. You tend to be more sheltered when you're working as a research associate. You don't think about this whole game of grantsmanship, and about working with students, and things like that.

So I probably would have matured more rapidly professionally. It was really a practical decision.

COLE: Of course, you have had and continue to have a position in really one of the best institutions in the world. The question is, have you ever felt, because you haven't had a faculty professorship, a kind of status deprivation?

DUPREE: I went through a phase where I was very concerned about that, thinking I should have a faculty position. The trouble is, then I looked and the Astronomy Department at Harvard was one of the weakest in the country, although we have one of the strongest research complexes overall. But the Department of Astronomy, which has only something like 6 full-time professors out of 120 scientists who are there, is not one of the strongest departments. So maybe I've just accommodated myself.

Perhaps I would have more prestige if I were called "Professor." But I look at everything else I have and I'm very happy. I'm getting a salary that's substantially in excess of most professors, except for the super-stars at the State Universities of New York and the University of Texas, for instance. I have travel funds. I have research funds.

ZUCKERMAN: And you have the equivalent of tenure.

DUPREE: No question. I have the equivalent of tenure; and it's a 12-month salary.

COLE: You said earlier that you ran into two problems with salary that were finally rectified. Have you run into other gender-related discrepancies in the way you and men would be treated?

DUPREE: I'm sure I have and I don't know about them. Salary is the one issue that comes to mind, because that can be codified and quantified. But there are a lot of other subtle activities that are important as well.

There are things like membership in committees. There are activities, invited papers, invited lectures, colloquia, scientific recognition from the community, that women statistically don't participate in to the same extent that men do.

The salary issue is easy because that's one to identify and compare. But I'm sure there are other things that I don't know about. I remember once writing a proposal for the National Science Foundation. When it came back, thrown in our face, it was suggested, why don't you have this professor at Harvard be the principal investigator, and why don't you go work for him? Well, that professor hadn't worked in that area in 15 years, and so we were outraged about that.

COLE: What about the length of time it took to promote you?

DUPREE: That was painful at the time. It was a very long period of time, and I felt that I had to really exemplify top-notch, first-class, international research on all levels. As a matter of fact, my group had repeatedly won more time on this particular telescope than all of Europe combined. Europeans were competing from all different countries, and my group got more time. And we were making discoveries in a number of fields with this telescope. We were clearly successful there.

I suppose, looking back over it, I could say, well, my mentor certainly didn't promote me or push me for a particular position, not by any means. He still acted as if he were amazed. He once said to a reporter doing a feature story on me, "You'd better find out what makes her tick, because she's such a strange animal."

COLE: Was there just a kind of insensitivity to issues of prompt promotion among people at Harvard or the Smithsonian?

DUPREE: That certainly is true. Just because a supervisor is a good scientist doesn't mean that he or she is a good administrator or very good with personnel problems. I see examples of that all the time. So that may be part of it.

Frankly, I have stopped sitting and agonizing over past injustices, because they may be there or they may not be there. Some of them certainly were.

COLE: Do you think that women at the Smithsonian Observatory are now treated the same as men, in terms of salary, promotion, and other aspects of responsibility?

DUPREE: I think they're treated 70–80 percent as well as the men at this point. Salaries are open. Everyone knows what the salaries are and what the levels are. I have found that the women who are at higher levels, who have succeeded or are commensurate with men, are women who have gone and demanded. When I went over to the Smithsonian, to the government, I said, look, I'm coming in at a certain level. They said, Oh, no, that's going to be a huge pay raise for you. I said, that's because my pay was so low when I was at Harvard. This is crazy. Look at what I'm doing. I demand to be at that level. And I got it.

The women that I see now are successful. They are articulate about their needs. But in terms of getting resources (and I'm mentally going through a handful of 10 women), they have to demand more. They have to be noisier about their needs and their wants than what I see happening to many of the men in similar situations.

COLE: What are some of the things that are part of the 20 or 30 percent differential between men and women?

DUPREE: Resources such as data aides, assistants, travel funds, computer funds.

ZUCKERMAN: Earlier, you said that after all, what scientists really want is recognition from their scientific peers. How about when it comes to recognition: do women get the right amount of recognition relative to their contribution?

DUPREE: Definitely not. You can see that in the distribution of women in faculty. As it happens, we were just looking through the percentage of women that gave invited lectureships at the American Astronomical Society. You know, some very outstanding senior women have never been invited. I'm now in charge of inviting the lecturers for the next two years. Well, we are going to be sure that we have women at every one of our meetings.

Or look at the prizes, which in our Astronomical Society tend, unfortunately, to go either to the very young or the very old. And among the very young there are very few women. I have my eye on some women to sponsor for the next few times around, but there's a real dearth of younger women. So I don't think women are getting their just representation at all.

ZUCKERMAN: All of this is a little puzzling. Women have told us that they don't think that men judge the quality and importance of work according to the sex of the person that does it.

DUPREE: Sure.

ZUCKERMAN: And yet, even though the work is judged on more or less gender-blind criteria, the rewards are not there. Do you have any hunches about what's going on?

DUPREE: First of all, men tend to invite their friends to do things. I'm thinking in particular of invited lecturers. I'm thinking of committees. I'm thinking of meetings and colloquia. You tend to think of your friends. And men predominantly have men as other friends. The more I know, the more I see that it's a very tight-knit little group. In various subfields of astronomy, you'll find the same little group of 10 people going from meeting to meeting and giving all the standard colloquia, going from conference to conference, or just emerging and re-emerging again in different combinations. So I think that's one reason for the differential.

Also, when you are making nominations, you're identifying the young Turks. And I think it's very hard for many men to project women into that role or to identify women as being young Turks.

And there's still a lot of the senior people—those in their fifties and sixties who provide a lot of the nominations, who sit on a lot of prize committees—who are still very uncomfortable with women. They are uncomfortable simply in the sense of lacking social graces. They think, well, if she's married, she may have a child, or why hasn't she had a child if she's already married? If she's not married, why isn't she married? I mean, these things really occupy their time. It just outrages me. I've heard a lot of this because I talk a lot with these senior people, and when talking about junior women, this will come up over and over again. They really aren't used to dealing with these women.

Maybe when the group that's 35 or 40 years old moves on up, they will be thinking of women as their colleagues and be gender blind in appreciation of their work and in giving recognition.

COLE: Have you observed among people who work for you or with you that a rejection of a paper or a grant proposal leads women and men to react at all differently?

DUPREE: I haven't given the women a chance to react differently. I told them they ought to go in and they ought to resubmit it and they ought to redo it, or they ought to go do something else and use this good proposal. I tell them to revise it and send it in somewhere else.

COLE: Do they take more time to do that?

DUPREE: They take more time to do it. But these are such small statistics. I'm also talking about two women in particular, outside of myself. And so they take a lot of time. They handcraft and work over the proposal. They polish it. They have resubmitted, and in most cases, they have won.

But I also have men working with me, too, who meet rejection. I have some men who just say, Look, I'm here, I'm happy not to think about anything else. Give me my office and leave me alone.

ZUCKERMAN: Is there any indication that when women submit papers for publication, there's a difference in the rate of acceptance or rejection?

DUPREE: Our major journal in astronomy accepts about 95 percent of submitted papers. Some people say, if you hang in there and stay long enough with arguing, eventually everything will get in.

Now apparently this is different in other fields. I was just checking with my mother this morning about a medical journal. She's on the faculty at Harvard and she said, yes, there are all kinds of gender-based problems with publication. But in astronomy, I just don't see it. I don't know, but I wonder whether the women are not as driven as the men. I'm only speculating.

COLE: For some of the same reasons you were speculating about before, in terms of having other possibilities or alternatives?

DUPREE: They may have other options.

COLE: Switching subjects a bit, you observed earlier that principal investigators have only limited time (six months) during which they have complete sole access to astronomical data generated from government-sponsored research. Can that actually lead to the proliferation of papers that clutter the literature or are not sufficiently thought through? Do you see that changing at all?

DUPREE: No question that's the case. It may be changing as more observational material becomes available. This six-month policy came as a result of a NASA decision to release the data from a government-sponsored satellite. The policy is to release it six months after delivering the data to a researcher. Also, this release date is combined with a one-year cycle for proposal funding and observing time.

Therefore, it's obviously prudent, when you have your data, to write it up quickly and get it into the literature, so that you can then show that you've achieved something and you can go on to the next cycle of funding.

COLE In your experience are women less likely to just march into print with all these things than men?

DUPREE No, they do it. At least in this particular instance, they're doing that just as quickly as the men are doing it. There's no question there.

COLE: Would you tell us a little bit about how marriage and motherhood were related to your scientific work and to your scientific productivity?

DUPREE: I continued to work full time after my first child was born. Three years later, after I had two children, I started working part time for two or three years. I think I started working three-fifths time, or something like three days a week and taking two days off.

I'm sure that also contributed to the perception that I wasn't really serious, that there were other things occupying my mind and that I wasn't putting out papers as fast or as prolifically as I had previously.

Frankly, I didn't look to see. I remember nursing my babies and reading the *Astrophysical Journal*. You know, science is done in all kinds of places. You do it in the kitchen, you do it in the shower, you think about it all the time. You really can't just turn it off and on again.

ZUCKERMAN: As I remember, a lot happened soon after your children were born that led to a tremendous amount of new work.

DUPREE: That's right, because a satellite I was involved with was launched in 1978. I had written all of these proposals to move away from Harvard, to try to get out and do something on my own. It was also in a different field, which was a real move for me. It wasn't working on solar problems. It was working on interstellar and other problems.

And I was just fed up with being one part of a large team where I was clearly a junior person and could never go anywhere. And I wanted to do something that was interesting, and this was challenging, interesting, and fun.

COLE: You had the resources, though, to have help when your children were young.

DUPREE: That's right; we've always had someone come in. I've never had anyone live in, but I've had someone coming all the time to the house, 40 hours a week or something like that.

Having children certainly did inhibit my traveling. I wouldn't go places to give colloquia. But I definitely would go to certain meetings. I remember going to a big international meeting in England two months after my first child was born, because I just couldn't stay home. I thought, this would be such a defeat if I stayed home nursing this baby and missing this big meeting. And so that was a goal, to have this baby and then go to this big international meeting.

But seriously, I don't remember that it was a hindrance, any more than it was having to stay in the same place and not being able to take a professorship somewhere else.

I enjoyed the children. I still enjoy the children. They are a very important part of my life.

ZUCKERMAN: You also said earlier that it's terribly important to get out on the circuit, to establish your ideas and establish that you're responsible for them. Do you think that not being able to travel freely interfered with putting your stamp on the work as yours?

DUPREE: It probably put a crimp in it. That may be part of why I didn't get a permanent position until 11 years out, instead of, say, 8 years out.

ZUCKERMAN: At what point did you find that you really could travel as much as you wanted to?

DUPREE: Never. The problem is, I could literally travel three weeks out of every four, in terms of meetings. Perhaps all the time. But then, no research would get done. Astronomy is very international, and now for instance we're talking about making a grand tour, to look at satellite and space development all over the world. So I'd have to go to Moscow, London, Paris, Italy, China, Japan.

COLE: You made the observation, and others have as well, that in astronomy people's careers tend to peak later than, let's say, with physics.

DUPREE: It's going to get better and better. It really is!

COLE: So that 11 years doesn't make that much of a difference in astronomy as it might, for example, in physics. The differential between 11 years and 7 or 8 years in terms of the career and when people make their reputations may not have as dramatic an effect as in physics or math.

ZUCKERMAN: Are you saying that being slow doesn't somehow rule you out all together?

DUPREE: I'm not sure of that. Maybe some people would say I'm ruled out all together. I don't have a professorship at Princeton or a professorship at Caltech. For that, being smart is the first prerequisite, and many people qualify. Then, you would have to start out at the gate running, publish before you complete your thesis, finish your degree in a hurry, get out and start publishing and do post-docs at various places and get around to meetings, and really not see your spouse or your children or your dog very much at all. Then you're making your reputation early. You're getting some of the prizes that the astronomers like to give to their young people. And then you move into one of these positions.

But there are some differences. Look at the Nobel Prize or the Vetlesen Prize, or different kinds of prizes that are given for the equivalent of Nobel efforts. As you know, in physics and mathematics, they tend to come at a very young age. But astronomy is more a synthesis of what's going on, and significant contributions tend to be made at later ages than mathematics and physics.

ZUCKERMAN: Do you feel your own powers are growing?

DUPREE: Oh, definitely. And I also think it's more exciting now. I have more freedom now to go try new things. We've just gotten time on the multi-mirror telescope for April, and we're doing something clearly different, on the structure of the galaxies and large-scale structure of the universe. We're really anxious to do that. I just think it's tremendously exciting. I'm looking forward to it.

I think life is going to get better and research is going to get better. We're going to have new tools and detectors.

COLE: How about resources?

DUPREE: That's difficult. I'm very fortunate, because at the Smithsonian Institution we're reasonably well funded, but even these resources are getting tight. But nationally, we may be in for a standstill for a while, because our equipment is getting more and more expensive. There are many demands for big science projects. But whether we, as astronomers, as a community, get new high technology telescopes, which are $80–100 million efforts, is not clear.

COLE: Do you think the whole game will change with the Hubble space telescope?

DUPREE: I hope not. As I alluded to earlier, there's a concern that these large instruments are fostering team competition and making team effort a more important component of research. Statements have been made to the effect that the only way you'll get time on the super-large space telescope will be if you form a team of 50 people and then you attack a big problem.

That's a different style of doing research, and as I said, a lot of people, myself included, are very uncomfortable with that.

5. Interview with Sandra Panem

SANDRA PANEM HAS FASHIONED an unconventional career in science. Beginning as an academic research scientist at the University of Chicago, she spent time at the Brookings Institution in Washington, worked for the federal government at the Environmental Protection Agency and then for a private foundation. Now, she is a venture capitalist specializing in biotechnology with a major investment banking firm. She is known for her work on tumor virology and also for her studies of science and public policy, genetic engineering, and the growth of industrial biotechnology.

Panem did her undergraduate work at the University of Chicago, went on to become a USPHS Predoctoral Fellow (1967–70), and received her Ph.D. there in 1970. She served as a Damon Runyon Memorial Fund Postdoctoral Fellow at l'Institut de Recherche sur les Maladies du Sang, Hospital St. Louis, Paris (1970–71) and returned to the University of Chicago to complete her post-doctoral studies. She became a Research Associate in the Department of Pathology at Chicago (1971–76) and was promoted to an assistant professorship in 1976, a post she held for eight years.

During her years at Chicago, Panem held a Leukemia Society of America Special Fellowship (1975–77) and was a Leukemia Society Scholar from 1978 to 1983. Her growing interest in the connections between science and its social contexts led her to apply for and receive a Kellogg National Fellowship (1981–84), and to become a Science and Public Policy Fellow at the Brookings Institution in Washington, D.C. (1982–83). This was followed by a two-year stint as a Senior Science Advisor at the U.S. Environmental Protection Agency. She returned to the Brookings as a Visiting Scholar in 1985–86 and went on to serve as a Program Officer at the Alfred P. Sloan Foundation in New York.

Panem is the author of many research papers. She has also published two books for the general public: *The Interferon Crusade* (1985) and *The AIDS Bureaucracy* (1988), an examination of the federal government's failure to develop a national policy for dealing with AIDS.

Much of the interview centers on the chain of events which led to her being denied tenure at the University of Chicago and her reflections on the tenure process in universities.

ZUCKERMAN: Could you tell us how you saw yourself as a candidate for tenure, by describing your research achievements, your teaching, and what your record was like when the tenure issue came up.

PANEM: Let me start by making certain observations. I am struck by the parallel between discussing this issue and doing any kind of fundamental experimental research. There you are engaged with a problem which is far greater than your contribution and solutions will ever illuminate. One continually asks the same question, but gets different answers.

In this case, what you're going to get from me are different sets of interpretations based on the experiences that I've had and my distance from those experiences. Whereas, in the laboratory, one would have a series of interpretations based on the quality of information and experimental tools.

My comments are formed by three different sets of experiences. One, my own personal experiences. Second, my experiences as part of a group which was analyzing tenure procedures at the time, and third, my observations after the fact. Perhaps a fourth; my thoughts on how I'd view tenure if I returned to the academic milieu. That's relevant, because I've had a number of discussions about offers within the past few years, and there's always been a tenure component. The way in which I now think about tenure is very different from my view at the time.

Like many people, when my tenure review came up, I had no personal experience to draw on. Any information was strictly anecdotal or folkloric. It was perhaps compounded by the fact that I had been serving on a committee to examine tenure procedures at the University of Chicago. At the time when I was undergoing the review and thinking about those questions, the whole concept of tenure in American universities was undergoing examination for many reasons, but not the least one concerning economic issues.

My views on the subject are both detached and very intimately involved with my own situation.

I was a young assistant professor. My work had high visibility in my area, though it had not been fully accepted. I had been encouraged to believe by my supporters that the work which I had been doing was interesting and of high quality and should not be problematic in the tenure decision.

Although there was always discussion about the controversial nature of my work, I was never really led to believe that there would be a problem with my staying at the university. Obviously, if you're doing "forefront research," there will always be controversy. If you're doing something safe, you'll be minimized. There was a sense, which I think still persists at the great research universities, that researchers should be engaged in the questions which lead to controversy.

Retrospectively, I would say that the American research universities with which I'm most familiar tend to be extraordinarily conservative. One of the things that I did not understand at the time is the distinction between what individuals in institutions *say* that they want and what they *really* want: what they are going to support and accept.

This observation of tenure as a political exercise can not be restricted to tenure in the sciences.

COLE: Could you tell us about the ways in which your work was controversial?

PANEM: As you know, within the context of the current AIDS epidemic, no one disputes the fact that there are viruses which are the agents that are associated with the causation of this disease. The viruses that are the causative agents of AIDS are among a group of viruses that are called retroviruses, in this instance human retroviruses, because they're engaged in the causation of human disease.

At the time I was working, in the middle 1970s, our aim was to examine whether human retroviruses indeed existed. Although there was a literature showing that these kinds of viruses occurred in different species, it had yet to be shown definitively that such viruses occurred in man and that they actually participated in the causation of disease.

In fact, this is really a wonderful story, because it shows how quickly the intellectual climate can change. Recently, I talked with a young woman who was a research associate in my laboratory at the time that I was going through my tenure decision. She subsequently took a job in industry doing genetic engineering. A few months ago she called and said, do you remember this gentleman who was on the faculty when we were at the University of Chicago? I said, yes. She said, he came to visit our company the other day, soliciting for this great university. And so she went up to him and introduced herself and said, perhaps you will remember me, I used to work with Sandra

Panem. And he said, oh yes, she really was prescient. She was before her time.

That's an amusing story, not necessarily about me, but about a group of academic scientists who, within the context of their own comfort, reevaluate the information because they are no longer involved in something that might turn out to be wrong. The information which those people had to judge at the time has not changed, but they would look at it now in a different light. This comfort with established "fact" would allow this man to say something like that, whereas, at the time, it was very uncomfortable to deal with something which either was or was not going to be true.

ZUCKERMAN: Was your work a particular focus of controversy, or was all the work in that area?

PANEM: The latter, but we can't distinguish one from the other. To answer that question directly is to recognize that the field in which I worked was cancer research. We were not looking for viruses which we thought would cause a disease like AIDS—rather, we were looking for agents which we thought would cause certain kinds of human tumors.

That entire area was highly emotional and highly politicized. We all understand the symbolic importance of cancer in our society and the aura that surrounded the people who were doing that work, and the potential rewards for doing something of major import in that area.

The idea of looking for human tumor viruses was not unique. People have been doing it since the turn of the century. The National Cancer Act mandated, in fact, that one is going to go and find this agent. And the issue was less whether a particular experiment in a laboratory was important or controversial as much as the issue of putative accolades that would accompany it. I did not understand that, although at the time I thought I was very politically sophisticated. I certainly didn't understand it within the context of a conservative university structure.

In addition, there was an independent set of political questions which were far more relevant to the tenure situation than the work, regardless of whether it was correct or not and whether it was controversial in the larger scientific community. And those reflected internal academic politics within the community.

ZUCKERMAN: Do you think now that the controversial character of your research was not really the important factor in the tenure decision?

PANEM: It was important, but it was only one of several different factors.

Had I been working in a less flamboyant area, it might not have had the emotional connotations; I doubt, however, that it would have made a difference to the politics within that university.

But there are two independent sets of political issues. The flamboyance of the area in which I worked, and to which I was sensitive, and what I focused on in my discussions of the politics of science. Perhaps I would have been better advised to have focused on internal politics at that institution.

As a woman scientist at Harvard once described it, "Tenure is a Byzantine process." And I think that the ability to navigate through such a Byzantine process is one that should be described in oral history. But because the raw data and experiences are held as confidential, one only learns about them anecdotally and through others' tutelage. The mentoring process, in this regard, is extraordinarily important.

Retrospectively, I didn't understand the process as well as I should have. And had I done a different style of investigation perhaps I could have understood it better.

I should have investigated it more carefully and understood it better. But I had bought into what I thought was the appropriate political stance: that I was dependent upon the actions of those people who were presumably my supporters. There was little that I could do until after the fact to understand whether they would provide the support that I anticipated. Also, I had no way to evaluate their political sophistication. Retrospectively, I bet on the wrong horse. But that's something I could not have known at the time.

COLE: Consider the following quasi-controlled experimental question. First, about the issue of controversy. Were there any men who were roughly comparable in age, in comparable situations at Chicago, working in controversial areas, who did get tenure? And second, do you think that women are more naive about university politics because of a certain kind of condescending protectionism at those institutions? "They shouldn't be involved in such political processes, we'll take care of them"; that kind of thing. In other words, do men tend to learn this Byzantine process and how it operates better than women do?

PANEM: I first want to relate an anecdote, because I think it leads into the question.

On receipt of your letter, I had dinner with a colleague of mine—a man who is in the same generation as my "mentors" in Chicago and the people who decided my tenure. I have known him for 25 years, as a colleague, and now as a friend. When I started in biology, it was a very small area, so it was not hard to know most of the people who were the professional tumor biologists.

This is an interesting guy, because he runs a science department, and he was telling me his views on his women faculty.

We talked a little bit about what had transpired in my career. He remembered a conversation with the man who was my major professor. They discussed what was going to happen. My friend said that given the nature of the kind of work I was doing then, my problem was a political problem. The political problem was that my greatest supporter was a lightweight in the area in which I was working, and the area where he was a heavyweight was not the area I was working in. And the person in the university who was the heavyweight in the area most allied to my work was highly unsympathetic to the kind of research I was doing. Moreover, this opponent's academic political agenda was not particularly suited to my own academic agenda.

To answer your questions directly. First, at the time were there others in a comparable situation who were men, and who received tenure when I was denied it? Secondly, were women less likely to learn the institutional ropes, and more protected from the politics than men?

At the time when I stood for tenure, there were seven people in my department who in a two-year period were going to have decisions made about their tenure. I was the only person who was denied tenure, without the recourse of revisitation. There was one individual who was given standard academic tenure and promotion. The other five individuals, all of whom were denied tenure at the same time, were given term commitments with the opportunity to revisit the tenure decisions.

They were all men. Should one give a gender-specific interpretation, or alternatively another very legitimate interpretation, which might be that all of them were doing work of much higher quality than I was—by whatever criteria. Or, one could say that I was doing more controversial, flamboyant work.

There's another element which makes it a very difficult situation for me to comment on, which is an example of how anecdotal discussion is very hard to avoid. I was in a clinical department, and of those seven people, I was the only person who couldn't do service. That is, as a Ph.D., I was not running a clinical laboratory. I couldn't bring money into the department. Each and every one of the other people involved spent time doing clinical service.

That's a liability for someone in a department where a faculty member may have to do more than just teach and do research; although that's what the university may argue is its primary mission in a department which also does some form of clinical service.

I don't know how to weigh these matters, to argue one way or the other. I don't know whom I was comparable to.

The other question concerned mentoring. Perhaps it is valid to argue that a male who has a man as his mentor may have a rapport that lends itself to more candid discussion than a female who has a man for her mentor.

But I would prefer to say that it is the quality of the relationship between the mentor and the younger individual and the validity of the advice that is given, which is really important in terms of whether the mentor is wrong, both in managing the tenure experience and in helping one to cope with it before and after.

Knowing other people who had the same mentor, I would say that this man did a poor job of mentoring everyone. I don't think it was gender-specific. That was just the nature of the personality in question. In my opinion this man was an ineffective mentor for both men and women.

ZUCKERMAN: What about your perception of the riskiness of your work?

You implied that you thought that scientists ought to be in the forefront—that the forefront was risky, but that you didn't see any severe risk to your career. You thought it would probably be good that you were doing flamboyant research.

PANEM: Yes. Retrospectively, I was guilty of those things and I certainly hope, in the future, that I will be even more guilty of them. Which is to say, if your intuition is correct, you follow it up. You persist. This enthusiasm is necessary: even though the field out there says something isn't the case, you want to do those experiments anyway.

I don't think that science goes forward without this kind of enthusiasm. Retrospectively, my political error was to assume that other people would be as interested in supporting that kind of activity.

So I don't view the decision to work in a controversial, and perhaps high-risk area, as incorrect. In fact, I would encourage people to do so.

COLE: Do you think that belief in risk is part of your personality?

PANEM: It's definitely part of my personality. But it's also part of the personality of a lot of people who manage to make things happen in science.

There are large numbers of people doing experiments, and there are many styles. Some individuals' style leads to very safe work, which is not particularly attractive to others. Then there are those who do things which are "on the fringe," who are not doing, perhaps, the safest work. Clearly, that is personality dependent.

Which type of work do I think is important for science? Both are important. What do I think great research institutions should be doing? They should foster risk taking, because I think the gain is there.

Remember that one may have a mentor who helps them through that part of the research process, but one may also be, as I was at the time, a mentor to younger people. I had my own students who were doing their degrees, my post-docs and my research associates, so I was playing both roles at the same time. Certainly, while I was enthusiastic about this work in the lab, I wouldn't let my students work only on that because I knew that it was important for them to finish their degrees. In order to do that, they needed to contribute a certain style of information. It was very easy for me both to plan for them and to explain to them that when you do research, you have to do it within the context of your goals. As a student, your goal is to get an education, be well trained, learn how to think, and to get a degree, so that you have the credentials that allow you to move along. Part of getting that credential is to get it in a timely fashion, to be able to plan something, and to recognize that this was a training exercise, and consequently, it was unnecessary and perhaps inappropriate for the student to work on a problem which might be the most attractive and exciting but also too risky.

Being able to articulate that to the students during the time that I was going through my experiences indicates that I was aware of the ramifications of risk taking. How did I feel that it would work for me in my tenure decision? I knew there were elements of riskiness, and I analyzed virtually everything I did in my career from a political standpoint. I played it wrong in a number of ways, but that's not to negate the fact that the analysis was done, whether it was faulty or correct.

For example, just prior to the tenure decision, it became clear that some of the work we were doing was clearly not going to be well accepted. We had described the phenomenon and the community was not convinced that the phenomenon was as we had described it.

What do you do under those circumstances? Presumably, you design other approaches to bolster your position, because ultimately science provides the opportunity to present and marshall arguments.

I feel very strongly, looking back on my experience as well as prior experiences with other people, that at that stage most of the problems I had were caused by limitations in technology. It became clear that the data which we were presenting and which were not being well accepted had limitations, and that perhaps the time had come to marshall new technological approaches in order to argue the point.

So I decided to take a sabbatical and to learn techniques with which I was not familiar, in order to bring them back to the laboratory and improve the case for my work.

By doing this, I thought, I was going to improve my chances for tenure at the University of Chicago. Retrospectively, my choice of venue for that sabbatical was interesting. Because there I was seduced, intellectually, from what I was doing into what I currently do.

I've told you this story as a way of indicating how my analysis was done. But I got very poor advice on the implementation of the correct political assessment that I needed to apply different methodologies to my research.

COLE: Did your mentor say, this risky stuff is exciting, but you also have to do other work which is less risky, which is going to give you something to fall back on if the risky stuff doesn't pan out?

PANEM: No. Part of the reason is that my mentor really didn't know enough about the community that was providing the resistance.

And again, these discussions are very dependent on my individual case, in which a complicated new process—which I hadn't known about previously—was used. Unusually, committees from several academic units were involved.

COLE: Did you choose this man as a mentor, or did he choose you?

PANEM: I chose him. I do think there is something more happening in that selection for women in science that our discussion hasn't indicated. The tenure decision in my case may well be instructive, because it may force us to look at some other questions which are generally interesting, independent of the one of tenure.

Women of my generation had lower expectations about jobs. For example, when I was finishing my degree, I received an offer to run the lab of the man I got my degree with. So I came back as a research associate. It seemed to me a very good deal at the time. I would never recommend that to a student of mine now, and I would be horrified if someone now were to tell me that they had taken such an offer, because I see it as a classic example of a woman taking the second track as opposed to the first track in academics.

I say that now because I have a much clearer picture of careers in academia than I had at the time. I had no understanding then of the fact that this position would be the surest way not to succeed. That is to put yourself into a position in which you will never be allowed membership in the club. It's not a perspective I had at the time. And it was long in coming for me to understand that a position as a research associate, with parenthetic rank, is really not what I should have looked for. I would have been better advised to go to another university and have a full faculty appointment, as an initial appointment after my post-doctoral training, than to take the kind of employment that I did. It is strictly a reflection of what I believed at the time—that it was better to have a lesser position at a prominent university than a regular position at a less prominent university.

COLE: You once told us that you believe that you made every mistake possible. What are some of your other mistakes?

PANEM: There is a natural tendency to believe that those who are helping a woman's career are doing it with her best interests at heart. But no one ever has a strictly altruistic viewpoint. I don't think it's human. People have multiple agendas. When you're young, it's hard to see that.

It's often difficult to have the confidence to go with your own agenda. What in a man is called ambition is seen as aggression from a woman. And it's a hard thing for a woman to hold to a balance between the two.

COLE: What about staying at the same university?

PANEM: A major mistake. I did it for the same reasons that lots of people do it. It's easy to be comfortable someplace where you know the climate. It's less risky to be in a familiar environment.

But it's a terrible mistake to take such a job, for many reasons.

It's a mistake from the political point of view, because it forces you to always be viewed as someone who is affiliated with the person who has supported you. You never develop your independence or identity. I suffered from this.

Intellectually, it's a mistake, because what's good for anyone in the research mode is to be challenged and to be exposed to new approaches. Academics, in general, tend to be parochial. Staying in the same place with the same group makes it easier to remain narrow and not to have to defend one's position so strictly from an intellectual and training point of view. It's a mistake.

ZUCKERMAN: As I remember, you were also personally involved with someone who was in Chicago at the time and you didn't want to leave for that reason.

PANEM: At the time, that was a factor. But by the same token, there are probably many circumstances where people would say they made a decision to leave because there were personal reasons to go. I think that's independent of the analysis.

ZUCKERMAN: Some say that women, more often than men, make career decisions on grounds that are essentially not career-related.

PANEM: I don't have any way of evaluating that. From my own and my friends' experiences, I don't see differences between men and women. Certainly, within my generation, the decisions that my male peers were making were as much to do with personal factors as those decisions made by women.

COLE: What you're suggesting is that you don't believe that your mobility was constrained because of your marriage, any more than it would be for an equivalently situated man.

PANEM: I don't know. I was making my own professional decisions. One factor was my own personal social situation. I don't know of a difference between my male and female peers who were also factoring many things into decisions. I don't know how to evaluate whether someone weighed the ratio between personal and professional factors differently, or whether that was gender associated.

COLE: But just from your own point of view, did your husband's location in Chicago affect your thinking about job opportunities outside of Chicago?

PANEM: Initially, it wasn't an issue because my best offers seemed to be in Chicago.

Had I had no social reason to be in Chicago, would I have made a greater effort to look elsewhere? I don't know. But the majority of the professional decisions which I made were formed more by my desire for a familiar situation.

Was I encouraged by anyone to look elsewhere? Was I advised to go someplace else because it would be good for my career? No, I was never well advised.

If I had been well advised, would I then have opted for the domestic situation? I don't know, but I can tell you that I was not well advised in that regard.

COLE: Did anyone ever tell you that you establish your market value at your own university by entertaining offers from other places?

PANEM: Not really. Did I tell my students that? Yes.

It's true throughout academia. In fact, it's true throughout life, and I think that this is part of my more general statement, about the difference between what somebody says they want, what they really want, and what they're really willing to support. In my experience, in the four different sectors of science in which I have worked, it's

equally true. In academic science, in the Washington policy community, with the government, and with the foundations—this distinction
is so pervasive that I don't think it should be restricted to an academic
analysis.

ZUCKERMAN: You said earlier that one of the really damaging aspects of
the tenure process was that you didn't see early enough that it might
go bad, and you didn't seek jobs elsewhere until it was too late.

No one said to you, look, who knows how it's going to turn out?
You probably ought to be on the market anyway.

PANEM: In order to be scrupulously fair, there was discussion that it wasn't
a fait accompli. However, I was led to believe that the worst-case
scenario would be that I would be able to revisit the tenure decision.
Given that, the issue was not whether one would eventually have to
look elsewhere, but when one would have to look elsewhere. That
was the decision.

I do not want to leave you with the impression that the scientists
who have taken risks and had that risk pay off in ways that they didn't
anticipate were so naive as to think that there wouldn't be costs to that
risk.

So I never felt that the tenure situation was a fait accompli, nor
was it presented as such. What was presented was that if it didn't go
through the first time, then it would be revisited.

Retrospectively, the denial of tenure shouldn't have been surprising. But my reconstruction and my understanding of what I see as
academic-political theater was only formed after the fact by conversations with people who were either participants or observers.

At the time, I did not have a really good understanding of the
events, and it's curious that I didn't have a better understanding.
Retrospectively, I think that the reason for this naiveté is that I was
willing to believe the counsel and the presentation of events by my
supporter, who I now see as not being as politically sophisticated as I
felt he was at the time.

There were issues of allocation of limited resources, having to do
with laboratory facilities. There were issues of others within the institution who wanted to create a cohesive group of researchers. I was
doing work which no one else on the campus was doing and which no
one was terribly interested in.

In coming back to my earlier point that I made every mistake in
the book, I think it's the nature of someone that's a risk taker, that
they often put themselves in situations which are risky. To be a Ph.D.
scientist in the pathology department involves risks. My research was

not directly related to anything else that was going on in the department; and I wanted to have students who weren't even primarily associated with that department. These were all things which weighed against me.

When I look at that situation retrospectively, it's a clear pattern for failure.

COLE: Was the pathology department, as opposed to the department of biology, a logical place for you to be?

PANEM: Appointments at the University of Chicago were not made reasonably. And from what I know about most universities, everything is so idiosyncratic and so dependent upon people's interests, that I don't necessarily know whether there's a right or a wrong place.

ZUCKERMAN: Should you really have been in some other department?

PANEM: No, not in that institution.

COLE: Were you on the periphery in that department?

PANEM: Retrospectively, recognizing the limitations of the work I was doing, the limitations of my ability to do everything that needed to be done to have moved those observations the next step, recognizing that there were things that could have been done strictly from the scientific point of view, that weren't done—if one asks whether I was working in the best place, the answer is absolutely no. But there were other universities in other settings which would have been more supportive than this department.

In terms of some of the other mistakes I made, perhaps the most critical one was an inability to understand the importance of collaboration in science, as opposed to independent discovery.

I say that because it is clear that in order to overcome my problems with methodology and technology, there were two solutions.

One was to acquire in my own laboratory that technical capability, to then utilize it and hopefully to document the phenomenon. The other would have been to treat it collaboratively. As a more mature individual, I would argue that it would have been better science and better politics to have done it collaboratively.

That's where I should have been given some advice. Instead, what I elected to do was to go off and try to acquire that technical capability myself. I was not successful, partly because I just couldn't do those things well enough. Retrospectively, I don't think it was possible because I didn't understand the technical dimension—it would have been much better done collaboratively, and that was a critical question.

COLE: Were the linkages for a collaboration there?

PANEM: Not at that institution, at that time. I did a lot of collaborative work, but it was always more biological.

I have to be more explicit. That is, I was able to develop excellent collaborations with a variety of biological clinicians at a variety of sites and that fostered the debates which we had in biology. We did an excellent job there. But I didn't present the more molecular data, which would have been necessary to convince the other part of the community.

That is where I elected to acquire the technology on my own. That's exactly where I should have developed the same kind of collaborations that I had with my clinical colleagues. That was the major error.

A part of it was youthful enthusiasm, where I felt that I should have been able to do it myself. Part of it was being caught up in this enthusiasm, where a scientist describes a phenomenon and wants to hold onto it, and foster it by himself. And part of it was getting poor advice.

Memory is highly selective. I don't want to leave you with the idea that no one I knew provided me with the correct advice. Rather, these were not alternatives that I seriously considered.

Whether people may have given me the advice and I didn't understand it or I rejected it, I can't tell you. But I can certainly tell you that the message didn't come through.

ZUCKERMAN: Have you found that collaborative relationships are good not just for science but for creating bonds between investigators in different fields that can then be useful in terms of careers?

PANEM: Without question, particularly in the tenure process, from the point of view of what goes on behind closed doors (and we're not privy to that in this particular case, so we can only talk about it in a broader sense). But if we postulate that part of academic success, as manifested by grants and tenure, it is the generation of a dossier in which your peers who are broadly dispersed through the subdiscipline speak well of you. Anything that fosters that network is, of course, positive. Being involved in any kind of extensive collaboration supports that.

I recognized this at the time. When I went off on that sabbatical, I genuinely believed I was going to work on the problem and that the people in whose laboratory I was doing my sabbatical were then going to become my collaborators.

So it wasn't as if I didn't know the importance of collaboration and didn't have that as part of the plan. As it turned out, the nature of the sabbatical and the people with whom I worked did not allow for that. It opened up an entirely new vista, which I then found as interesting as what I was doing, if not more so.

ZUCKERMAN: One of the things that is important to add here is the fact that you were funded all along. You had a track record. It is obvious there was peer review of your work and people thought it worthwhile.

PANEM: Absolutely.

ZUCKERMAN: So these are subtle observations about your situation. There weren't gross questions of your not having outside support.

PANEM: Absolutely; I was always very well funded. When I finally decided to leave laboratory science, I returned the funding to the funding agencies. I have no beef with that system. All the criticism of the work we did was very justified, and I think there really are times in science when there are limitations to what you can do in a given system at a given time. And there were real technological limitations.

Retrospectively, I think there were ways where we could have brought more evidence to bear on the situation, in a way that would perhaps have made my colleagues in the community more sympathetic to what we were doing, and I didn't do that.

But I was always well funded.

ZUCKERMAN: Did you ever have papers rejected for publication?

PANEM: Sure, but not any more or less than anyone else, and we had the kind of experience where, if something were rejected, than we would have redone it and eventually it would be published.

In the areas in which I worked, it was more rare to find someone who never had a paper rejected than to find someone who had. It's part of the nature of that kind of experiment.

COLE: Did other researchers see the next step that you saw, and then go on and do the work at the molecular level?

PANEM: They did, and that's why they were eventually successful.

COLE: So your work was being picked up.

PANEM: Yes and no; I was not working in an isolated situation. There were many people doing the same kinds of things. I think that our work was understood; people knew about it and they read it.

One of the things that I am more and more convinced of is that there are times when ideas are acceptable: when their time has come. Certainly, I'm also a great believer in the importance of lots of people approaching the same question at the same time from very different areas. The actual "viral" isolate which we worked with had difficulties with it. It was not picked up in the sense of lots of laboratories working with it, but they were working with the same intellectual theme.

People trying the same approach had comparable experiences, and eventually other technological advances occurred which allowed scientists to move on, far beyond anything we were doing.

What I want to leave you with is not that what we were doing was so unique, but that we were persistent. We were seeing what other people were seeing and we were just a little bit more stubborn, trying to work on a system that other people thought wasn't the most fruitful.

That doesn't make us right or smarter. It was just the way that work was being done.

COLE: So you weren't working on a far-out problem, but were really in the center, in a central university, working at the frontiers of knowledge which were being worked on by other people, and your work was visible and being attended to, whether it be critically or not. You were a young, visible scientist.

PANEM: Absolutely.

COLE: In talking about the tenure process, you used the metaphor of "theater." Tell us about a few of the other players at the central administration level, and the relationships between players that didn't even involve you but were part of this process.

PANEM: First, all of this was hearsay, which is one of the interesting elements. Again, it was my academic life. It was my professional life. And yet everything that was going on was hearsay.

I did not have any primary knowledge of it and so I am hesitant to talk about anything in terms of people's names. But I'm happy to tell you my own perception. And while I see this as trivial academic theater, that community is the major component.

ZUCKERMAN: But the general point is nonetheless true, that these decisions almost, for everybody concerned, had a "dramaturgical" quality.

PANEM: Absolutely; a ritual quality. And we could make a case for there being rituals in the life cycle of an academic, in science. But I don't see it specific to science. We could talk about a graduate degree being virtually qualifying exams and then a defense of the thesis, in this case also as a rite of passage.

So who were the other characters? Some of them were internal to the institution. Some of them were external. It was an administrative process. For example, I can recall discussions about deciding who one asks outside the institution to write for the candidate.

I mentioned earlier that as far as I know, never before and never after was the situation handled in the way mine was. I retrospectively view that as the reflection of an internal university political drama, of which I was unapprised, and which to this day I don't understand.

There was the marshalling of the dossier in my department, which is in fact the unit in which I was presumably being considered for tenure, and then subsequent to all of that, an independent committee revisited that whole question, all of which I did not know about.

I subsequently discovered that there were people whose opinions about me had been solicited, about whom I would have said that I had either political or professional conflicts with and would not consider them independent observers; I would have viewed their being consulted as a conflict of interest. But I was not consulted and I was unapprised of it. That made an unusual situation, and I would hope for others this would not be the case. Because I don't know of anyone who has gone for academic tenure who hasn't been concerned with the individuals that would be asked about them. I know of very few institutions where the candidate is not a participant in discussing whether given people really are qualified to do the evaluation. This type of participation was denied me, in one of these two Byzantine procedures.

That is my comment on the external players. In fact, they are often selected for their effect, either positive or negative. In my case, I had examples of both. And I view that as a dramatic kind of event, one which requires a certain kind of strategy.

Secondly, there are of course the people who sit on these committees. How those committees are formed is probably different from university to university, and probably varies within different divisions of universities; but certainly those people are players as well, very much as in a courtroom drama.

In general, the chairman of the department is presumably arguing for a member of his faculty. And in cases where there is beforehand a sense that one needs to develop a constituency, either for or against the candidate, there has to be an individual who wants to build that constituency.

Do I know who those people were in my case? No. Did I do a really thorough, after-the-fact investigation to find out what happened? No. Was it because I was so traumatized that I didn't want to know? No, I didn't see any purpose in it. As far as I was concerned, this situation was closed.

ZUCKERMAN: You said that you didn't press further on it because you didn't see any point in wanting to be a member of the club.

PANEM: Absolutely. But in addition, it didn't serve any purpose. What should I have done? It would have depended on what my objectives were.

COLE: You made the observation that after the experience at Chicago, when you went around for different job opportunities, that in some sense, they said, this woman has given a brilliant talk, she's done very interesting work, why did the University of Chicago not give her tenure? What you were describing is stigmatization. Now, if that is a result, why not fight it?

PANEM: I'm happy to answer that. I think that it depended upon my academic objectives. I did not understand the stigmatization which does exist, nor could I have understood the personal effects which that kind of situation might have. I was subsequently offered not one but several good jobs, jobs which, compared to what I was being offered in terms of real opportunity to do science, were far better than anything I would have had at Chicago. Several of these jobs came with tenure. So it was not a situation which precluded an academic career. I had every opportunity to continue.

However, the question of Chicago continually arose. It really was a stigma, and perhaps the person who is the least capable of answering the question is the person whom it was about.

You are, by no means, a disinterested party. If you say it wasn't my fault, because there were factors beyond my control, who is going to look at that situation and say anything but? As an interested party, it's very hard to admit that you're not, perhaps, up to snuff. It's very hard to admit that you are the person who's most responsible for what happens to you.

So any defense is discounted in advance. In addition, because of the Byzantine process and the fact that I did not have access to the discussions, there was no way to know what the actual reasons were or what was to be done.

Consequently, the individual who isn't granted the tenure, who then has to defend or explain that position subsequently, may be in the least favored position to be well informed.

In addition, and this is something that's really tough, if you have an experience where you are led to believe one thing and the outcome is different, that tends to promote a kind of skepticism and a discounting of anything which you are told after the fact.

I am frankly unsure about exactly what transpired. I am also confident that I would not have been able to have gotten information that would have had an effect.

Also, my political analysis was that it wasn't worth my time. After all of this had happened, it was suggested to me that I might bring a sex discrimination suit against the university. My analysis was that it certainly wasn't going to do me any good and I certainly wasn't going to be a tool for somebody else's agenda. I did not see anything of value in it.

COLE: Do you, in fact, see any sex discrimination in the process at Chicago?

PANEM: In my case, I do not. There is sex discrimination in academic science. I think all the data bear that out. It's very subtle. Especially at the major research institutions, it's very hard to pinpoint. There is nothing in my life that doesn't have a gender component, because I happen to be a woman. Can I identify, in this particular episode, aspects which I think are discriminatory? No. That's not to say they're not there, but I can't identify them. So I do not see it as sex discrimination.

ZUCKERMAN: You told us earlier that you were not apt to see episodes as involving discrimination, because you didn't want to admit that such criteria could be brought to bear in judging you.

PANEM: I'm sure that's part of it, but even now, it would be hard for me to do that, especially within the context of the people in that institution, at that time. A number of my peers, not in my department but people doing biology at the university, were appointed and were women. I think in those cases it was more of a matter of the individual players vis-à-vis the politics, with different mentors.

That's not to say that I don't think discrimination existed. It does exist, and everything bears it out.

COLE: You mentioned in a prior discussion your early interest in mathematics. You also indicated that there was always a sense that you might also like to do other things. And you also had a Kellogg fellowship before the whole tenure process played itself out. Did anyone think that it indicated that you were less committed to laboratory science than you should have been?

PANEM: At the time, this kind of interdisciplinary fellowship was such a queer bird that very few people had any idea of what it was. It was so peculiar that my colleagues didn't understand it as well as they ought have, and I doubt that they would have been concerned with what its potential impact was. In fact, one of my colleagues in Chicago, when I subsequently took the position at Brookings, asked whether they had laboratories. That's just an indication of the breadth of the scientific community that I was a part of.

Subsequently I've learned that the Kellogg program, which is a very interesting experiment in interdisciplinary activities and is now going into its eighth year, has a serious concern with its effects on recipients. They have collected anecdotal information which indicates that the receipt of such a fellowship prior to the granting of tenure is a negative factor to a number of people. They are discussing whether it would be better for an academic to already have tenure before they would grant that kind of thing.

In my case, given the parochial nature of many of my colleagues and the newness of that award, it would have been so poorly understood that it would not have been clearly interpreted as a factor.

Given the fact that I had opportunities that were within the traditional structure of the science, as well as opportunities in other areas, why did I elect to leave laboratory research? I would say that even applying for the Kellogg was a clear indication that I had interests beyond laboratory research, way before a tenure decision was ever made.

I only know this from hearsay, but the folks at the Kellogg Foundation told me that, at the time I applied for the fellowship, there were very few fundamental scientists who applied. And one of their concerns was that it would not be viewed appropriately within the university community, vis-à-vis my tenure decision. They, on their own, investigated, and were told that nobody thought I would have any trouble. Again, that's hearsay. It's after the fact. I don't think it had any effect—but with all these dramas going on, there's a confluence of events; so who knows?

ZUCKERMAN: One of the themes that runs through this is that the primary actor in the drama knows least about what's going on.

PANEM: I believe that's the case. Even if you think you know, you don't know. You're the least capable of understanding because you're so emotionally invested in it.

ZUCKERMAN: Also part of this peculiar business is not knowing anything about what others have experienced. You've said that only after you went through tenure process did people begin to tell you what had happened to them. So it's only after the fact that you get the information.

PANEM: Exactly. Again, I can only speak from my own experience. At the University of Chicago, people would often volunteer success stories and would rarely bring up their failures.

When the tenure decision did not go in my favor, many people, all of whom I only knew as very successful tenured individuals at Chicago, told me their experiences of having been denied tenure at other institutions and the circumstances of their moving to Chicago.

As a young assistant professor, I was both appalled that no one had told me stories like these before—and comforted that I was not the only one.

But what was interesting was that this information had never been provided beforehand. Perhaps people felt that it wasn't relevant. I prefer the interpretation that within the academic culture, one tends not to speak of one's negative experiences.

COLE: Before we switch to the issue of what led to the decision to finally move out of the laboratory, but still stay in science in a different capacity, I was wondering whether you might comment on the impact of being married, per se (and married to a scientist).

PANEM: There are two implicit questions. One, is it negative, positive, or neutral to be married if you want to pursue an academic career in science, and second, is it negative, positive, or neutral if you are married to a scientist, in that maybe there is something alien about the commitment or the hours or the nature of doing science, that would be either supportive or not supportive.

I think my husband's profession had little relevance to my own career. I'm sure that what you want me to comment on is whether that makes it easier or more difficult to be able to put in the time and have the commitment to the work that's required.

To accomplish anything as an experimental scientist you have to work full time, all the time, and then overtime. One of my colleagues in Chicago used to say, to be really successful, it wasn't enough to be very bright, it wasn't enough to really work overtime, but you had to be very lucky. And that's all very true.

But I don't think that my own personal domestic situation got in the way at all, because I really was able to put in enough time. I put in the maximum. His profession was neither a plus nor a minus.

COLE: Did it adversely affect in any way your relationship with your husband?

PANEM: My only reaction to that is I only have one life. I don't think the science side of it had an adverse effect, not that I could identify. If I had been writing books at the time, I would have been just as intensely engaged in that as I was in doing experimentation.

ZUCKERMAN: You did say that there was one positive aspect to it, which is that you were relieved of the time, inclination, and so forth, that would be required to have other social relationships or to find them.

PANEM: Yes.

COLE: Tell us about how you made the decision to go into a different type of work in science, away from the laboratory, and what the major forces were which determined that.

PANEM: As you know, I started out as an experimental scientist, and I elected to go off in the late 1970s to become a gene cloner, because I felt that the technology was necessary for me to make any progress in my research.

You also know that I had prior interests in issues independent of science and research, as evidenced by my Kellogg Fellowship. In fact, at the time that I went to do this sabbatical, I had decided that I really needed a break.

That's an early indication of clear interest in the direction which I eventually elected to take, which is more an understanding of the environment in which science is done, in addition to straight scientific research.

During my sabbatical, I did an experiment that I didn't know in advance I was going to do. Not only was I participating in learning the nuts and bolts of a new technology, I was a participant in the development of commercial biotechnology and the struggle at that time concerning the entrepreneurial nature of fundamental biomedical research and what the correct role of the academic scientist in the commercial enterprise should be. I found myself in one of the laboratories which was engaged in the original foundation of a major genetic engineering company.

When I returned to my lab, I found that I began to practice the new trade. But in addition, I discovered that the circumstances in which I had found myself led me to believe that science, as I knew it, no longer existed. I found that the questions of who controlled the scientific agenda, what kind of research would be done in which setting, the nature of relationships between the government, industry, and the university, were now all very different from when I had been trained. That was something I was really interested in, as well as looking at the power of the new technologies. I felt very strongly that if you were really interested in science and in the generation of new knowledge and its eventual practical application, then there were many different points in the scientific enterprise where you could participate and contribute to that growth. And I was clearly interested in several of them.

I felt that my own skills as an empirical researcher were perhaps not as acute, given the new setting, as they might have been under other circumstances. And given my political experiences, I felt that I was perhaps better suited for several other areas in which I might apply my interest in science and the scientific enterprise.

Many of the academic questions which I was fundamentally interested in, I felt, could be approached by others and would quickly be resolved. In fact, I was correct.

I also had a clear sense that the scientific future was large laboratory science, as opposed to small laboratory science. I was pretty sure that I did not want to participate as a bench scientist in that kind of enterprise.

I also felt that there was going to be an increasing role, not yet well expressed, for people who were interested in the science policy, in para-science questions. I thought that I might have a real talent for that. So when I was offered the resources of the Kellogg Fellowship to write about what I saw as a fundamental change in this profession, and then when I was subsequently offered the opportunity to go to Brookings and work on it, it was an opportunity to take a look at a potentially attractive and important profession, one in which perhaps I might have more to contribute than I would on the laboratory side.

COLE: Would this be a fair statement about you? It's important for me to be in the top 2 percent of people, in whatever field I'm working in, and I would rather be in a slightly different field than I am right now, and be in the top 2 percent of that, doing what I think is important work, than being in, let's say, the top 20 percent or top 15 percent of the other field.

PANEM: I don't think I'd do the analysis in that way. I wouldn't disagree with it, but it's not necessarily the way I would characterize it. I'm not sure that that's the motivating factor.

COLE: Let me be a little bit more explicit. You had a sense of multiple talents that could be directed in a variety of different ways, and in the field of science policy you thought that you could be right up there in the top 2 percent, which might not have been the case in academic science.

PANEM: I guess that's part of my personality.

ZUCKERMAN: But you're implying, too, that the character of the research that you would have gone back to was a somewhat different research.

PANEM: Absolutely. The character of doing research had changed. Whenever you work in an area which is moving and shows some growth because of an accumulation of information, the research project always changes. Whereas perhaps you had very few people initially, success means that you are going to have a larger number of people down the road.

More fundamentally, the character of the profession and the character of the environment in which that work was done, had changed.

This is a totally different discussion, but consider, for example, our discussion of the growing number of women in academic positions. Now, the optimist says, that's because finally women are going into science. The realist says that's because more men are going into industry. So the character of the profession changed.

Also, as someone who was more capable of understanding the professional environment in which I was operating, I had a different sense of what it could do and it is absolutely plausible, given the entire tenure discussion, that in a sense I was preparing for an alternative profession. So I had inclinations and interests in doing something other than staying as a strict bench scientist in this new environment.

ZUCKERMAN: You didn't see yourself changing your role and being the chief of a major laboratory, in which you wouldn't be the bench scientist?

PANEM: I already had that. I had a lab group of close to 17 people, at the time when the lab was largest. So I'd had that experience.

The most direct way to make the point is to say that the decision to change careers was not a response to the lack of academic tenure at the university, and that the seeds of interest and different directions were there long before the tenure decision was made.

ZUCKERMAN: Well, the fact that you didn't take one of the several job offers later on (which offered tenure and good scientific opportunities) is a test of that.

COLE: Would you classify yourself, as others might, as a dropout from science? Or would you say you're still in science, but you're in a different aspect of it?

PANEM: I would say the latter. Certainly I no longer do laboratory research, but I feel that I am as much a participant in the scientific enterprise as I ever was.

COLE: Those who might be classified as dropouts are frequently people who have done things very similar to what you have done; switched careers into some other aspect of science, where in some ways they may be making many more fundamental contributions to science, as an enterprise, than they were in the laboratory.

PANEM: I don't see myself as a dropout, but there are lots of people who do, because I've had those conversations with people who knew me when I was a bench scientist. They genuinely do not understand why I'm doing what I'm doing, and they genuinely do not understand how it might be interesting. I believe that they are uncertain as to how it relates to what they do.

ZUCKERMAN: If you were to make a judgment about where you could have the most impact, what's your sense about where it would be greatest?

PANEM: It's impossible to project from my current vantage point. Clearly, the effect is the amount of leverage I have where I am.

If someone stays in science and makes a revolutionary discovery, that's a greater effect from a leverage point of view. Do I think I would have done something like that? I doubt it.

I think the element of luck in that is so great that there's no way to discuss it.

COLE: What do you consider to be important policy issues for women in science, both for their entry into science and for maintaining careers?

PANEM: My own feeling is that one has to be clear about what they feel the eventual objective goal ought to be, of having a policy for women in science. And I consider myself fundamentally feminist. Ultimately what I'd want to see is a parity both of opportunity and of any criteria that would be used at all stages of the enterprise.

Needless to say, within the context of this society, where to be a successful woman you have to be twice as smart and work twice as hard and be twice as lucky, that will be a long time coming. So when one speaks about contemporary objective goals in policy for women in science, I think one certainly needs to talk about parity at the entry level.

Parity within the hierarchies at major universities, in major corporations, where scientific R & D is the business of the corporation, is something which is going to require two things. One, it's going to require a change more generally in society, and that's going to be a long time coming. And second, it will reflect a pool of people coming out of the training programs.

So the only objective policy that is correct has to focus initially on the capabilities of those entering the training programs and on the maintenance of interest and support and those incentives that can provide opportunities for women professionals across the board.

It's important that it's across the board, because success in seeing more women professors, at the expense of seeing women in higher positions in the industrial sector, is very short sighted and is a Pyrrhic victory because one doesn't want to see academic science become a handmaiden to where the big boys are playing.

The only way you get there is when you see 50 percent admissions in medical schools, women and men, and 50 percent Ph.D.s being women and men, and then 50 percent of those who are hired into different sectors.

And short of that, we have to have a discussion on different points, where I can be more concrete.

ZUCKERMAN: As I'm sure you know, there has been discussion about whether women do science differently from men; that is, whether they choose their problems using different criteria; whether they choose systems to work on that are different from those men choose; whether they're better at some things than men are. Does that correspond to your experiences?

PANEM: Probably not. There are a lot of women in the discipline from which I came. So I have had a lot of contact with peers and students who were both men and women and were comparable. The difference is that in the generations older than mine, there were proportionately so few women that it's hard to talk. But in my peer group, there are a lot of women, certainly among younger people in increasing numbers.

From what I've seen, I don't think it's gender-specific.

Do women select different questions? My feeling is that training in the graduate and postdoctoral periods have a longlasting effect on what someone does in science.

In my experience and observations, it's rare that a student really selects his or her project. The student is largely formed by the professor, and if you want to start to see gender differences, you have to look at that relationship.

ZUCKERMAN: Do you have the sense that women students are given different problems?

PANEM: Not in the settings in which I have worked. That is because these settings were small group enterprises, where the general theme of study was really set by the professor and everyone worked on elements of it.

In terms of how young scientists select questions, it's partly a reflection of where they've been working and partly a political decision in terms of where they feel they can both be supported as well as have a little bit of time to be able to establish themselves.

And my sense is that that reflects less one's scientific intuition than other factors.

I'm familiar with studies where people talk about whether there are gender differences in the way people conceptualize problems. For example, I know that if I had any talent as an experimentalist, I would have to have biological intuition. I know a lot of guys who have it, and a lot of guys who don't; I also know a lot of women who don't have particularly good biological intuition.

So I do not see gender-specific differences, where I would say, now, that's the kind of experiment I would give a woman.

I do have colleagues who say things like that.

COLE: What you're saying is that in your experience there is variability both within sexes and between them.

PANEM: Yes, exactly.

COLE: Now, what if this colleague of yours were to realize that he or she sees similarities because women with particularly female characteristics are sorted out by the whole process. The only people who ever get through all the barriers created by male science are people who think like men. In an open science, we would have a lot more women who use a different approach to problem solving.

PANEM: I would say that's an experiment that's going on. But only when we get to the point of parity will we be able to address that hypothesis.

There is no question that science is currently practiced in American universities as very much of an apprenticeship profession, and there's no question that students reflect the experience and the prejudices of their mentors.

And when we see breakthroughs, it's less because the process has changed, and more because some constellation of events has occurred that allows people to move a little forward. That hasn't changed the process or the structure.

I think that may change; I really do think that the change from small to big science in the life sciences is going to mean very different things. In fact, my earlier comment, that it is inappropriate for someone to be a research associate, may well change in the future. It may well be, if we see a coalescence of large laboratory groups, that I might counsel someone who really loved the experimental stuff, who did not want to get involved with any of the administrative part, to take that kind of position.

I genuinely believe that career paths in science are very different now than they were before. And my comments and my advice really have to be taken within the context of the kind of system that I was operating in.

ZUCKERMAN: That's a very important point, because a young person listening to your story might take away the erroneous impression that the only way to do it is the way that was appropriate ten years ago, even though the structure of opportunities may have shifted.

PANEM: Exactly. It may be very different.

ZUCKERMAN: Certainly what people have told us about what life is like being an experimental high energy physicist, is very different than it was 30 years ago. Those fellows are very rarely prime contributors to the experiment. They work in very large groups.

PANEM: In fact, one of the interesting things one might learn from the interviews in this volume is that the structure may be changing.

I wouldn't be surprised, from the point of view of career models, that for someone coming up in the life sciences today, regardless of whether a man or a woman, the model of, say, physics 15 years ago may be a more appropriate model than that of biology 10, 15 years ago.

I don't know how you would go about identifying such parallels, but I wouldn't be surprised. But there is a terrifically fundamental change that's occurring, and it's occurring in terms of the speed at which information can be generated and in terms of the potential for an individual to have even an incremental effect independently.

Most people will talk about the change in small to big science as being associated with costly instrumentation and facilities. I don't think that's the important part of it at all. It's more the speed at which information can be generated and changed, and the requirement for a critical mass of people to do anything unique and to be able to move something forward.

As exciting as the life sciences are at the moment, they are stalled. There will be a conceptual revolution in the life sciences in the next few years, I am confident, very possibly falling out of all of the mass of information that is being accumulated now with technology which may look very spiffy, but is really 10, 15 years, 20 years out of date from an intellectual point of view. This will change even further the way the life sciences are done and undoubtedly will move more toward big science in terms of the group-work models we have in physics.

This will be such a fundamental difference that many of the former models and the kinds of things you've been talking to me and these other women about may well lose their relevance.

II

ARE WOMEN LESS PRODUCTIVE SCIENTISTS?

JONATHAN R. COLE / HARRIET ZUCKERMAN

6. Marriage, Motherhood, and Research Performance in Science

STUDIES OF scientists' research performance, as gauged by their published productivity, find that women generally publish fewer papers throughout their careers than male scientists matched for age, doctoral institutions, and field (see Figure 1.4). Various explanations have been proposed to account for this disparity in scientific publication, ranging from systematic gender discrimination to biological differences in scientific aptitude.

One frequent explanation holds that women, far more than men, bear the burdens of marriage and child care and that this fact of social life best accounts for gender differences in scientific publication. Whether or not this is true, the belief that it is so affects women's career opportunities, their decisions, and the way they are actually treated.

We decided (as part of a larger investigation of the careers of American men and women scientists) to test the counter claim, made in earlier studies, that marriage and motherhood have no effect on women's research performance. We did so by assessing the dynamic relation between family life and women's research throughout their careers—an approach not taken by earlier investigators, who simply correlated the number of papers published with current marital and parental status. Our study draws on interviews with 120 scientists: 73 women and 47 men. We wanted to know whether scientists (both male and female) believe marriage and motherhood to be generally incompatible with a scientific career, whether this had been the case for them in particular, and what quantifiable effects (measured in numbers of publications) marriage and motherhood have actually had on the research performance of women scientists. Since men traditionally have not had primary responsibility for child care, we focused here almost entirely on women, comparing publication rates for those who are married and those who are single, those who are mothers with those who are childless.

Publication counts are, to be sure, an imperfect indicator of scientists' contributions. Yet such counts are highly correlated with better measures such as peer evaluations; moreover, the extent to which scientists publish is consequential for their careers. We therefore took the extent of publication as a rough but serviceable gauge of research performance. We recognized that women scientists are, in a sense, "survivors." They have passed through the rigors of graduate training, have earned doctorate degrees, and are employed in science. We did not seek to examine the impact of cultural expectations on women's chances of running this gauntlet; that would have called for investigating the processes by which women are winnowed out of scientific careers.

We chose subjects for this study by a stratified selection process that took into account gender, professional age, field of expertise, and scientific standing. To compare the effects of marriage and motherhood on women who took degrees in different historical periods, we divided the scientists into three age groups: those who received their doctorates between 1920 and 1959, before the advent of the women's movement; those during the decades from 1960 through 1969, and from 1970 through 1979, decades when the movement was getting under way and then becoming widespread. Eighty percent of them were drawn from mathematics and the physical and biological sciences; the remainder from economics and psychology, with the same 4:1 ratio being applied in each age group.

Scientists were further divided according to their peer recognition in relation to others of roughly the same professional age. The top tier of scientists (those we designated "eminent") in the oldest group were members of the National Academy of Sciences or the American Academy of Arts and Sciences or were full professors in departments ranked in the top ten in each field by national surveys of the quality of doctoral programs. In the intermediate age group, Guggenheim fellows or tenured professors in the top ten departments were classed as eminent. Younger scientists were considered eminent if they had held Guggenheim Fellowships or were assistant or associate professors in a top ten department.

Scientists not meeting these exacting criteria were designated as rank and file. They were randomly selected from lists of the faculty and research staff at accredited four-year colleges and universities located in the same geographical regions as the eminent scientists. They were also matched to these scientists for professional age and scientific field. Although these scientists were selected randomly, the criteria we used, their small numbers, and our exclusion of some important groups of scientists (such as those working in industry) mean that this is decidedly not a true random sample of all U.S. scientists.

Our subjects were asked about their research and publication histories and for comments on graphs we prepared that showed the number of papers they had published each year along with important events in their careers and personal lives.

Both men and women scientists report having come up against the belief that marriage and motherhood cannot be meshed with demanding scientific careers. Not surprisingly, the oldest group of scientists encountered this belief most often. Before (and even shortly after) World War II, the proper priorities for women were widely held to be marriage and motherhood first and science, second, and good science was believed to be all-consuming. The notion that women could simultaneously be traditional wives, traditional mothers, and productive scientists seemed patently absurd.

These beliefs were articulated by a woman zoologist who reported that "if one had children and a working husband it was not part of the psychology to suppose that one's job was anything more than a secondary consideration." Many men and women scientists at the time shared these views. They believed most women could no longer be serious scientists once they were married. A distinguished woman biologist who is now in her seventies said her woman laboratory chief was appalled by the idea that her protégé would marry: "She threw me out of the lab the minute she heard I was going to get married because that was treason against women." For one male chemist, marriage meant that women scientists were "finished"; for a male physicist, "As soon as women got into domestic life, that was the end of it for all of them."

This climate of opinion meant that women determined to have serious research careers often did not marry. In the words of one biologist now in her seventies, "Marrying was not considered the thing to do [for women scientists]. In science, you're dedicated. You go into a shroud, you don't wear normal clothes . . . you shouldn't get married; you shouldn't have children."

Not all women scientists accepted these views, of course, and some did marry, have children, and continue to work. Yet until the end of World War II at the earliest, women scientists were few in number and fewer than half were married. Married women with children were nearly invisible in

American science. Such women were viewed, by many, as violating the prevailing family norms.

Although social attitudes concerning the roles of wives and mothers have changed significantly, even the youngest women scientists report that many people still consider marriage and motherhood to be incompatible with a scientific career. One young chemist said, "When I got pregnant I was written off as a serious scientist . . . by lots of people." When those occupying positions of power and authority act in accord with these beliefs, they severely limit the opportunities and careers available to married women.

To assess the actual impact of marriage and motherhood on women in science we needed answers to four questions: Are married women, as a group, less productive researchers than single women? Among married women, do those with children publish fewer papers than those who are childless? Is there a drop in women's published research performance after childbirth? Does the number of children women scientists have affect their research performance?

The publication and career histories of eminent scientists provided the first clues that marriage and children do not generally affect scientific productivity. On average, these eminent married women (and eminent women are just as apt to marry and have children as their rank-and-file counterparts) published slightly more over their careers—not less—than eminent single women: an average of 3.0 papers per year compared with 2.2. Among the eminent married women, those with children publish 2.9 papers annually and childless women publish 3.3. Moreover, during the three-year period preceding and following the births of first children, the annual published productivity of eminent women rises from 1.5 to 2.7 papers annually. Finally, rate of publication of these scientists is unrelated to the number of children they have.

These statistical findings are plainly counterintuitive, and yet they are consistent with earlier cross-sectional studies. How well do they correspond with the subjective reports of women in the interviews? Do they think that marrying and having children really is unrelated to the amount of research they publish, and if they do, how is one to account for this?

The publication histories of two older, eminent women scientists (see Figures 6.1a and b), one with four children and the other with three, illustrate one general pattern. Such scientists published less when they were young and had young children; there is a marked upward trend in the number of papers they published after the first decade of their careers. There are also year-to-year fluctuations—peaks and valleys—within the general upward trend. All these women do, to be sure, acknowledge that children take up a great deal of time. They are "a definite time commit-

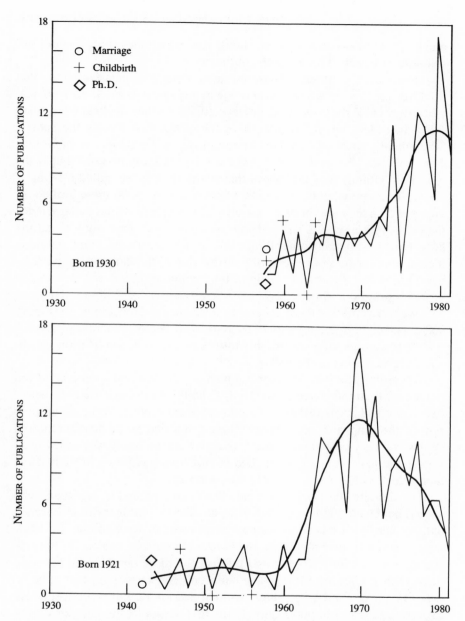

6.1a *(top)*, **61.b** *(bottom)* **Publication histories of two eminent women physical scientists who had married and had children.**

Note: Eminent women scientists, during the first three decades of their career, typically show a general upward trend in the number of papers published annually. The publication histories of two married women, one with four children *(top)* and the other with three children *(bottom)*, conform to this pattern; they show no lasting negative effects of marriage or children. The jagged line indicates the number of papers published in each year; the curved line is calculated to indicate the general trend.

ment. That means that you are doing less with other things"—but not scientific research. The research continues.

How can it continue? These eminent women emphasize, first, that thinking about science goes on at home as well as at work. It does not end when they close the doors to their laboratories. "When the kids were small . . . I had ideas when I was washing the dishes and nursing the babies. Scientifically speaking, I did my best work during the period when the kids were coming." Second, if they have a scientist husband (and that is the typical situation), they talk about their work during "so-called off-time." Third, professional obligations other than research are far more limited for younger than they are for older scientists. "I spent more time doing science then than I do now . . . the calls on my time [then] were my job and my kids. Now it's so many other things." Fourth, lower rates of publication in the early years are not necessarily attributable to the demands of mother-hood but rather are characteristic of the beginning phase of a developing research program. As one physical scientist observed, "In the first years . . . we were building those enormous instruments. Developing the theory and experiment [took up] a lot of time so there weren't that many papers." Another, commenting on the downturns in her own publication graph, "You are very busy in the valley."

According to these eminent scientists, marriage and motherhood did not reduce their published productivity. Should one believe their retrospec-tive accounts? Perhaps their perceptions are not correct. After all, inspec-tion of the graphs of these older eminent married women with children shows that they indeed were less productive scientifically when they were young and had young children. Did having young children in truth affect their rate of publication at least in the short run?

We thought the publication patterns of two groups of scientists who should not be affected by marriage or parenthood—eminent single women and eminent men whose wives took responsibility for looking after chil-dren—would further illuminate these counterintuitive results. Do the pre-sumably unencumbered scientists publish at a more rapid rate in the early years than women who had young children? The answer appears to be no (see Figures 6.2 and 6.3). Single women and married men are just as apt to show low levels of publication in the first decade of their careers. They are also as likely to show oscillations and an overall rising slope of publi-cation with time. The fact that the early publication patterns of these two groups do not differ much from those of married women with children lends credibility to the married women's observations.

Another question comes up however. The eminent older women say that marriage and childbearing did not reduce their scientific productivity.

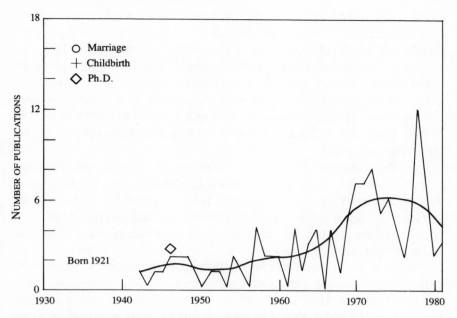

6.2 Publication history of an eminent woman biologist who never married.

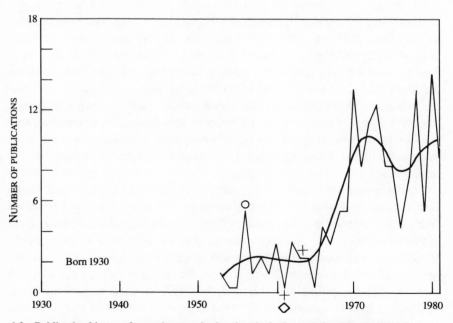

6.3 Publication history of an eminent male chemist who had married and had children.

If that is so, why do their publication rates increase as they pass beyond the child-care age—particularly in view of the additional distractions and responsibilities that they say come with professional maturity and a higher degree of recognition? Part of the answer is that the opportunities for collaborative research increase as one's career progresses. Beginning scientists do most or all of the benchwork themselves; more established ones often assume major administrative roles and oversee the work that goes on in their laboratories. Their publication records reflect this upsurge in collaborative research.

We should mention that the publication record of some eminent women scientists does not exhibit the typical rising slope with time. These women, too, say that marriage and motherhood had little bearing on their scientific productivity (see Figures 6.4a, 6.4b). Moreover, this pattern of comparatively constant productivity appears as often among married women as among single ones.

Our data seem to indicate then that older, eminent women with children generally published as much early in their careers as their unmarried counterparts. Could we have made a critical error, however, in compiling and interpreting the data? Could it be that women who have children and remain scientifically productive are "self-selected," that is, are simply more talented scientists than those who choose to remain childless?

Although strict comparisons of scientific ability cannot be made, we can compare the publication rates of older eminent scientists who did and did not have children, focusing on the years before motherhood. To this end, we matched the two groups of women roughly by their birth dates. The publication rate during the three years before women with children had their first child was compared with "equivalent years" in the life span for women without children. We found similar early histories; approximately 1.3 papers annually for women who subsequently had children and 1.6 for those who did not. In other words, older eminent women who eventually had children published inconsequentially fewer papers initially than women who never had children.

More important, might we be mistaken by concentrating on the histories of eminent women scientists instead of on those who might be more likely to experience the debilitating effects of marriage and motherhood on publication rates? The eminent women, after all, are successful scientists. If marriage and motherhood take a toll in their case, such women would presumably not have been able to achieve the recognition they actually did achieve. Do the publication histories of other women scientists, those designated as rank-and-file, testify to the negative impact of marriage and child care?

Rank-and-file men and women scientists do, of course, in general, publish less than their eminent colleagues. Within the rank and file in our sample, married women did publish slightly fewer papers than single ones (an average of 1.1 a year as compared with 1.7). But married women with children publish no fewer papers than married women without children; both groups average about one paper per year. As in the case of eminent women, their publication rates did not decline after children were born. Rank-and-file women averaged well under one paper per year (.2) in the

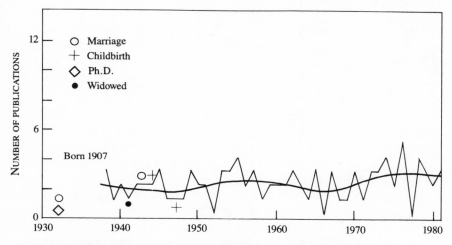

6.4a Publication history of an eminent woman biologist who had married and had children.

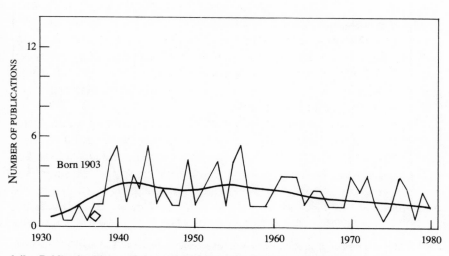

6.4b Publication history of an eminent woman biologist who never married.

three-year period prior to the birth of their first child and they averaged just under a paper per year (.8) in the three years following the birth (Figure 6.5).

Much the same impression is conveyed by these scientists' own testimony: having children did not significantly affect their records of research and publication. As one relatively unproductive behavioral scientist observed, "It didn't occur to me to stop working when I had (a child). . . . In

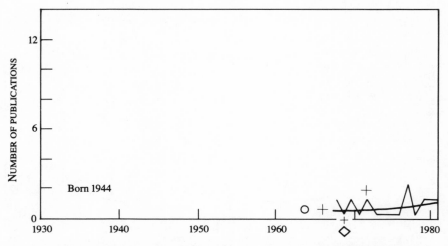

6.5   Publication history of a rank-and-file woman chemist who had married and had children.

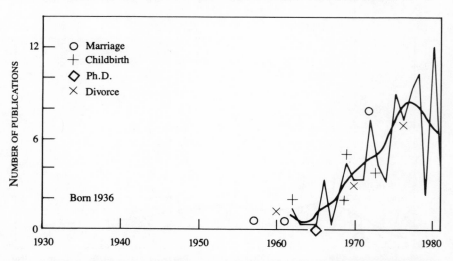

6.6   Publication history of an eminent woman behavioral scientist who had married four times and had four children.

fact, right after she was born I wrote one paper and started to work on the next one . . . so if anything . . . I seemed to work better," that is, more efficiently if under greater pressure.

Her account is consistent with accounts of other women, such as a young biochemist who asserted that her publication rate was unaffected by family obligations: "It's just fortuitous. . . . The kinetics of me as a parent and me as a researcher don't have a direct relationship. . . . One has not interfered with the other." Contrary to expectation then, such women are no more likely than older eminent ones to say that marriage and family responsibilities account for their rate of publication, and the statistical data we have in hand bear out what they say.

Would we find a similar pattern among younger women? Because marriage and childbearing usually come early in a woman's career, the records for younger scientists should show their effects on published productivity, at least in the short term.

A behavioral scientist who is now a full professor in a high-ranking department and who had just had a child suggested that motherhood was unrelated to the pace of publication. "Having a child is draining in many ways, but not in terms of having affected my work, especially when I look at how much I've . . . done this year. No, it's really movies, social life and things like that [that go]. . . . I feel chronically slow and behind and this year I'm blaming it on the baby, but I realize that has nothing to do with it." In the past, it was "too many graduate students, . . . [a] grant reviewing committee, [an] editorship, [and now a] baby. So I have always had some kind of baby to blame it on."

Perhaps the limiting case is a woman scientist who has an endowed chair in a major department. She has been married four times and divorced three times, and she has had four children by three different husbands. If marriage and motherhood should bring a scientist's career to a halt, they should have done so here. Yet her published output has risen throughout this complex history. Ironically, the largest dip in her pattern came in 1979, one of the few years in which she did not get married, have a child, or get divorced (see Figure 6.6). Asked about her publication pattern, she replied, "Suddenly you're ready to report on three different projects and therefore [the papers] roll off the presses. The ups and downs [have] nothing to do with the rest of my life."

Even so, pregnancy and its aftermath did interfere temporarily with research in the case of 3 of the 37 women in our sample who had children. A woman biologist told us "I was one of those women who said I'm so well organized, I'll just drop a child and that will be it—but I didn't realize that hormones could do such a job on a person." Asked to explain why her publication rate declined only temporarily and not by much, she said, "I was lucky because by then I had people in my lab. They were working [and] productive. . . . But I found that for a whole year my mind wasn't functioning."

These longitudinal data indicate that marriage and children are not inimical to published productivity of women in the aggregate. Although a few people would question that marriage and motherhood impose formidable responsibilities, apparently many women scientists can manage a career and family obligations simultaneously. How do they do it? And how, when conflicts arise between home and career, do women scientists deal with them?

The answer can be found, in part, in how women scientists manage their "status set," that is, the array of social positions each of them occupies (such as professor, laboratory director, wife, mother, and citizen). We focused on three interconnected aspects of status sets: size (the number of positions held simultaneously), congruence (the extent to which various obligations are consistent rather than in conflict), and the timing of the addition and deletion of status obligations.

At the extreme, several women, convinced that marriage is incompatible with scientific work, chose not to marry, in effect, limiting their status sets. Yet three-fourths of the women in our sample did marry—a proportion that appears to be typical of women scientists in general now. For the majority of them, the fundamental question about marriage was one of timing. As a young, woman economist said, it "would also be a big career disadvantage for me [to get married]. Once I have tenure and am more settled down in a university, it would be a little bit easier."

Two-thirds of the married women had children. Timing their arrival, many women say, helps to maintain a research career. A renowned physical scientist delayed having her first child for nine years after marrying in order to prove herself as a valuable professional. Many younger women said they were delaying motherhood until they received tenure. Doubting that she could have a child and maintain the level of performance needed for tenure, a young biochemist noted, "My ideal scenario is to get a tenured position, and then have a child or two."

All told, eminent and rank-and-file women had about the same number of children: an average of two, with none exceeding four. Our data show that annual rates of publication are virtually the same for women with one child and for those with two or more.

There are aspects of marriage and motherhood other than timing that can make for congruent status sets. Close to four-fifths of the married women we interviewed were married to scientists (again, a proportion typical of married women scientists generally). Such "assortative" or selective mating apparently gives these women (and men, too) a variety of benefits, including ready understanding of their professional obligations and way of life. A molecular biologist observed that her husband could scarcely be upset when she came home late because "he knew that no matter how well I had planned something, experiments do get delayed. I think that's made it a lot easier." Women scientists married to scientists publish, on average, 40 percent more than women married to men in other occupations, and this is so both for eminent and for other women scientists. This difference in publication rates may result from self-selection, congruence of values, or the flexibilty of academic schedules.

Women scientists may also achieve congruence of status obligations by compartmentalizing their lives. But compartmentalizing, they report, is not always feasible. In fact, many find it harder to keep their minds off work when at home than to keep their minds off children when at work. Moreover, every woman with children emphasized that they relied on some form of child care or household help—necessary arrangements but fragile at best. The illness of a spouse, child, or housekeeper can throw the entire scheme awry.

In view of these difficulties, is it possible that women and men scientists manage to continue research only by neglecting spouses and children? Our study was not designed to answer this question. This much we do know. The divorce rate for these women and men is unrelated to published productivity.

Married women scientists with children do pay a price to remain scientifically productive. They report having had to eliminate almost everything but work and family, particularly when their children were young. As an eminent psychologist observed, what goes first is "discretionary time. I think I can only work effectively . . . 50 hours a week. . . . If I didn't have children, I'd probably read more novels . . . or go to more movies."

Loss of discretionary time not only affects leisure pursuits but also sometimes has serious consequences for women's research and careers, even if they have no significant effects on their rate of publication. Women scientists who adhered to rigid family schedules lost the flexibility to stay late in the laboratory to work on an interesting problem; they report not feeling part of "the club," not having time for informal discussions with colleagues.

Other investigators have shown that only some 12 percent of women scientists stop work after getting their Ph.D. Surely some of them do so because of the intense conflicts that arise between science and parenthood.

One woman who had left a promising research career for a less demanding job said in a supplementary interview, "I was only in the lab the hours that the children were in school. . . . I was working with really bright people who ground out the publications at a rate I couldn't keep up with. . . . It was too frustrating." A small subset of women, then, find that science and motherhood do not mix and alter their careers to give more time to their families. For some, science and motherhood do not mix.

Our study shows, however, that for most of these women, science and motherhood do mix. Women scientists who marry and have families publish as many papers each year, on the average, as single women. Managing the simultaneous demands of research careers, marriage, and motherhood is not easy; it requires organization and an elaborate set of personal adaptations.

The results of this study should not be interpreted as meaning that marriage and children have no effect on the careers of women scientists. They do, but they generally do not take their toll on women's research performance. How then can the disparity in rate of publication between men and women scientists be explained? Why do men publish substantially more papers on average over the course of their careers as women of comparable backgrounds? This difference remains a puzzle requiring further comparative inquiry into the research careers of both men and women scientists.

WILLIAM T. BIELBY

7. Sex Differences in Careers: Is Science a Special Case?

IN NEARLY ALL industrialized countries, men's and women's patterns of labor force participation are converging (Mincer 1985). Despite this trend, there has been little change in occupational segregation and earnings disparities by sex. In the United States, the work roles of men and women were about as differentiated in 1970 as they were in 1900, and there has been only a modest decline in the level of occupational segregation by sex over the past two decades (Beller 1982, 1984). When men and women do pursue similar occupational roles, they are often segregated across organizations or by job titles within organizations (Bielby and Baron 1984, 1986). Over the past several decades, women in the United States working full time, year round have earned about two-thirds as much as men, and the discrepancy is almost as large when comparisons are made between men and women in the same occupation (O'Neill 1985; Reskin and Hartmann 1986).

Men and women bring different skills, aptitudes, interests, and work orientations to the workplace, and this accounts for some of the discrepancy in their career outcomes (Marini and Brinton 1984). Women's family responsibilities also affect the choices they make concerning work roles outside the home (Becker 1974; Polachek 1981). But these "supply-side" factors alone do not account for the persistence of sex segregation and earnings disparities. Structural factors include arrangements and practices in schools, workplaces, and other institutions that cause men and women to be treated differently even when they have the same qualifications and work orientations (Roos and Reskin 1984). Finally, there is a cognitive or social psychological basis to sex differences in the workplace; because of widely held cultural stereotypes about gender, the qualifications and performances of men and women are perceived differently. These stereotypes affect the perceptions and actions of teachers, counselors, employers, co-workers, policymakers, and even the ways in which men and women perceive themselves (Ashmore and Del Boca 1986).

After two decades of empirical research by behavioral scientists, a consensus is emerging on how supply-side, structural, and social-psychological factors combine to produce different career outcomes for men and women (for reviews, see Deaux 1985; and Reskin and Hartmann 1986). The existence of cultural stereotypes, structural barriers, or differences in "human capital" investments are no longer at issue, although the relative contributions of these three factors are still debated. While some neoclassical economists argue that supply-side factors account for most occupational segregation and nearly all of the earnings disparity between men and women (e.g. Polachek 1984), most social scientists agree that structural barriers and cultural stereotypes are at least as important. Moreover, the debate has become less ideological and more scientific as the empirical research by economists, sociologists, and psychologists grows increasingly sophisticated.

In this chapter, I examine the implications of research on gender and work for the study of sex differences in the careers of scientists. At issue is whether the distinctive cognitive, normative, and institutional structure of science (Hagstrom 1965; Merton [1942] 1973), attenuates the extent to which structural barriers and cultural stereotypes shape the careers of women scientists. Jonathan Cole (1979), among others, has presented research results in support of the hypothesis that the reward system in science is essentially meritocratic and not biased against women. At the other extreme, the case has been made that the very content of science itself has been constructed from a masculine approach to the natural world (Keller 1985).

Sex Differences in Scientists' Career Outcomes: A Familiar Pattern

Sex differences in career outcomes in science are strikingly similar to those in other institutional sectors. As in other occupations, the representation of women in science has increased markedly over the past three decades. At the same time, women remain segregated by academic field, by institution, by job, and by rank. In 1981, nearly half of the female Ph.D. recipients obtained their doctorates in the fields of psychology and education, compared to about a fourth of the men. Nearly a fourth of the men received Ph.D.s in physical sciences and engineering, compared to less than 5 percent of the women (Fox 1984). The index of dissimilarity, a widely accepted measure of the extent of segregation, shows that the level of segregation across 24 fields for new Ph.D.s actually increased slightly between 1971 and 1981. In the same period, however, the level of segregation by field at the baccalaureate level decreased substantially.[1]

Women faculty in the United States are also segregated by institution. They are more likely than men to find employment in two- and four-year colleges and less likely to be employed in universities (Fox 1984). Female Ph.D.s are also less likely than men to find jobs outside of academia (National Research Council 1979). Female Ph.D.s are also about twice as likely as men to be unemployed (National Science Foundation 1986, but unemployment rates are low for both).

Even when men and women scientists work in the same institutions, they are segregated by rank. In 1981, women constituted nearly half of non-tenure-track faculty but only 25 percent of full-time ladder faculty (National Center for Education Statistics, 1981). Among cohorts of Ph.D.s receiving degrees in the same year, men are more likely than women to be represented in the senior ranks of university faculty (National Research Council 1981; Bruer 1983).

Finally, salary discrepancies between men and women scientists are similar to earnings differences by sex in other institutional sectors. In 1984, women engineers and scientists earned about 71 percent as much as their male counterparts (National Science Foundation 1986), roughly the same as within-occupation wage differentials estimated from national labor force data (Treiman and Hartmann 1981). Nearly all of the earnings differential can be attributed to segregation by field, institution, and rank (Fox 1981; Zuckerman and Cole 1985).

In short, sex segregation is about as pervasive in science as it is outside of science, with predictable consequences for differences in men's and women's career outcomes. On the one hand, this might suggest that the structural barriers and cultural stereotypes that create inequities in other institutional realms also harm the careers of women scientists. On the other hand, science is a highly stratified institution, and even small differences between men and women in skills, training, or work orientation can lead to large differences in career trajectories (Cole 1979). The segregation described above may largely reflect individual choices men and women have made to specialize in different fields and in different kinds of careers. The debate over sex discrimination in science in large part centers on the extent to which differences in career outcomes of men and women are the product of structural barriers on the one hand versus achievement-related "supply-side" factors on the other, that is, factors or attributes of men and women relevant to job performance but which result from decisions they made long before entering the job market.

Fair Science? The Case for Science as a Special Case

More so than in other institutional sectors, scientists share a belief in an institutionally sanctioned goal and a normative principle governing actions in pursuit of that goal. The goal is the production of certified knowledge, and the normative principle is universalism (Hagstrom 1965; Merton [1942] 1973). The norm of universalism prescribes that "rewards and resources are to be allocated on criteria of scientific merit and that judgements of scientific merit are to be made only in terms of contributions to knowledge" (Zuckerman and Cole 1985: 1; Merton [1942] 1973). Sex, race, and other ascriptive criteria for allocating resources and rewards are functionally irrelevant and illegitimate within the normative structure of science.

Moreover, there is widespread agreement among scientists that the extent and quality of research publication is the most important measure of an individual's contribution to the production of knowledge. With such a visible and objective measure of productivity, deviation from the norm of universalism should be easily detected and negatively sanctioned. As in other institutional sectors, science as actually practiced does depart somewhat from its normative principles. But the meritocratic model is so widely accepted that it is the criterion by which scientists evaluate their own actions.

Most scientists who study discrimination in science also subscribe to the meritocratic model as an ideal type. Discrimination is measured as a departure from a base-line model that represents the distribution of rewards and resources as a function of the productivity of individual scientists (Reskin 1978; Reskin and Hargens 1978; Cole 1979). Functionally irrelevant criteria are presumed to operate when the meritocratic model fails to account fully for sex differences in the career outcomes of scientists. However, structural, cultural, and other sources of discrimination are rarely studied directly, especially in quantitative studies of scientific careers. Instead, they are typically inferred from the residual net difference between men and women after controlling achievement-related variables in the meritocratic base-line model. As a result, we know much more about the extent to which "supply-side" factors do and do not account for reward and productivity differences between men and women scientists than we do about the impact of specific structural barriers and cultural stereotypes.

There is, of course, another view of gender issues in science, one that is less prone to treating science as a special case. Historical and organizational case studies are more likely to take a societal view of structural and cultural forces that shape women's careers and then examine how those same forces operate in the scientific community (e.g. Fox 1981; Rossiter 1982; Szafran 1983). I return to that line of research below, after discussing the research that builds more directly on the view as science as the meritocratic special case.

"Supply-Side" Factors: Human Capital Investments and Differential Productivity

Neoclassical economists argue that differences in human capital investments account for much of the disparity between men and women in labor market outcomes. Because of intermittent labor force participation, women accumulate less experience and on-the-job training than men, which in turn leads to lower earnings (Mincer and Polachek 1974). Occupational segregation comes about because women and men choose to specialize in different kinds of jobs. Unlike men, women tend to choose jobs that are easy to reconcile with family demands and do not impose a wage penalty on workers who drop out and re-enter the labor force (Polachek 1981). Moreover, compared to men with similar labor market experience, women are argued to be less productive because they allocate more effort to household tasks and less to job tasks (Becker 1985). Labor force studies show that differences in human capital investment account for at least some of the differences in men's and women's work experiences, although the extent to which the economists' hypotheses are supported by empirical data is the subject of vigorous debate (England 1984; Bielby and Bielby 1988). Nevertheless, the human capital model offers a comprehensive explanation for sex segregation and differences in the career trajectories of men and women—an explanation that might be especially applicable to careers in science.

PRODUCTIVITY OF SCIENTISTS AND FAMILY DEMANDS

In fact, research on men and women scientists provides mixed support for the human capital model. Productivity, as measured by number of publications, is lower for women than for men (Cole and Zuckerman

1984). However, sex differences in productivity fail to account completely for differences in the career outcomes of men and women scientists, especially with respect to academic rank (Cole 1979; Zuckerman and Cole 1985).

Moreover, the impact of family demands implied by the human capital model is not supported by research on men and women scientists. Marriage and family status appear to have no impact on the research productivity of female scientists (Reskin 1978; Cole and Zuckerman 1984, 1987). Married female scientists with children appear as committed to their careers as their male counterparts (Reagan 1975; Rosenfeld 1984). On the other hand, marriage and the problems of reconciling the constraints of dual careers do appear to limit the job mobility of women scientists (Marwell, Rosenfeld and Spilerman 1979).

As noted above, male and female scientists differ in their fields of specialization, but family demands probably do not explain segregation by field. Although little (if any) research has been done on the topic, it is probably no more difficult to reconcile work and family demands in the physical and mathematical sciences than it is in the social sciences. Since labor force participation of women scientists is much less intermittent than it is for women in other sectors, differences across specialties in rates of skill atrophy and wage penalties for dropping out are unlikely to explain segregation by field.

EARLY SOCIALIZATION AND SPECIALIZATION IN QUANTITATIVE FIELDS

There is one obvious factor that is associated with sex differences in fields of specialization. Women are much less likely than men to earn advanced degrees and then to work in the more quantitative disciplines. Although not related to family roles, these are "supply-side" factors insofar as they reflect different kinds of choices made by men and women long before they begin their research careers. Sex differences in the extent of interest in science emerge prior to high school, and those interests have a strong impact on the propensity to obtain quantitative training in high school and college. Thus, perhaps because of early childhood socialization, many young women self-select out of the pool of individuals capable of pursuing research careers in highly quantitative fields (Berryman 1983; Matyas 1985).

Even if science were based on universalistic principles, the career outcomes of men and women could well be unequal if men and women differed in their human capital investments and other "supply-side" factors they bring to their jobs. Differences between men and women scientists in their household and family responsibilities appear not to affect the kind of investments they make in a scientific career. To the extent that family demands limit the mobility of women scientists, they may be less able than men to take advantage of opportunities that would enhance their careers. Because of early socialization, girls express less interest in science than boys, and as a result they are less likely to get the kind of training that will prepare them for careers in science, especially in highly quantitative specializations. Differential childhood socialization is a "supply-side" factor insofar as it can generate sex differences in careers even when the allocation of resources and rewards in science itself are based solely on meritocratic criteria. On the other hand, sex differences in interest in science among children and adolescents may be shaped by structural barriers and cultural stereotypes similar to those discussed below (Kahle 1985).

Unfair Science? Structural Barriers and Organizational Dynamics
SCIENCE IN AN ORGANIZATIONAL CONTEXT

Science, like other forms of work, is done in organizations. The careers of scientists are influenced by both the rules and procedures of the organizations in which they are employed and the decisions made about how to formulate and apply those rules and procedures. In addition, decision making in organizations is an inherently political process, influenced by the distribution of power among groups with differing and often conflicting interests (Pfeffer 1981). It is also shaped by the organization's history and by constraints imposed by its environment (Pfeffer and Salancik 1978). In short, the course of scientists' careers may be influenced as much or more so by the dynamics of the organization in which they work than by the "normative principles" or "goals" of science as an institution.

The point is readily acknowledged in historical studies of the careers of men and women scientists. Historians have documented the impact of policies that kept women out of graduate programs of institutions like Harvard and Princeton, excluded them from faculty positions in coeducational institutions, and channeled them into subordinate staff roles in research laboratories (Rossiter 1982). Some quantitative studies note the impact of past structural barriers on the careers of women scientists who began their careers early in the century (e.g. Cole 1979). However, blatantly exclusionary policies and practices have diminished; today's barriers are much more subtle. Unfortunately, studies of contemporary structural barriers and organizational dynamics are relatively rare. Nevertheless, the limited research that has been done suggests that structural factors in science affect women's careers in much the same way as they do in other realms.

Decisions on personnel matters in university settings are largely delegated to academic departments. This is perhaps the most distinctive feature of the organizational decision making which governs the careers of scientists (Szafran 1984). Of course, universities differ in the amount of autonomy they cede to departments, and decisions on the allocation of positions and resources across departments are usually made in centralized administrative units, often in consultation with a faculty legislative body.

As in any organization, individuals form informal alliances or coalitions based on common interests. Trying to influence the outcome of decisions is just one way that coalitions attempt to pursue their interests. They also attempt to shape the goals of the institutions and the agendas that determine which decisions are to be made. So far, research work on the politics of organizational decision making has been more theoretical than empirical (Cyert and March 1963; Zald 1970; Pfeffer 1981), but anyone who has ever participated in an academic personnel case is well aware of how politics are played out in a university setting. Although debates are often explicitly couched in terms of universalistic principles and the institutional goals of science, the participants typically include many individuals whose own contributions to the production of certified knowledge are marginal at best. The underlying issues can range from conflicts over methodological and substantive preferences, teaching versus research, and the interests of younger versus more established scholars, or social and personality issues even more remotely connected to the cognitive content of the scientific enterprise (for a vivid illustration, see Schumer's [1981] description of a controversial tenure decision at Harvard's sociology department). Although the normative principles and goals of science are certainly relevant to personnel decisions, more proximate political factors can be just as decisive in shaping the careers of individual scientists.

Universities, like other organizations, face external pressures as they attempt to maintain access to the material and symbolic resources necessary for their survival (Pfeffer and Salancik 1978). Public universities depend on state support for financial resources; their fortunes are linked to general economic conditions and to the degree of public support for higher education. Increasingly, major research institutions rely on both government and private funding. It is particularly important that a university's external constituencies accept the legitimacy of the organization's goals and methods. A university's actions are likely to be especially subject to outside scrutiny—by public agencies, the mass media, alumni, professional groups, etc.—during periods when that legitimacy is in question (Meyer and Rowan 1977; DiMaggio and Powell 1983).

External pressures can produce inequities in the career prospects of men and women scientists. Direct pressure to improve the career prospects of women can be applied by government agencies and political constituencies, as can pressure to dismantle programs that are perceived to offer "preferential treatment" to disadvantaged groups. Policies aimed at supporting higher education for military veterans can indirectly place women at a disadvantage, as can financial aid programs that require full-time study. This seems subtle; but the impact of external constraints can be even more indirect. For example, volatile economic conditions may lead universities to increase their use of part-time, temporary positions. Hiring women disproportionately for such positions would reinforce sex segregation between tenure-track and off-track positions. In short, a range of factors external to the organization can shape the way rules and procedures affecting scientists' careers are formulated and applied.

BEYOND THE MERITOCRATIC BASE-LINE MODEL

In sum, examining scientific careers from an organizational perspective provides an alternative to the conventional approach to studying gender inequities in science. The conventional view emphasizes the normative principles and goals of science as an institution, and it abstracts scientists' careers from the organizational settings in which they are accomplished. Research focused on individual scientists examines discrimination in terms of departures from a base-line meritocratic model. In contrast, organizational approaches directly examine the structures and processes that can create barriers to equity in the careers of men and women scientists. Starting with the premise that decisions are the outcome of a political process and are subject to external constraints, an organizational approach suggests that the link between decision making regarding scientists' careers and the normative principles of science is almost certainly not nearly as strong as, say, those principles and editorial decisions regarding the publication of scientific research. As a result, there is considerable room for the emergence and persistence of the kinds of the structural and social psychological barriers suggested below.

RESEARCH ON STRUCTURAL BARRIERS AND SEGREGATION

The few existing organizational studies of scientists' careers consistently show greater sex stratification in elite institutions. For example, Cole (1979) found that disparities by sex in academic rank were greater at more prestigious institutions, even after controlling for differential productivity. Szafran's (1984) analyses of the 1969 Carnegie Commission data revealed similar patterns. In addition, he found a negative relationship between financial resources (measured as expenditures per student) and propensity to recruit female faculty. Szafran suggests that to protect their reputations, highly regarded institutions in the 1960s may have been using slack resources to attract and support eminent male scientists. The association between institutional resources and segregation in universities is consistent with findings from Bielby and Baron's (1984) study of sex segregation in other institutional realms in the late 1960s and early 1970s. In a detailed examination of organizational arrangements and job segregation by sex in a cross-section of California firms, they found that large, bureaucratic organizations were among the most segregated. To the extent that segregation introduces inefficiencies in the allocation of human resources, those were the firms that could afford a segregated workforce. In addition, highly structured firms have more dimensions along which to potentially segregate women from men—by department, division, geographic location, job title, and so on—and almost every large, bureaucratic firm in their study was virtually completely segregated by sex along one or more of those dimensions.

Research by Fox (1981, 1984) shows how sex segregation affected the careers of men and women scientists at one large, prestigious public university in the early 1970s. Her studies document the segregation of men and women by rank, administrative location, and field of specialization within the university and how that segregation sustained a dual reward structure. She found that for both men and women, achievement-related variables (rank, degrees, experience) were the most important determinants of salary. However, the *payoff* for specific kinds of achievements was consistently lower for women than for men. Moreover, her analyses show that both men and women earned less when they were employed in an administrative location dominated by the opposite sex. Fox argues that segregation by sex makes the dual reward system less visible. Given patterns of segregation at institutions similar to the one she studied, the most salient salary differentials perceived by any individual are likely to be among the same-sex colleagues employed in that individual's administrative location. Since those salary differences are largely related to achievements, they are likely to be perceived as equitable. Much less salient (and visible) to an individual would be the salary differences among opposite-sex employees in other administrative locations. Segregation therefore allows an achievement ideology to persist alongside a compensation system that rewards both men and women for their accomplishments but offers them different rates of return. Of course, we cannot tell from Fox's studies whether the same dual reward system operated at other universities in the early 1970s. However, patterns of segregation were probably similar at other large, prestigious universities, and the mechanisms she describes suggest a way that the inequities detected in aggregate studies such as Cole's (1979) and Szafran's (1984) may have come about.

The studies of job segregation and structural barriers described above are limited in two ways. First, they are all based on data from the early 1960s and late 1970s, before the era of vigorous enforcement of equal employment opportunity laws and regulations. We do not know the extent to which the interventions of the 1970s changed patterns of segregation. Second, those kinds of organizational studies do not examine the decision-making processes that lead to the implementation of the organizational arrangements that create structural barriers.

ORGANIZATIONAL RESPONSE TO INTERNAL AND EXTERNAL PRESSURE

Evidence from Szafran's (1984) study suggests that a sizable female constituency within a university can affect its employment practices. In his data from 1969, universities with a greater percentage of women in administrative roles had a better record on recruiting women to faculty positions. Similarly, universities exhibited greater equity in assigning men and women to off-ladder positions when women were better represented in both administrative and faculty positions. These results parallel findings from Bielby and Baron's (1984) study of segregation in a general sample of organizations in California, already noted. They found lower levels of job segregation by sex in firms with a sizable female workforce. In addition, the few organizations that exhibited significant reduction in levels of segregation over a four-to-seven—year period were ones that initially had a high percentage of female employees.

Evidence that universities have changed practices in response to pressure from federal and state agencies is mixed. Szafran (1984) found no relationship between level of government funding and measures of equity in faculty employment. However, his data are from 1969, and it was not until the mid-1970s that universities were subject to regulations on equal employment for government contractors. Szafran did find that public universities had a slightly better record in terms of parity between men and women in salary and rank, and this may be a function of the closer scrutiny of personnel practices in public institutions by outside agencies and political constituencies.

The formal procedures used in recruitment and promotion have certainly changed over the years since universities have been subject to Federal regulation of affirmative action. Most universities have an affirmative action office that monitors all personnel actions. Faculty openings at all levels are widely advertised through professional channels. Records are kept on applicant pools, candidates interviewed, and reasons why one candidate was chosen over another. Similarly, criteria for promotion and the procedures used to decide promotion cases have been formalized. While such efforts at reform by no means guarantee that personnel decisions are now made on universalistic criteria alone, there is considerable evidence that progress has been made since the early 1970s. Zuckerman and Cole (1985) cite recent research showing that women no longer face disadvantages in admission to graduate programs and are now no more likely than men to obtain first jobs in off-ladder positions. Overall, recent cohorts of female Ph.D.s appear to begin their careers in positions comparable in quality to those obtained by men. On the other hand, the same studies cited by Zuckerman and Cole show that, even among recent cohorts, women are somewhat less likely than men to be promoted to tenure and move much more slowly through the more senior faculty ranks. In short, it appears easier to reduce and essentially eliminate biases in procedures used to recruit and screen candidates for new faculty positions than it is to remove the variety of structural and social psychological barriers women confront once their careers within the university are under way.

It is perhaps too early to tell whether the weakening in enforcement of equal employment opportunity laws and regulations in recent years is eroding the gains of the early 1970s. However, recent statistics on female enrollments in scientific fields show that the trends of the past decade are not irreversible. A recent report of the Commission on Professionals in Science and Technology (Vetter and Babco 1986) shows that in 1984 the proportion of women declined in the cohort of first-year engineering students, among those obtaining undergraduate degrees in geology and earth sciences, and among those receiving advanced degrees in mathematics and biological sciences. These trends and the persistent sex segregation by rank among university faculty suggest that some of the barriers faced by women scientists are either so subtle or so deeply entrenched that they are difficult to dismantle, even when there is an organizational commitment to do so. Recent research on the social psychology of prejudice, summarized below, indicates that the biases that arise in day-to-day interpersonal interaction may be even more difficult to eradicate than the structural barriers built into the rules and procedures of organizations.

Unfair Science? Stereotypes and
Social-Psychological Barriers

In 1936, physicist Robert A. Millikan advised the president of Duke University against hiring a woman for a faculty position in the university's physics department. In his letter to President Few, Millikan asserted; "I should, therefore, expect to go farther in influence and get more for my expenditure if in introducing young blood in a department of physics I picked one or two of the outstanding younger men, rather than if I filled one of my openings with a woman" (Rossiter 1982: 192). He based his assertion on several assumptions about the competency of women physicists and the potential negative impact women faculty might have on a department's prestige. In regard to teaching, he argued: "In a coeducational institution where there are many women students it is undoubtedly also desirable to have for pedagogical purposes women instructors, but only in very exceptional cases would I think that the advance of graduate work would be as well promoted by a woman as by a man" (193). No doubt some male scientists still hold attitudes like these about the capabilities of women, but they are almost certainly in the minority today. Those who do, are at least more cautious about articulating them. Public attitudes about women's careers have become much more liberal over the past 50 years (Thornton, Alwin and Canburn 1983), and no doubt scientists' attitudes have moved in the same direction.

Few male scientists today would openly admit to being influenced by prejudice in their interactions with female scientists. On the other hand, narrative accounts by female scientists consistently report situations in which their careers have been affected by the prejudicial actions and attitudes of male teachers and colleagues (Associaton of American Colleges 1982; Clark and Corcoran 1986). Recent social-psychological research seeks to explain how dominant and subordinate groups can have such different perceptions of their social interactions. Both laboratory and field studies suggest that stereotypes and prejudices are more subtle and less visible today than they were in Millikan's era, that they influence the behavior of both men and women, and that most of the time we are unaware of how they influence our behavior.

Social psychologists view stereotypes as an essential feature of human perception, a cognitive shorthand invoked to achieve economy in processing, storing, and retrieving information (Ashmore and Del Boca 1986). Gender stereotypes are constellations of traits and behaviors attributed to the categories "male" and "female." Research on the content of those stereotypes consistently finds two clusters: "instrumental" traits ascribed to men and "expressive" traits ascribed to women (Broverman et al. 1972; Spence, Helmreich and Stapp 1975). Thus, traits and behaviors commonly associated with the scientist's role conform to cultural stereotypes of "maleness."

Stereotypes influence the attributions individuals make about the causes of their own behavior and the behavior of others. Laboratory studies show that men typically attribute success in performing "masculine" tasks to ability, whereas women tend to attribute their own success on the same tasks to luck or effort. Similarly, in evaluating others, performance consistent with stereotypical expectations tends to be attributed to stable, internal causes, whereas performance inconsistent with expectations is attributed to transient, contextual factors (Deaux 1985). Thus, a successful woman scientist might tend to minimize the extent to which her abilities contributed to an extraordinary research accomplishment. At the same time, her peers might attribute her performance to situational factors instead of talent (cf. Feldman 1974). As a result, the "Matthew effect," the tendency for advantages to accumulate for visible and successful scientists (Merton 1968), might be less strong for women than for men, especially in the most highly sex-typed fields.

Stereotypes affect actual behavior, not just attributions about what causes behavior. In a study of racial stereotypes, Ward, Zanna and Cooper (1974) discovered that white subjects exhibited more "low immediacy" behaviors—less eye contact, shorter verbal interactions, less expressive nonverbal behavior—when interviewing blacks than when interviewing whites. In another part of the same study, interviewers were trained to exhibit either low or high immediacy behavior. Interviewees randomly assigned to the low immediacy interviewers were rated to have performed less effectively by naive raters. Similar processes may explain the "chilly" climate in classrooms and laboratories that is often invisible to men but widely perceived by women (Association of American Colleges 1982).

Research on "expectancy confirmation sequences" shows how stereotypes can lead to self-fulfilling prophecies (Merton [1948] 1968): individuals subject to stereotypic expectations may eventually act in ways that play out the stereotype (Darley and Fazio 1980). In a laboratory study, Skrypnek and Snyder (1982) asked pairs of subjects, one male and one female, to decide how to allocate a list of tasks, some of which were male-typed and others female-typed. Subjects were blind to the actual sex of their partner. A male subject led to believe his partner was female typically allocated fewer feminine tasks to himself than when he believed his partner to be male. The female subject was also more likely to choose feminine tasks if her partner assumed she was female, even though she was not given information about her partner's expectations. If the same processes operate in scientific settings, then women scientists might adapt their behaviors to the stereotyped expectations of colleagues without even being aware of the ways in which they are being influenced.

In scientific work contexts, women are likely to find themselves interacting in social groups that are predominantly male, especially among networks of the scientific elite. Unfortunately, gender stereotypes are especially difficult to undermine in groups with highly skewed sex ratios. Individuals tend to retain information that confirms stereotypes and to ignore information that fails to fit their expectations. One or two exceptional cases can either be ignored or categorized into subtypes (Kanter 1977; Hamilton 1981; Deaux 1985). As a result, the social-psychological barriers faced by women scientists are likely to be greatest in the top ranks of scientific fields.

Pettigrew and Martin (1987) review the social-psychological processes described above as they pertain to barriers facing minority groups in organizations and conclude that they constitute a new kind of "modern racial prejudice." On the one hand, majority group members eschew both gross global stereotypes and blatant discrimination. On the other hand, members of the majority group are unaware of the subtle ways stereotypes influence both perceptions and behavior in newly integrated work settings. Women scientists may be similarly affected by what might be called "modern social prejudice." Few male scientists, for example, believe that women are biologically less capable of contributing to science or favor policies that would exclude women from university faculties or scientific laboratories. At the same time, however, scientists, like the subjects in the social-psychological experiments described above, are largely unaware of the more subtle stereotypes that affect their perceptions, expectations, and behaviors when interacting with members of the opposite sex.

Social-psychological research therefore suggests that the disadvantages facing female scientists are not exclusively structural. They are also embedded in the subtle nuances of everyday interaction between the male majority and female minority. Structural interventions that change the rules and procedures used by universities and scientific laboratories to hire and promote scientists have little effect on subtle and largely invisible social-psychological processes. Moreover, scientists subscribe to an ideology of achievement which denies that bias and prejudice have any role in the pursuit of certified knowledge. Many may appeal to that ideology in the face of any charge that their actions as scientists are implicitly sex-biased. As a result, identifying and eliminating social-psychological barriers are likely to be even more difficult than dismantling the exclusionary practices that have historically kept women from fully participating in science.

Conclusions

A decade of "supply-side" research on stratification in science has provided some important insights into the links between gender, careers, and the productivity of scientists. That research tradition has also produced important findings that it cannot explain. We still do not know why it is that women move more slowly through the highest ranks in the profession than do equally productive men, nor do we understand why it is that women tend to publish less than men. An adequate understanding of sex differences in scientists' careers requires more than simply documenting the extent to which there is a net association between sex and career outcomes after controlling for productivity-related characteristics of individual scientists. Regardless of "who owns the null" when it comes to assessing the extent to which functionally irrelevant criteria influence the status attainments of scientists (White 1982), the "supply-side" research designs that still dominate research on sex, productivity, and career accomplishments in science are simply incapable of determining why science is more fair to one sex than to the other. Models of scientists' careers that are abstracted from the actual settings in which the work is done cannot possibly explain how the structural and interactional features of those settings affect career prospects.

Over the past decade, the "supply-side" approach to the study of occupational status attainment has been supplanted by a research agenda that considers how the organization of work and its structural context shape careers (Baron and Bielby 1980). But the structuralist approach has yet to gain any widespread acceptance among those who study the careers of scientists. Perhaps we are comfortable with the "supply-side" approach because it allows us to measure discrimination as "deviance" from an otherwise meritocratic system based on universalistic principles. As scientists ourselves we subscribe to the ideology of achievement and the notion of science as a special case.

I suspect that there is much to be learned about sex differences in the careers of scientists by relinquishing the notion of science as a "special case." We certainly have nothing to lose by attempting to apply insights gained from recent social science research into the ways structural arrangements and social-psychological dynamics shape the careers of men and women. Existing studies that do examine the impact of structural arrangements on the careers of men and women scientists are limited to a few case studies of university faculty that are not easily generalizable and large-scale comparative studies using gross measures of organizational characteristics. Moreover, nearly all are based on data collected before universities became subject to equal employment opportunity laws and regulations. Over the past fifteen years there has been considerable variation in the labor market for new Ph.D.s, in the vigor with which equal employment opportunity policies have been enforced, and in organizational policies regarding the hiring and advancement of scientists, yet we have hardly any systematic knowledge of the ways in which these factors have affected the career prospects of men and women scientists.

Nor is there any existing systematic research on social-psychological barriers faced by women scientists. For example, while research studies continue to document the existence of social-psychological barriers with carefully designed randomized experiments using *student* subjects, I am unaware of any study which attempts to isolate the same processes using college *faculty* as subjects.

At the 1985 Macy Foundation symposium on gender discrimination in science, it was suggested that the nature of the scientist's work is undergoing a transformation (Bruer 1985). Increasingly, scientists are involved in the tasks of securing resources, managing teams of people, engaging in public relations, and other aspects of the management of science. If so, this could have important implications for both the structural and social-psychological barriers facing female scientists. More women scientists may confront the kind of structural barriers that have been documented in research on business settings, in addition to the arrangements that have traditionally contributed to segregation in scientific settings. In addition to stereotypes about women as scientists, prospects for career advancement may also be affected by stereotypes about women as managers (Donnel and Hall 1980; Sanders and Schmidt 1980). At present, we can only speculate on the possibility of such trends. Until sociologists of science undertake a systematic research agenda to examine directly the structural and social-psychological dynamics of the scientist's workplace, our understanding of the origins of sex differences in the careers of scientists will be limited to the realm of speculation.

MARY FRANK FOX

8. Gender, Environmental Milieu, and Productivity in Science

LIKE OTHER WORK in contemporary society, science and scholarship are largely organizational work. The notion of scientists in spontaneous intellectual exchange outside of administrative or organizational frameworks is simply illusory (Blau 1973). Scientists work in ordered organizational environments that limit, constrain, and facilitate both the development of their ideas and their actions. Further, they work in a larger community of science that can enable or disable their performance.

The community of science includes those in both academic and non-academic settings, with and without doctorates. Here, however, I will concentrate on *doctoral-level* scientists (that is, Ph.D.s, M.D.s, and other holders of doctorates) in academic institutions. In 1983, 60 percent of all doctoral-level scientists were employed in academia (National Science Board 1985). These academic scientists are a particularly appropriate group for the chapter's focus upon published productivity.

In explaining publication productivity, personal traits and dispositions such as motivation, stamina, and perceptual and intellectual style play a part (Fox 1983). But these characteristics do not exist in a vacuum; and such factors alone do not account for published productivity. Studies show no direct relationship between productivity and measured creative ability [1] or intelligence (Andrews 1976; Cole and Cole 1973), for example, and suggest that social and organizational variables interact with and affect the manifestation of individual attributes.

Thus it is difficult to separate the performance of the individual scientist

This chapter was written with support of a grant from the National Science Foundation, SES-850851. For their observations and detailed comments on an earlier version of this manuscript, I thank Jonathan Cole and Harriet Zuckerman.

from his or her social and organizational context. Work is done within organizational policies and procedures; it relies upon the cooperation of others; it requires human and material resources. Further, the scope and complexity of research and the use of advanced technology have heightened reliance on facilities, funds, apparatus, and teamwork. Performance is tied to the environment of work—the signals, priorities, resources, and reward schemes that provide the ways and means of research.

In the association of institutional environment and productivity, it is true, of course, that selectivity factors can operate so that more productive scientists are recruited into more resourceful settings. However, longitudinal studies, which have monitored the publication histories of scientists between locations and over time, indicate a stronger causal effect of location upon productivity than vice versa.

Among these studies, Long (1978) reports that while the effect of publication upon prestige of location is weak, the effect of location upon productivity is strong. For academics moving into a first position, publication is not immediately affected by current location; rather, it is affected by early, predoctoral location. However, after the third year on the job, productivity is more strongly related to prestige of department than to previous predoctoral publication. Specifically, those in prestigious departments increase their publication and those in less prestigious settings publish less.

In a subsequent study, Long and McGinnis (1981) extend these analyses beyond the prestige of academic department to the effects of larger organizational contexts—the research university, nonresearch university or four-year college, and nonacademic or industrial sectors. They report that the chance of obtaining employment in a context is unrelated initially to publication level. Once in the job, however, publication comes to conform to the context—but only after three years in the location. The fact that it takes some time for the new location to take effect suggests that productivity levels are not simply a result of changes in the individuals' goals or global barriers to publication in some settings that affect everyone working in them.

In accounting for gender differences in publication, some point then to the disparate locations of men and women: men are more likely to be located in higher ranks and major universities with favorable teaching loads and resources for research, and women in lower ranks and regional schools or four-year colleges with heavier teaching loads and limited resources to support research productivity. Yet even after controlling for institutional affiliation, gender differences in publication exist (Cole 1979).

This is not, in consequence, to suggest that women's depressed status in science is a function of personal deficits or that rewards accrue equitably to performance. Frequently, they do not. The correlation between publication and salary levels, for example, is much higher for men and women (Tuckman 1976). And further, even after controlling for quantity and quality of publication, women lag in rank (Cole 1979). Still—given that (1) women publish about half as many articles as men[2] (Cole 1979; Cole and Zuckerman 1984; Helmreich et al. 1980), and (2) that performance is a valued basis of reward in science, we cannot understand gender inequalities until we understand the productivity differences.

In assessing those gender differences, we need to look at social and organizational processes. First, we need to consider how factors such as collegial interaction, work climate, and collaborative opportunities affect the productivity of women and men. Second, and this is more subtle, the same institutional setting (major research university, minor university, or liberal arts college) may, in fact, offer different constraints and opportunities for one gender compared to the other (Fox 1985).

It is wrong, however, to presume that an environment has a uniform effect: that is presents the same conditions for all social categories, gender or other groups. I argue that environment does not operate uniformly, neutrally, or androgynously. Thus, although research opportunities for both men and women scientists are often more limited in liberal arts colleges than universities, the women may have heavier teaching loads, fewer collaborative opportunities, and lower claims on administrative favors than the men. Such matters of organizational membership and participation are central to science because research is a *social* process: a system of communication, interaction, and exchange. The data are not yet available to verify more conclusively the ways in which environment operates in explaining productivity among men and women.[3] In this chapter, I draw upon evidence to date and discuss both causes and implications of patterns of gender, environment, and productivity in science.

To understand the climate in academic institutions, the major employer of scientists, one must recognize conditions in the institutions' external environment. The external environment fuels a struggle over scarce resources in academia. Limited resources result from economic stringency, the outcome of a leveling of external federal support for science, a reduction in support from private foundations as the economy slowed and portfolios shrunk during the late 1970s and early 1980s, disenchantment of legislators with higher education following the protests of the 1960s and early 1970s, and, of course, the demographics of fewer college-age students in the population (Drew 1985:16).

Filled with caution and foreboding, colleges and universities have cut not only what appear to be frills but also basic programs, salaries and benefits, teacher-to-student ratios, secretarial services, laboratory and grading assistance, equipment, and the maintenance of plants and facilities (Bowen and Schuster 1985). The real earnings of faculty (adjusted for inflation) fell between 1972 and 1985. And with administrators seeking to keep costs down and avoid long-term commitments, the ratio of full- to part-time faculty declined from 3.5 to 1 in 1970–71 to 2.1 to 1 in 1982–83 (Bowen and Schuster 1985:5)—a dramatic decrease in a single decade.

From interviews with over 500 academics[4] on 38 campuses between 1983 and 1984, Bowen and Schuster (1985) report a mood that is "glum" and a faculty that is "apprehensive and discontent." Faculty expressed strong misgivings about their work environment with complaints of "too little office and laboratory space, woefully obsolete equipment, inadequate supplies, and far too little clerical support, . . . and travel funds preposterously low" (p. 156). Likewise, a survey of department chairs by the National Science Foundation reported that 46 percent of the chairs rated their departments' equipment inadequate for the pursuit of major research interests (National Science Board 1985).

Scientists have heavy workloads and experience strain in their composite roles as teachers, researchers, and administrators. As they and their units and departments deal with pressures of high institutional costs and limited aid and revenue, their focus is frequently on "utilitarian orientations" (Etzioni 1961); they aim simply "to get theirs" (Austin and Gamson 1983:67). In the jockeying for resources and rewards, pools of "chips" for bargaining power are important. The position, rank, and status of the players are frequently determined by capricious, informal processes. All of this operates to the disdvantage of women.

At center of the problem is a particular normative standard applied to scholarly and scientific work. In this work, standards are both "absolute" and "subjective." Performance is measured against a standard of absolute excellence, which, in turn, is a subjective assessment. Thus, the evaluative criteria are vague; the process of appraisal is highly inferential; and the decisions for allocation of resources and reward are largely judgmental (Fox 1989). In such a context, gender-stereotyped and biased assessments abound. Studies indicate that the more loosely defined and subjective the criteria, the more likely it is that white males will be perceived as the superior candidates and that gender bias will operate (see Deaux and Emswiller 1974; Nieva and Gutek 1980; Pheterson, Kiesler and Goldberg, 1971; Rosen and Jerdee 1974).

In a study of psychologists, Fidell (1975) sent professional summaries of 10 Ph.D.s to 147 psychology chairpersons across the country. The summaries contained different combinations of publication records, teaching performance, departmental committee service, and comments on sociability and conscientiousness. For each questionnaire sent, female names were randomly assigned to four summaries; the rest were assigned male names. Asked to make hypothetical hiring decisions and assign academic rank on the basis of the summaries, most chairs recommended the rank of associate professor for the summaries containing male names and the rank of assistant professor for the same descriptions identified with a female name.

In another treatment of gender, rank, and location, Jonathan Cole (1979) hypothesized that "functionally irrelevant characteristics such as sex will be more quickly activated when there are no or few functionally relevant criteria on which to judge individual performance" (p. 75). In a test with a limited sample of male and female scientists with no publications to their credit, he found that the "silent" (nonproductive) men were more likely than nonproductive women to be in prestigious departments. He also found that for scientists at *each* level of productivity, women were less likely to receive promotions than men (p. 70). In both better and lesser departments, gender was significantly related to academic rank, after controlling for productivity and citation levels.

Ahern and Scott (1981) also document large and pervasive sex differences in academic rank. Among pairs of men and women matched on year in which Ph.D. was received, field of Ph.D., institution from which doctorate was awarded, and race, the men were 50 percent more likely than the women to have been promoted to full professor 10–19 years past the doctorate. Among the younger matched pairs who received their Ph.D.s in the period 1970–74, women lagged behind men in promotion to associate professor, independent of marital status, the presence of children, or whether their work orientation was primarily research or teaching. Unlike Cole's study, however, Ahern and Scott's analyses did not control for publication productivity.[5]

When such biased practices are operating, high achievement is not enough and performance may be no guarantee of reward. Rather, as an experienced scientist or scholar comes to recognize, there are stock procedures people use to create and justify a negative assessment: finding "inconsistencies" between parts of the work; distorting the argument; suggesting that the work "lacks complexity"; asserting that it fails to include some significant piece of literature or aspect of the subject (Lieberman 1981:5). People who have the critical vocabulary and know the lingo can always find a way to justify a predetermined negative judgment.

Some argue that such biased judgments of work and other capricious and particularistic practices may be more characteristic of fields with lower levels of consensus on goals and methods, such as sociology or political science, than of fields with higher levels of consensus, such as physics or chemistry. Studies of editorial processes, for example, suggest that particularistic factors play more of a role in editorial policies and practices within social compared to physical and natural sciences. Thus, Yoels (1974) reports that in the selection of editorial board members, the shared doctoral origin of the editor-in-chief and appointee is a more important factor in the selection of editorial board members within the social sciences. Likewise, a survey of editorial practices in four fields (Beyer 1978) reports that editors in sociology and political science, compared to chemistry and physics, were more likely to use particularistic criteria (personal knowledge of the person, position in the professional association, institutional affiliation) in choosing editorial board members.

Looking more broadly to status attainment in science, Hargens and Hagstrom (1982) also show positive associations between the consensus levels of fields and status attainment based upon performance compared to nonperformance factors. However, it was only larger differences in consensus—differences between political science compared to chemistry, physics, mathematics, and biology, rather than differences between the latter four scientific fields—that resulted in performance-based differences in the career outcomes and prestige of first position and citations to work. Thus Hargens and Hagstrom conclude that large variations in consensus produce fairly small differences in status attainment patterns.

Although suggestive of disciplinary patterns of particularism and non-performance-based practices and criteria, none of these studies addresses gender explicitly. It may be that biased practices based upon gender differences vary with consensus of field or that they operate more uniformly across fields. Gender may be sufficiently different from particularistic factors such as shared doctoral origin that it influences more unilaterally across fields. This remains to be determined.

The confidentiality of decisions about allocation of resources and rewards within departments may also widen the margin for gender-biased judgments, insofar as such decisions allow for considerable administrative discretion. Accordingly, a study of scientists at the University of Michigan revealed that men appointed as assistant professors received higher levels of research support than women: more start-up monies for their labs, better physical facilities, or better placement within existing projects with funds and laboratory equipment (Feldt 1986). What is particularly notable is that this initial research support was subject to a "flexible" standard of allocation and the levels of support, a matter of "negotiation" with the administration.

Under such conditions, a key resource becomes the ability to build and manipulate a professional network in and out of the organization. This, however, depends upon an understanding of the criteria of membership and prestige and the ability to gain acceptance in the network and prominence within it (Collins 1979). Both the understandings and abilities come from previous group traditions, or they may to some extent be created within the organization (Collins 1979). But it is membership in this shared culture that becomes a critical factor in professional placement, performance, and reward.

As a consequence of men's numbers and dominance, however, the culture of the scientific world is a male milieu—both within the home department and in the larger community of science. In science, as in management (Kanter 1977), law (Epstein 1983), and other professions (Epstein 1970), the men share traditions, styles, and understandings about rules of competing, bartering, and succeeding. They accept one another, support one another, and promote one another (Fox 1989). As outsiders to this milieu and its bartered resources, shared influence, and conferred self-confidence, women are shut out of ways and means to participate and perform. As one woman in science put it:

Sooner or later, I and those around me knew I was not one of the boys. Professional identification in the absence of collegial identification gets you only so far. When somewhere along the line the boys start boosting their buddies in a kind of "quid pro quo" competition, you realize you've had it. You don't have the "quo" for the "quid." You don't have any coattails, and everyone knows it. The shift may be subtle or it may come as a jolt, but it *does* comes. (Tidball 1974:56–57)

A recent survey of the Great Lakes College Association supports these observations. In a survey of the female faculty (in arts and sciences) at 12 colleges, the greatest unanimity occurred with respect to the statement: "On my campus, there exists an 'old boys' network." Eighty-five percent agreed. Further, the same proportion disagreed that "women faculty relate well to the 'old boys' network." And 82 percent believed that "male faculty treat male and female colleagues differently" (Loring 1985).[6]

These matters of professional culture, organizational membership, and patterns of inclusion and exclusion are central to science because research is a social process. Scientific progress, more fundamentally than other types of creativity, "relates to, builds upon, extends, and revises existing knowledge" (Garvey 1979:14). The productive scientist must continually seek information pertinent to his or her research, shape and reshape the course of the work, update its potential relevance, and test its significance. This takes place not just formally through conference presentations or publications but more so, interactively, in conversations in the lab, the lunchroom, departmental corridors, and after-hour gatherings at professional meetings. In this sense, the work environment of scientists includes the larger scientific community as well as the organizational environment of home department and institution.

Yet, as Jonathan Cole put it, although women have moved into science, they are not *of* the community of science (1981:390). More often than men, they remain outside the heated discussions, inner cadres, and social networks in which scientific ideas are aired, exchanged, and evaluated. As indicators, we find that women are less likely than men to have professional connections as editors, officers of professional associations, reviewers of grants, and journal referees (Cameron 1978; Kashket et al. 1974); to appear on programs at national meetings (Morlock 1973); to be invited to lecture or consult outside of their institutions (Kashket et al. 1974); or to spend time off campus in professional activities (Bayer 1973).

Such exclusion limits the possibility not simply to participate in a social circle but rather to do research, to publish, to be cited—to show the very marks of productivity in science. First, face-to-face interaction with colleagues both within and outside of the one's own institution generates and supports research activity. Ongoing informal discussion among colleagues about research experiences—problems encountered, progress made—helps activate interests, test ideas, and reinforce the work (Blau 1973; Reskin 1978a; Pelz and Andrews 1976). Such informal exchange stimulates work because one cannot be a peer in these collegial groups without involvement in a research program oneself (Blau 1973). Through informal membership and discussion, a scientist can then generate ideas and evaluate problems, variables, concepts, and methods. In a study of 200 research efforts in psychology, Garvey (1979) found that less than 15 percent of initial ideas for projects originated from formal sources such as journal articles or presentations at professional meetings. Rather, the germ for the projects originated in informal networks of information.

Compared to formal communication, informal exchange also provides more room for speculation, retraction, and the sharing of failures as well as successes. Additionally, the flexibility of informal channels allows a scientist to direct the conversation and select specific information (Garvey and Griffith 1967). This provides immediate feedback and, in the process, alerts one to the prospects of being anticipated or, conversely, helps to establish the priority and importance of work before it is published.

Understandably, then, empirical studies have reported that professional connections and collegial interaction are associated with productivity in science. In an analysis of hard sciences in 18 disciplinary areas,

Biglan (1973) found that level of social connectedness (measured as numbers of collaborators, numbers of persons worked with on research, teaching, and administration, and sources of influence on research goals and teaching procedures) was related to publication productivity. Similarly, with a sample of faculty in a private research university and two private colleges, Finkelstein (1982) reports that faculty with strong collegial ties both on and off campus had the highest publication rates.

Ironically, it may be that while women are more often excluded from professional networks and connections, the collegiality is of even greater consequence for them than for men. As Barbara Reskin (1978a) argued, collegial expectations and exchange surrounding research may be especially important to those scientists who face conflicting demands from nonresearch roles in the institution (such as teaching and service) or from roles outside the institution (such as parenting). As a group, women are more likely to be concentrated in colleges compared to universities, in appointments with heavier teaching loads, and in demanding domestic situations, as single parents, for example—in roles, that is, with demands that conflict with research. Social psychological literature suggests that social support mediates role overload, role conflict, and stress and enhances well-being and satisfaction among various populations (House 1981). Likewise, social—and collegial—support may mediate role overload and enhance performance among scientists and among female scientists subject to role conflict, more particularly.

Along with exchange through informal discussion, the sharing of research findings and developments also occurs through preprints. Publication, the public portion of scientific communication, verifies information, legitimizes authorship, formally establishes priority, and functions as a primary basis of recognition and reward. But 90 percent of material published in journal articles has been disseminated previously (Garvey 1979:40). Active researchers say that they gain the information most relevant to their work through such informally circulated materials (Ziman 1968). Prepublished reports are not merely means of circulating information; they are mechanisms used by scientists to develop, refine, and test their work (Garvey and Gottfredson 1979). Garvey reports that over 60 percent of the authors who distribute preprints actually receive feedback and comment on their work which then leads them to modify their manuscripts (Garvey 1979:138).

But not all researchers are so connected to the informal community of science that they can distribute or receive prepublished work. Some evidence indicates that women depend upon published literature more than do men (Kaplan and Storer 1968; Reskin 1978a). This means that they may get information later, when, in fact, it may be "old"; and it means that they are deprived of feedback and comment on their own work.

Those located outside of circles of communication, interaction, and exchange are denied important means of testing and developing ideas and interpretations. This is true for both sexes. Men and women in science are not separate homogeneous groups, with all men inside and all women outside the informal scientific community. But as discussed, numbers of indicators point to women's greater marginality to the social system of science—with consequences for productivity and performance.

In assessment of these patterns of collegiality and productivity in science, a further issue is collaboration. During the nineteenth century, collaborative research grew slowly from 2 percent of all research in 1800 to 7 percent in 1900. Then, at the beginning of the twentieth century, a significant upward increase in collaboration occurred, with a slowing of the pace during World War I, followed by a continued increase, so that by 1920, 20 percent of research in biology, chemistry, and physics was collaborative (Beaver and Rosen 1978).[7] In the next decades, the growth in collaboration was exponential, so that 50 percent of the abstracted research was collaborative by 1950 and the majority, 60 percent, of research collaborative in the next decade (Beaver 1986; Beaver and Rosen, 1978).

In recent years, the upsurge in collaboration has continued. The Institute for Scientific Information, which indexes 2,800 journals, reports that the average number of authors per article rose from 1.67 to 2.58 between 1960 and 1980, and in some areas and journals such as the *New England Journal of Medicine,* papers average five coauthors (Broad 1981). The long-term trend toward collaboration has been called "one of the most violent transitions" that can be measured in patterns of scientific performance (Price 1963:89).

The long-term increase in collaboration owes to a number of factors, including patterns of funded research (Heffner 1979; Subramanyam and Stephens 1982), specialized research technology (Meadows 1974), the maturation of disciplines (Maanten 1970), and the increasing professionalization of science (Beaver and Rosen 1978). However, because collaboration typically grows out of informal collegial contact (Hagstrom 1965), women's restriction from collegial circles may limit their access to collaborators (Reskin 1978a).

In an examination of authors of articles in 14 sociology journals, Mackie (1977) reports that women are more likely than men to be single authors. Likewise, Chubin (1974) reports that women sociologists are less likely to collaborate. On the other hand, in their analyses of the publications of a matched sample of men and women who received Ph.D.s in 1970 in six scientific fields, Cole and Zuckerman (1984) find that the women are as likely as men to publish jointly authored papers. However, even when women publish jointly, it may be that they have more difficulty finding collaborators and have fewer possible collaborative partners available to them. An interesting analysis of publication among academics in sociology, psychology, and English indicates that this may be the case (Cameron 1978). Cameron reports while women are as likely as men to collaborate, the men have a significantly higher number of different collaborators. The men collaborate more broadly, while the women collaborate with a few persons.

Access to collaborators and patterns of collaboration are critical because solo authors are handicapped in their rates of publication and, probably, rewards. With the growth of increasingly technical and specialized teamwork, isolated research is difficult to initiate, fund, and sustain. And collaborative work seems to fare better in the publication process. In a study of the editorial fate of 1,859 papers submitted to a leading astronomy journal, for example, Gordon (1980) reported that coauthored papers were more likely to be accepted than were single authored papers. This may owe in part to the greater likelihood of coauthored papers to result from funded compared to nonfunded projects (Heffner 1979). It may also owe to the greater likelihood of coauthored papers to be empirical rather than theoretical pieces (O'Connor 1969; Meadows 1974), which are, in turn, easier to review and evaluate and, perhaps, accept.

Beyond this, however, coauthored work provides checks in research and helps avoid outright error. Thus, in his analysis of papers submitted to a social psychology journal, Presser (1980) found that coauthored papers were rejected less often than single authored papers and more often received invitations to revise rather than outright rejections. Differences in rates of initial acceptance, however, were minor for the two types of papers. This suggests that "collaboration leads less to producing very good papers and more to avoiding bad ones" (p. 96).

Correspondingly, analyses of other disciplines indicate that collaborative papers are more likely to be cited (Beaver 1986), to be of higher quality (Lawani 1986), and thus to be more visible. This pattern of collaboration and citation varies by field, however, and is less characteristic of social compared to natural sciences (Bayer 1982; Smart and Bayer 1986).

To the extent that women are excluded from collegial channels and collaborative opportunities, their productivity can suffer. Two studies tracking the accomplishments and tenure outcomes of assistant professors at the University of Michigan point to such patterns. Although these are case studies of limited numbers of scientists at one research university, they shed light on the problem of collaboration, collegiality, and productivity.

The first study (Feldt 1985, 1986) tracked 29 assistant professors appointed between 1973 and 1978 in one college of the university.[8] By focusing upon one time period in one college, the study was able to relate qualifications, achievements, and career outcomes in the given unit with consistent stated promotion criteria. Of these assistant professors, 13 percent of the women and 43 percent of the men were promoted to associate. The entering qualifications of the women were comparable to men in quality of schools attended and number of publications prior to appointment as assistant professor. Once in place, however, the women did not publish at the same rate as the men. Marital and parental statuses did not explain productivity differences. Collegiality and collaboration patterns provided the best clue. Only 25 percent of the women compared to 52 percent of the men had coauthored papers with senior professors. Such collaboration was closely associated with promotion: of those who collaborated, 62 percent were promoted, compared to 13 percent of those who had not.

A second study (Feldt 1986), called the Cohort Study, corroborates these patterns with a group of 23 women and 97 men appointed as assistant professor between 1974 and 1980 in the University's School of Medicine. For this group, as well, publication of peer-reviewed articles was strongly related to promotion as associate professor. And more pertinently, collegial interactions—specifically, collaboration in coauthoring authors and mentoring relationships—related to both productivity and promotion rates.

Collaboration is both a means to, and an indication of, involvement and integration into the research community. In this latter case study, women were more productive when they received research funding from their departments for a lab or were placed in an established lab, had a mentor, and coauthored papers with senior colleagues. These factors were connected, and their incidence and relative influences are difficult to isolate.

While post-doctoral collaboration is important to productivity, so too is earlier, pre-doctoral collaboration. In fact, in their analyses of career processes in science, Long and McGinnis (1985) conclude that "collaboration in publication appears to be the most influential act that mentors can perform on behalf of the careers of their graduate students" (p. 278). Specifically, they report that collaboration with a mentor affects pre-doctoral productivity and job placement, which, in turn, influence later productivity. But the mentor also influences productivity independent of these indirect effects. For those who collaborate with a mentor, the mentor continues to affect the student's productivity. This probably reflects the effects of research training and experience.

The significance of such training and collaboration lies not simply in its consequence for technical knowledge and skill, but more so, in consequence for taste, style, and confidence in research. In a classic study of the scientific elite, Zuckerman (1977) explains how apprenticeship cultivates norms, values, and opportunities. Socialization into the culture of science involves transmission of exacting standards for judging work and assessing performance, criteria for distinguishing between important and trivial problems, and elegant compared to common solutions, and the capacity to cope with difficulties, setbacks, and failures in research (Zuckerman 1977).

In assessing women's compared to men's graduate school background, we find small gender differences in certain measures of attainment and support. First, with some variation by discipline, men and women are similar in the proportions of each group who received their doctorates in top-ranked scientific departments between 1970 and 1980 (National Research Council 1983, table 2.2). The most substantial gender difference is in mathematics: 46 percent of the men compared to 37 percent of the women received Ph.D.s in departments rated as "strong" or "distinguished" (based upon ratings of Roose and Anderson 1970). In physics, women were somewhat less likely to have graduated from a top-ranking department, and in psychology and microbiology more likely to have done so. But overall the pattern is one of similarity for men and women.

Likewise, in certain rates of financial support for graduate training, gender differences are small. Among those who received Ph.D.s in 1950, 1960, and 1968, women in the social and physical sciences were slightly more likely than men to receive a research assistantship sometime during their graduate education, and in the biological sciences women were less likely to receive such an assistantship. In each of these disciplines, the women were as likely as men to have been awarded some form of a fellowship or scholarship for part of their doctoral training (Centra 1974, table 7.1). Data from more recent, 1970–1981, recipients of Ph.D.s in science and engineering also show little difference in the percentages of men compared to women who had held research or teaching assistantships (National Research Council 1983:2.9). None of these data, however, indicate the level of financial support or duration of support.

Despite these documented similarities, the quality of men's and women's graduate school experiences differ. Here, again, environment does not operate neutrally or uniformly. Thus both men and women acknowledge problems such as impersonality and lack of faculty commitment in graduate school (Centra 1974), but women report more isolation and marginality in their training. Women see faculty members and research advisors less frequently than do men (Holmstrom and Holmstrom 1974; Kjerulff and Blood 1973). And their interactions with faculty are less relaxed and egalitarian (ibid.). Women tend to regard themselves as students rather than colleagues of faculty and they say that they are taken less seriously than men (Berg and Ferber 1983; Holmstrom and Holmstrom 1974). In short, women report that they are more marginal—especially outside of formal classroom proceedings. This has consequences for their opportunities to join research projects, to collaborate, and to gain access to the culture of science.

Yet, there is a negative underside to collaboration and connection with senior scientists in the case of the research associate. In the pre–World War II decades, the options for women in science were basically an appointment in a woman's college or a position as research associate. The research associateship at least provided a chance to do systematic scientific work. But while research associates executed experiments, solved problems, and managed labs, they did so without status, and without the opportunity to control and design their own work. Usually, they had no chance to obtain funding and establish their own labs, and often, no right to be named on publications.

In 1938, about half of the academic women in biochemistry, astronomy, and anthropology, and nearly half of those in microbiology, worked in research institutes where the funding was channeled through a few established investigators, who hired assistants and expanded their domains (Rossiter 1982:204–205). Female research associates represented a good investment. They were skilled, low cost, and grateful for the work. While the male research associates were able to use the position as an advanced post-doctoral fellowship and then move on to their own institutes or labs, the women could not. Although the research associateships were essentially temporary, dependent positions, tied to the funding of a particular investigator, women remained in these posts throughout their careers (ibid., chap. 7).

Fewer research associate positions now exist, but those that do remain disproportionately female.[9] Contrary to the general argument about collaboration and productivity, in these positions, collaboration infrequently brings visibility or recognition to the junior associate.

Because women are more likely to be outside of informal channels, they may need to depend upon other ways and means—frequently upon the more inexpedient formal processes—to support their work. I have observed, for example, that with less access to pre-print networks, women may depend for information upon published articles, which frequently contain dated research. Furthermore, women may depend upon formal processes to reinforce their work. Located in less research-oriented institutions and outside of collegial ties within their institutions, it is likely that women receive less informal feedback, comment, or validation of their research. Thus formal citation can become more important in reinforcing women's compared to men's subsequent productivity, as Reskin (1978b) reports. Specifically, Reskin finds that citation is more important in reinforcing the productivity of chemists outside of universities, particularly for women. In interaction terms, the effect of citations upon productivity of chemists located in universities compared to elsewhere is greater for women than for men (ibid.).

Without tests for such interactions of gender and location, Cole and Zuckerman (1984) report, however, that women's productivity is less apt than men's to increase with citation but more apt than men's to decrease without the reinforcement. At the same time, Cole and Zuckerman (1984) suggest disciplinary differences in the effect of reinforcement, since when they confined their tests to chemists, they obtained results similar to Reskin's.

To the extent that women lack informal contacts and connections to funding agencies (Sandler and Hall 1986), they may depend also upon formal, published announcements, with their frequently incomplete information on the agency's program objectives, needs, and priorities, and their appearance relatively late in the agency's planning. In addition, to the extent that women have fewer collaborative ties before and after receiving the doctorate, they are restricted in their socialization into the entrepreneurial culture of science—experience in proposal writing, communication with program officers, modification of research projects to meet the review of the agency. These are processes cited in a study of National Science Foundation's program to stimulate research as the "weak points" in the research activities of fledgling scientists (Drew 1985).

Agencies—NSF and NIH (see Zuckerman 1987:138)—have reported that women and men receive grants proportional to the numbers of proposals submitted by each gender group. However, these agencies also report that women do not apply for grants at the same rate as do men. The critical issue then becomes what enables scientists in submitting proposals? What is the role of informal contacts and connections in the process? To what extent are women disadvantaged in learning the entrepreneurial skills of science? And how does this affect levels of productivity and performance?

Those on the margins of the social and organizational milieu of science are disadvantaged in access to resources, claims to significant membership, patterns of communication, and possibilities for collaboration. When restricted in social and organizational resources, people rely upon what is left for them—and for women this may be their own personal resources. Astin and Davis (1985) asked a sample of productive researchers to identify the factors that enabled their productivity. The factors identified are those the researchers perceive to be important. The perceptions may not be equivalent to that which actually enables their productivity. Yet, with that caution, we find interestingly that compared to men, women were more likely to cite personal variables such as "hard work, being motivated, being interested in the research topic, and possessing the necessary skills to do the work" (p. 150). Men, on the other hand, were more likely to cite organizational factors such as funds, assistantships, and institutional support. One might argue that women are under-using organizational resources. But it is more likely that women draw upon what is available to them—and that may simply be themselves. Yet with science tied to the complexities of funds, facilities, and teamwork, personal resources provide only a part of the requirements for research.

Finally, in this assessment of gender, environment, and productivity in science, there is another factor: the cognitive authority of the dominant culture. Many maintain that research is not gender-specific, that women do not do science differently than men. Another, smaller group argues that the experience of women—as women—is special, unique, palpable; that women can ask different questions, develop different methods, and combine elements more intuitively and creatively (see profiles in Gornick 1983). From this latter perspective, the male scientists are imbued with a cognitive authority that takes their understandings and interpretations as scientific truth, and many women's as peculiar, if not false.

In this sense, the very laws of nature are more than expressions of objective inquiry, more even than reflections of the general social and political processes accepted by those doing social studies of science. Rather, they are the result also of connections between mind, nature, and masculinity (Fox 1986). This view is represented by Evelyn Fox Keller. In a series of essays and books, Keller (1982, 1983, 1985) maintains that science is inherently masculine in character, not just male dominated by numbers. The consequence is not simply an exclusion of women, but a schism between masculine and feminine, subjective and objective, understanding and control (Keller 1985). She argues, for example, that the tension between scientific explanations that stress hierarchies and unidirectional causal paths and those that stress multiple interactions and multidirectional paths stem from different views of nature, domination, and control (ibid.).

Illustrating the struggle of ideology and practice in the production and interpretation of science, Keller wrote the biography of Barbara Mc-

Clintock (Keller, 1983, 1985). Barbara McClintock has been a philosophical and methodological deviant. By challenging the "master molecule theory" of genetic control in a single molecule within a hierarchical organization, McClintock offered a different view of nature, a view of subject and object that were outside the mainstream and, Keller argues, outside of gender-defined definitions and boundaries. After being largely underappreciated, McClintock was awarded a Nobel Prize in her eighties for work begun 40 years earlier.

Barbara McClintock is a maverick redeemed. But, for the larger scientific community, the point is this: If the authority of a dominant scientific culture defines ways in which questions get posed, interpretations are made, and knowledge is established, and if those patterns are *gender-specific,* then women's research and productivity are constrained even beyond the limits of the human and material resources I have discussed here.

In sum, productivity in science is irrevocably tied to the environment of work: the signals, resources, and reward schemes of the institutional setting, and the networks of communication and exchange in the larger community of science. In assessing the gender differences in productivity, I have looked then to the ways in which social and organizational variables affect the productivity of women and men—through processes of appraisal, collegial interaction, work climate, and collaborative opportunities, among other factors. I argue that differential environment is not simply a matter of men and women being disproportionately located in different types of settings (more or less research oriented, more or less prestigious). As I have said, a given environment does not necessarily operate uniformly, neutrally, or androgynously. Within the same type of setting, women scientists can have fewer and different collaborative arrangements, claims to enabling administrative favors, collegial opportunities for testing and developing ideas, and entrees into the informal culture of science and scholarship. These issues are central to science because research is a social process of communication, interaction, and exchange. Exclusion from this process limits the possibility not simply to be part of a social circle but rather to do research, to publish, to be cited—to show the crucial marks of productivity in science.

STEPHEN COLE / ROBERT FIORENTINE

9. Discrimination Against Women in Science: The Confusion of Outcome with Process

WOMEN HAVE ACHIEVED LESS than men in science. This statement is true no matter how we choose to measure achievement. Since science is a relatively high prestige occupation, we might begin by simply examining the proportion of scientists who are women. If we define as a scientist all those who hold Ph.D.s, we will find that slightly less than 20 percent are women.[1] The proportion of women scientists in fields like physics, chemistry, and mathematics is substantially smaller. If we then consider achievement among those who are scientists, we will find that whether our indicator is receipt of a prestigious prize such as the Nobel (Zuckerman 1977), admission to prestigious societies such as the National Academy of Science, or virtually any other measure one could use, that male scientists have as a social *aggregate* achieved more than female scientists.

How can we explain the correlation between gender and achievement in science? Among many social scientists and members of the public in general, the answer to this question is sex discrimination. It has been assumed that women have not been afforded the same opportunities as men to enter science—or, once in science, to make important contributions. For example, a committee of the National Research Council issued a report

This research was supported by a grant from the Josiah Macy, Jr. Foundation. We thank John Bruer for the encouragement he gave to this research. We also thank Jonathan R. Cole, Norman A. Goodman, and Maria Misztal Cole for comments on earlier versions. Any errors are, of course, the responsibility of the authors.

in 1979 in which they blamed the discriminatory practices of universities for the relatively low proportion of women in science: "The very low rate of participation in graduate study by women following World War II is largely a result of well-documented overt sex discrimination practiced for many years in some graduate science departments" (CEEWISE 1979: 19).

A notable exception to the majority of sociologists who explain differences in achievement between men and women scientists by sex discrimination can be found in the work of two of the editors of this volume, Jonathan R. Cole and Harriet Zuckerman. In their substantial body of work on this subject they have indeed pointed out instances in which the evidence suggests that women may have been discriminated against; but they have also been careful to point out those cases in which the data do not support that conclusion. In particular, they have focused on the importance of taking role performance into account. Data collected by Cole (1979), by Cole and Zuckerman (1984), and by others show that women scientists publish significantly less than men, on average about half as much. In their contribution to this volume and in other papers, Cole and Zuckerman have presented both quantitative and qualitative data with the aim of understanding this "productivity puzzle."

This chapter has two primary goals. The first is methodological: Under what circumstances can we reach the conclusion that an observed inequality is a result of discrimination? Here we shall analyze what we believe to be a widespread logical error that is made when interpreting the causes of inequality between the sexes in science and other occupations. The same error is frequently made in interpreting the causes of all types of social inequality: between men and women, different racial groups, or different social classes. This error is to use inequality itself, an outcome, as evidence of a process: discrimination or "lack of equal opportunity." Our analysis is based upon the premise that discrimination or lack of equal opportunity is only one possible cause of any observed inequality between two or more different groups. The second goal is substantive: in recent years cultural explanations of inequality have become unfashionable. Our research to the contrary, however, suggests that culture is an important determinant of occupational inequality between the sexes. We will describe a theory, termed "normative alternatives," which offers a framework that can be used to explain much of this inequality.

The essay will begin with a discussion of the three basic types of processes which can create inequality. We will then show that it is common practice to assume, even without empirical justification, that inequality is a result of discrimination. Finally, we will describe a research program which will show how the processes creating inequality may be more directly studied and at the same time demonstrate the importance of cultural explanations for gender differences in occupational attainment.

Processes Leading to Inequality

If one wanted to classify the possible causes of inequality between men and women in an occupation like science, they would fall into three broad categories: biological, structural, and cultural. In any given case all three or any combination of these processes may be interacting to produce an observed level of inequality. It is sometimes difficult to determine which of the three processes is most influential. For example, if women consistently score lower than men on quantitative tests, it is difficult to determine whether this is a result of biologically based ability differences, culturally based socialization practices, or structural discrimination in the education system. Disparity in scores could be a result of any or *all* of these processes.

Biological differences between men and women could *potentially* explain the differences in their behavior which lead to eventual inequality. Many biological theories purporting to explain sex differences in achievement have been developed. There are, undoubtedly, significant biological differences between men and women. What has been debated, however, is the extent to which these differences cause the observed social inequalities.

An example of a biological theory of sex differences in achievement is Goldberg's "testosterone" hypothesis (1973). He argues that because of a different hormonal makeup, men are more aggressive than women and this aggression leads to greater achievement. Maccoby and Jacklin (1974), in a comprehensive review of the literature on sex differences, conclude that the evidence suggests that men and women do indeed differ in their levels of aggression.[2] They go on to point out that there is little evidence, however, to link lower levels of aggression with lower levels of achievement. To what extent is achievement dependent upon aggression in occupations like science? Given the current state of our knowledge on the influence of hormones on behavior, this must remain an open question.

The absence of any compelling evidence linking biological differences with differences in achievement in particular occupations plagues other biological theories as well. Some sociologists, including the sociobiologists, have become increasingly interested in using biological variables to explain human behavior; but thus far it has been impossible to disentangle the possible influence of biological factors from structural and cultural factors.

The paleontologist Stephen Jay Gould (1981) has presented a convincing argument that in the past, biological determinist theories were heavily influenced by the ideological biases of their supporters. He argues that biology may set very broad limits on what humans can achieve but that there is no evidence that any achievement differences between different groups of humans (blacks or whites, men or women) are a result of biological factors.

We have no evidence for biological change in brain size or structure since *Homo sapiens* appeared in the fossil record some fifty thousand years ago. . . . All that we have done since then—the greatest transformation in the shortest time that our planet has experienced since its crust solidified nearly four billion years ago—is the product of cultural evolution. (p. 324)

Thus, although it is possible that in the future some biological cause may be found for sex differences in achievement, currently this remains only a possibility with no solid empirical evidence demonstrating the linkage.

The second general process which could create inequality may be termed structural. Here, social structure refers to the way in which social institutions are actually organized and operated. Thus, the social structure of a university would be its formal organization and informal practices to the extent that these were regularized. A structural explanation of inequality would point to how members of two different groups, men and women, for example, are treated differently by particular organizations. If a particular organization utilizes the sex status of an individual to determine their position or their rewards, this could clearly explain differences in achievement between men and women in this organization.

Sex discrimination would thus be classified as a structural explanation of inequality between men and women scientists. Most social organizations have to decide who from among a group of applicants or eligibles should be admitted to particular positions or receive particular rewards. All the criteria used by organizations to make these choices may be called "social selection." The criteria used in social selection may be divided into two broad groups: those that are functionally relevant to performances and those that are functionally irrelevant. Discrimination is defined here as the utilization of functionally irrelevant criteria in making decisions as to who will be allowed entrance to a status or become the recipient of some scarce reward.[3]

Discrimination can be overt or covert. In overt discrimination, a functionally irrelevant criterion will be part of the official social selection system. Thus, if a medical school stated that applications from women would not be accepted, this would be overt discrimination. Such discriminatory practices were widespread in the past in the United States. In many traditional societies—in Arabic countries, for instance—women are still overtly prohibited from entering many occupations. In covert discrimination, the functionally irrelevant status will not be part of the official criteria but will be used nonetheless. In such cases the people controlling the social selection system, the "gatekeepers," may deny that they are using such criteria. Since overt discrimination on the basis of sex is now illegal in the United States, most of the discrimination we are likely to find is covert.

Functional relevance refers to the actual ability of an individual to perform the duties of the position or to utilize the reward. There is, for example, no necessary relationship between an individual's sex, race, religion, or any other so-called demographic characteristic and that person's ability to perform the duties of a scientist.[4] Therefore, the use of any of these statuses in evaluating applicants to graduate school, for example, would constitute discrimination. On the other hand, using scores on the Graduate Record Examination as criteria for admission to graduate school is not discrimination because there is some reason to believe that this criterion is functionally relevant, i.e., has some predictive strength as an indicator of the ability of the applicant to perform well in graduate school.[5]

The third general type of process which might explain inequality is culture. Cultural explanations are those which use beliefs, attitudes, norms, and values internalized by individuals through the socialization process to explain behavior.[6] Before an individual can attain a particular position or reward, that individual must decide that he or she is interested in that attainment. Thus, before an individual can become a scientist, that individual decides to pursue this career and apply to a graduate school. Before an individual can become a physician, that individual must decide that this is the occupation he or she wants to enter, take pre-med courses, and then apply to medical school. All of the processes influencing such decisions may be called "self-selection" or "cultural selection."

It might be possible to call inequality resulting from cultural selection "cultural discrimination." Some might say that if women are "discouraged" (either actively or passively) from pursuing a career in medicine or science, this represents discrimination against women. Thus, if women learn that medicine and science are not "appropriate" careers for them, this could represent a type of discrimination. If we were to define discrimination in this way, then any observed inequality, when not the result of biological differences, would have to be a result of some type of discrimination, either structural or cultural. It would not be a question of whether discrimination existed, but of the source of discrimination. Although we sympathize with the belief by some who place a high value on achievement orientation that cultural selection may be "unfair," we believe that our ability to understand the causes of inequality is not well served by lumping together different processes under the same term. We would like to use "discrimination" to refer to the distribution of rewards or active discouragement on the basis of functionally irrelevant criteria.

It should be evident that simply looking at an outcome, such as the number of women entering careers such as science or medicine, will not tell us whether this outcome is a result of social selection processes, cultural selection processes, or both. Yet some social scientists use these proportions as evidence of discrimination without any other systematic evidence. In part because the achievement of women has become an intense political issue, particularly in academic circles, the scientific analysis of this question has become intertwined with political values. The term "self-selection" is virtually a "dirty" word in the social sciences today. This is true for all cultural theories or any theory which questions whether discrimination (or, as discrimination is sometimes referred to, "structural barriers") is the prime determinant of inequality.[7] This is because the analyst employing cultural explanations or the concept of self-selection is viewed as utilizing a "blaming the victim" approach to explaining inequality. If inequality results from the unequal group having "voluntarily" decided that they were not interested in competing or attaining a particular social status rather than from discrimination against able and willing people, the victim is seen as being blamed for the inequality. This interpretation of cultural explanations and the concept of self-selection is in our opinion misguided because it underestimates the constraining power of culture.

If a group of people—women, for example,—grow up in a society where they have been socialized to believe that certain occupations are inappropriate for them to pursue and then act upon the basis of these internalized values, their behavior is not entirely voluntaristic. Individuals or "victims" have no more personal control over the culture they are influenced by than they do over the social structure they are influenced by. Especially in the past, young women have not aspired to be scientists, doctors, or lawyers because they grew up in a society in which it was believed that these were not appropriate careers for them. To say this is not to "blame" the women; it is to "blame" the culture which has influenced them.

Because culture works in a diffuse rather than a specific way, we sometimes overestimate the ability of individuals to overcome cultural constraints. Culture can be just as constraining as social structure. Once individuals internalize values, these values can have a constraining and lasting influence on individuals' behavior and their ability to achieve particular goals. In order to emphasize the constraining nature of culture, we use the term "cultural selection" to refer to those aspects of self-selection which are constrained by culture.

Confusing Outcome with Process

The assumption that outcomes are evidence for process is at this time thoroughly embedded in sociological thinking and the society in general.[8] A good example of how outcomes are taken as evidence of process in the general society can be found in a recent suit brought by the EEOC against Sears Roebuck for sex discrimination. The EEOC accused Sears of having systematically discriminated against women in hiring for the highest-paying sales positions, those in which so-called big-ticket items are sold and the salesperson receives a commission. The EEOC case was based exclusively on outcome data. Of all applicants for sales positions a disproportionately low number of women were hired for the commission sales positions. The EEOC did not have a single female witness who claimed to have been discriminated against; nor did it present any direct evidence that sex was taken into account in making hiring decisions. Their argument, and that of the academic witnesses who testified for the EEOC, was that the level of opportunity offered women by Sears could be determined by looking at the proportion of women in various positions.

Rosalind Rosenberg, a feminist historian testifying for Sears, claimed that the difference observed between men and women could possibly be explained by differing value systems of men and women. She argued that women have historically had differing attitudes toward work. Earning money has been less important to women and having a job which is compatible with their family role as mother and wife has been more important. To infer from an outcome (a smaller proportion of women in high-earning sales positions) that this outcome is a result of discrimination rather than cultural selection factors, Rosenberg testified, was simply unjustified. The proportion of women in various positions should be viewed as a result of both opportunity and cultural values.[9]

In his book *Fair Science* (1979), J. R. Cole gives examples of what he calls "naive residualism," or the practice of attributing any difference in rewards received by men and women scientists to discrimination. Although most sociologists today are aware that before a simple zero-order correlation between sex and some outcome can be called discrimination some controls must be applied, they frequently begin with the *assumption* that any differences *must* be the result of discrimination. Weitzman's (1979) book, *Sex Role Socialization,* is a good illustration of this tendency. After a comprehensive review of the literature on socialization, Weitzman concludes that although men and women are socialized differently, no evidence is available to link this socialization specifically to achievement differences. She then concludes, without examining any studies or data, that the achievement differences must be a result of discrimination:

When women are denied real opportunities for advancement and are discriminated against at every stage of the process leading to a professional position, it is not surprising that they have not "made it." Thus, the answer to this question [Why has the proportion of women in the professions and in other high-status occupations remained so low?] lies not so much in the socialization of women as in the structural opportunities available to them in our society. (Weitzman 1979: 86)

Examples of research in which there is an attempt to explain zero-order differences in achievement by controlling for relevant variables may be found in the status-attainment literature. Let us examine the typical procedure followed. A good example of this type of research is the article by Suter and Miller (1973) on income differences between men and women.

There is a significant difference in mean income between men and women. In the 1966 data analyzed by Suter and Miller, the mean income of men was $7,444, whereas the mean income of women was $2,875; a large difference of $4,569. In order to explain this difference the authors compute a regression equation which shows what the mean income of women would be if they were like men on certain characteristics known to influence income. Thus, they control for education, occupational status, full-time or part-time work, and lifetime work experience. (They do not have to control for variables like socioeconomic status of family of origin or IQ because there are no differences between men and women on these variables.) After controlling for all these variables, they conclude that if women were like men on all these variables there would still be a mean difference of $2,828 in annual income: "The inability of women to convert occupational status into income to the same extent as men suggests that much of the remaining unexplained difference in male/female earnings could be attributable to discrimination in payment for jobs with equal status" (p. 971).

In the status attainment literature this residual difference is frequently called the cost of "discrimination." This is a typical example of what J. R. Cole (1979) has called "sophisticated residualism." It is sophisticated because it realizes that *zero-order* relationships (that is, simple relationships between two variables) alone cannot be used to assess the impact of discrimination; it is residualism because all of the unexplained variance or residual is attributed to an unmeasured variable, "discrimination." To what extent is this procedure justified?

Residualism can be a weak ground on which to conclude that discrimination is operating because it assumes that all of the variance which cannot be explained is a result of a specific unmeasured variable. The logic behind this would be that if we have taken into consideration *all* of the other variables which could influence income, then any income differences would *have* to be a result of the unmeasured variable. The problem is that it is difficult and sometimes impossible to measure all or even the most important other variables that could cause the inequality. In the Suter and Miller analysis, for example, the authors control for occupational prestige. Within each category of occupational prestige there is wide variance in the income of various occupations. Occupational prestige, in fact, explains only a small amount of the variance on income (Jencks et al. 1972). Clearly, there is a large element of cultural selection in deciding which occupation to enter. It is possible that women, because of the way in which they have been socialized, may be less concerned with earning a high income and may therefore give income a lower priority when selecting an occupation. Thus, within each category of occupational prestige women may *on average* be more likely to *select* lower-paying jobs. The residual variance found by Suter and Miller could be significantly influenced by this uncontrolled "selectivity."[10]

Even within the same occupation, differences in income result from a wide variety of variables. Sociologists often seem to assume that everyone wants to maximize income and will do whatever they can to do so; any observed differences in this scheme must be a result of social selection not cultural selection. This assumption is clearly incorrect. Not everyone cares enough about money to make career decisions which will sacrifice their leisure, family time, control over their own working conditions, and fulfillment in their work.

Male salesmen, for example, might work harder than women and sell more goods, thus accounting for higher income with discrimination playing no role. Or consider a potential analysis of income differences among male and female scientists. If male scientists earn higher incomes than female scientists, even when number of publications is controlled, would we be justified on the basis of this evidence alone in concluding that the employers of the scientists have discriminated against the women? In fact, such arguments have been made by groups of women faculty members in bringing sex discrimination suits against universities.[11] These arguments ignore the possibility of cultural explanations of the observed difference. It is possible that differences in income between male and female scientists could result from the differing values that men and women have about money.

Surveys on the values of men and women have consistently shown that men are more likely than women to place a relatively high value on earning money both in occupational choice and in making occupational decisions once they enter an occupation (Rosenberg 1957; Lueptow 1981). Although in recent years money has become increasingly important to women, there is still a significant difference in the importance of this value to the two different groups (see below, p. 223). *If* male scientists are more concerned than female scientists about earning money, it is *possible* that they would be more likely than female scientists to engage in occupational behavior specifically aimed at increasing income. Such behaviors could include accepting extra teaching assignments during the academic year, teaching summer school, making an effort to obtain research grants which would supplement salaries, and seeking outside offers from other universities which could then be used to increase their salaries. We have no evidence that there indeed are differences in the extent to which men and women scientists engage in these practices; but certainly there is a possibility that these sex differences exist and that they could influence income. Before ruling out this possibility, it would be important to have some evidence that there are no significant sex differences between men and women scientists' behavior here.

To return to the Suter and Miller example, if women on average were to care less about earnings than men, they might choose different occupations and they might not have the same income as men even within the same occupation. To attribute this difference to discrimination—without taking into account the possibility of self-selectivity—would be an error. The error is in essence the same as that made by the naive residualists: the use of outcome data to infer that one among a number of possible processes is the cause of an inequality.

How could Suter and Miller (1973) handle our "potential self-selectivity" critique? With their data set it would have been impossible. It is conceivably possible to collect data sets on individuals in which one would attempt to control for self-selectivity. To our knowledge this has never been done in status-attainment research—and for good reason. How would one measure the variables causing self-selectivity? The only way to do so would be with a longitudinal study in which the influence of values at relatively early ages on occupational selections were traced and then income measured at a later point. The study would have to take into account other variables besides selectivity, such as role performance in work. This would be a massive study and one that is very unlikely to be conducted.[12]

How then can we prove that an observed inequality is a result of discrimination? Lieberson (1985) presents a convincing argument that in situations like this, in which the results can be influenced by uncontrolled selectivity, the control variable approach is unlikely to provide satisfactory answers. As long as the unexplained variance or residual can be explained both by discrimination and other variables such as either selectivity or role performance, the use of residual variance to assume discrimination represents the confusion of outcome with process.

This does not mean, of course, that the control variable approach is never valid. The utility of sophisticated residualism as a tool for assessing discrimination depends upon the nature of the dependent variable. Some dependent variables are clearly more or less influenced by selectivity. Consider some examples in the area of science. To the extent that the particular reward being studied is *not* influenced by self-selection, it can be adequately studied by sophisticated residualism provided that adequate indicators of role performance are included among the controls. For example, there are some awards such as the Nobel Prize and the MacArthur Foundation Fellowships for which it is impossible to apply. If individuals do not have to apply, then the potential influence of self-selectivity is greatly reduced and the procedure of sophisticated residualism makes more sense. Obviously, the problem in such a study would be to obtain adequate measures of role performance. At the level of the Nobel Prize and Mac-Arthur Fellowships there are many more scientists seemingly qualified than there are awards. Very fine distinctions in qualifications must be made at this level, so fine that they are hard to quantify.

An example of a dependent variable better suited for research on which there is potentially little self-selectivity is academic rank. Here it is possible to control for role performance in a large sample by looking at the number of publications and citations to them. Since at any given productivity level virtually everyone would prefer a higher rank, we can assume that self-selectivity does not play a major role in this process.[13] If, when the relevant controls are made, it is found that male scientists are more likely to have higher ranks than female scientists, it is reasonable to assume that the residual variance may be largely a result of sex discrimination. In fact, this is the one area in which Cole (1979) did find some evidence of sex discrimination in science.

If academic rank lends itself to a control variable approach, income and rank of department do not. We have already pointed out the selectivity factors which could operate in regard to income. Since assignment of individuals to particular departments is clearly a combination of social and self-selection processes, it would be an error to use the control variable approach to determine whether women are discriminated against in terms of the rank of their department. In order to study this problem, we would have to look at the social selection process directly. We would need data on applicants to various departments, their qualifications, and the decisions made. If women applicants with equal qualifications to male applicants are not given equal treatment, we can conclude that they are being discriminated against. Existing data on this topic (Cole 1979) indicated that there is not a significant zero-order correlation between sex and prestige rank of academic department; it is thus unnecessary actually to carry out a study of the hiring process.

The problems we have pointed out in the use of the control variable approach give a substantial advantage to the null hypothesis, i.e., there is no discrimination at work (White 1982). Essentially, using the control variable approach it is possible to prove that discrimination is *not* a cause, but in many cases virtually impossible to prove that discrimination *is* a cause. Discrimination is posited as an independent variable to explain an observed difference in outcome. If there is no difference on the dependent variable between two groups, then we may conclude that discrimination has not occurred. If there is no inequality between men and women in a particular case, then there is nothing to explain and no control variables need be utilized. Thus, if there is no difference in the prestige of the academic departments in which men and women scientists are employed, it is clear that no sex discrimination has been practiced in regard to this reward.[14]

If there is a difference on a particular dependent variable between two groups and if this difference can be "washed out" or explained by controlling for any of the variables we have data on, then this can be taken as evidence that the inequality was not a result of any *direct* discrimination. Thus, if men and women scientists do differ in the rank of their academic departments but this difference can be completely explained by measures of the quality and quantity of their publications, we would have no evidence that the hiring departments were directly discriminating. It would, of course, still be possible that discrimination prior to hiring created the conditions which enabled men to produce more and "better" papers; but such evidence would have to be gathered.

The problem in using the control variable approach occurs only when there is an observed inequality which is maintained even after important other variables are controlled or in a situation in which the null hypothesis must be rejected. Since inequality can be produced by many relevant variables not included in the model as well as by discrimination, assigning all of the unexplained variance to "discrimination" can be an error.

A potential solution to this problem is to adopt a different style of research. Instead of relying on anecdotes, qualitative examples, or precollected data sets which require the use of the control variable approach, we should look for strategic research sites in which we can focus on the process of social selection. In doing so, it is possible to rule out the potential influence of both self-selectivity and role performance and therefore to conclude that inequality is a result of discrimination.

Studying Process Rather than Outcome

Thus far we have been critical of sociologists who use data on inequality to make assumptions about the processes creating inequality. If inequality, even after controlling for significant variables such as measures of role performance, cannot be taken as evidence for the operation of structural factors such as discrimination, how then can the sociologist demonstrate that discrimination is the cause of inequality? Even though it is substantially more difficult, in order to disentangle the causes of inequality, the sociologist must focus his or her research on the processes creating the inequality. This must be done in a quantitative and systematic way rather than depending, as so many do, on qualitative observations and statements by participants that they have been discriminated against. Although qualitative work is generally useful for gaining insight into a process, it alone cannot be used to "prove" anything in sociology.[15]

The work currently underway by J. R. Cole and Zuckerman is an attempt to use both qualitative and quantitative data to study the process through which inequality is generated among men and women in science. We have been studying why women have been less likely than men to enter the occupation of physician. Our research strategies and the problems we have encountered are relevant for studying inequality between men and women scientists and, we believe, for studying most types of observed social inequality between groups. We describe this research program as an illustration of how it is possible to zero in on the processes through which inequality is actually created.

First, we began with the premise that in order to study how inequality is created, we must break the process down into its component parts, collect data on each part, and then put the parts back together. If there are fewer women doctors than male doctors, this could be a result of conditions that influence men and women from the time they are born until the time they graduate from medical school and begin the practice of medicine. If discrimination is a cause of this inequality, where and how does it work? Our research focused on the end stage of the process: application to medical school.[16] In studying how applicants are treated (acceptances and rejections), we are studying the social selection process directly—and at its most critical point. Also, since we wanted to study *only* those people who had already applied to medical school, *cultural selection could be ruled out as a cause of the outcome: admission to medical school.*

Social scientists and historians writing about women in medicine have sometimes claimed that the small number of women in medical schools resulted from quotas on female applicants and sex discrimination in admissions policies (Walsh 1977). We found systematic data on *all* applications made to American medical schools since 1929.[17] (For a complete analysis and description of these data, see Cole 1986.) With this data set we could conduct a direct test of whether discrimination was operating to restrict women from entering the medical profession. If medical schools were discriminating against women applicants, we should see lower proportions of women applicants admitted than male applicants. The data show that this is not true. Since 1929, women applicants have been admitted to medical schools in essentially the same proportion as have male applicants. Even during the period after World War II, when there was a substantial increase in male applicants (as soldiers going to college on the G.I. bill later applied to medical school), there was no evidence that women were discriminated against by medical schools. In 1949, a year in which the number of male applicants to medical school was more than double the pre-war rate and the proportion of male applicants admitted was at an all-time low (29 percent), we found exactly the same proportion of female applicants admitted (29 percent).[18]

This research illustrates both the serious errors that social scientists can make by taking outcome as evidence for process and how the study of the social selection system itself can provide more convincing evidence on the extent to which an observed inequality is the result of discrimination. From 1929 to today the outcome has been that more males than females have been admitted to medical school. If outcome is used to assume process, we can use these data, as Walsh (1977) and others have done, to assume that the medical schools were practicing direct sex discrimination against females. But when we study the process, looking at how medical schools treated applications from female students, we find that throughout the entire period the medical schools were just as likely to accept women applicants as male applicants. This means that it would be an error to assume that the outcome of having fewer women in medicine could be explained by discriminatory admissions policies. This part of the study did not deal with whether the differential rates of male and female application were a result of prior discrimination; but it did enable us to reach a conclusion on whether discrimination was being practiced by the medical school admitting committees.[19]

Since this study found that the relatively small number of women in medicine could not be explained by discrimination by medical schools, it suggested that in order to understand the inequality we would have to find out why women were less likely to apply to them. This required us to study the process of occupational choice. Perhaps women were less likely to apply to medical school because they had been discriminated against in high school or college by being steered into academic programs which would not allow them to apply to medical school. Perhaps women did less well in science and mathematics courses and thus decided that they did not have the qualifications to enter medical school.[20] Or perhaps, for cultural reasons, women were socialized to believe that medicine was not an appropriate career for them. In fact, all of these explanations could be true; but without systematic data, we cannot assume a priori that any are true.

In order to determine the best way to study the processes leading to fewer women than men applying to medical school, it was necessary for us to pinpoint the time in life at which this decision was made. If women were less likely than men to want to become doctors from a very early age, this would indicate that the phenomenon we were studying could probably be explained by early childhood socialization. We began our research by studying the occupational choices of contemporary college students. Why, even today, are there two men applying to medical school for every woman applicant?

We found, surprisingly, that among a sample of more than 250,000 college freshmen in 1984, women were just as likely as men to aspire to be doctors (Fiorentine 1986). This means that despite the fact that men and women may be socialized differently as children or in high school, these experiences are unlikely to explain the lower application rate of women to medical school. This does not mean that there are no differences in the way in which men and women are socialized. In fact, we will argue that there are substantial differences. It simply means that these differences do not affect the decisions that men and women make at the time of entering college as to whether or not to become a physician. These data also suggested that it would be useful to examine the process of occupational choice of college students. If in the freshman year men and women are equally likely to want to become doctors and if in the senior year men are twice as likely as women actually to apply to medical school, the crucial process affecting the decision to apply must be occurring during the four years of undergraduate education.

We must then study the processes influencing persistence in aspiration to become a physician during college. This "persistence gap" could be the result of both self- and social selection processes. The social selection processes could consist of irrelevant criteria such as sex (discrimination) or relevant criteria such as grades in pre-med courses. Again, it would be necessary to have systematic quantitative data to understand this phenomenon. We collected data on undergraduates who graduated from the State University of New York at Stony Brook in the years 1981–1983. Our sample consisted of all students who began their education at Stony Brook by wanting to become physicians. Since the only data we had were the students' college transcripts, we had to use an indirect measure of this aspiration. We included in our sample all students who took three or more of the required pre-med courses during their freshman year.[21]

For the sample of close to 700 pre-med students we obtained a complete transcript of their college record and information on whether or not they had applied to one or more medical schools. This sample showed patterns virtually identical to those in the national data referred to earlier. As freshmen, almost half of the pre-med students (48 percent) were women. As seniors, only a third of those who completed the pre-med program and applied to one or more medical schools were women. Thus, at Stony Brook, there was a persistence gap between men and women pre-med students.

The first hypothesis we tested was that the gap might be explained by superior academic performance of men in the pre-med courses (functionally relevant social selection). This turned out not to be the case. Although men did do slightly better in the pre-med courses (although not better in overall GPA), this difference explained only a small amount of the persistence gap. When grades were controlled, women were still significantly less likely than men to persist in their aspirations to be a physician. The analysis, however, revealed a notable specification. There was no persistence gap among men and women pre-med students who received the highest grades. Thus, among those students who received an A − or higher grade point average, women were just as likely as men to persist in their aspiration to be a physician. The persistence gap was large at grade levels below A − . This suggested that women were less likely than men to persist when they encountered difficulty (as evidenced by receiving a low grade). (This research is reported in detail in Fiorentine 1987.)

In order to understand this persistence gap better, we conducted a series of qualitative interviews with both persisting and "defecting" men and women pre-med students. We then replicated our original study by collecting transcript data on Stony Brook freshmen entering between 1982 and 1985 and completing surveys with 542 students who began as pre-med students. The data from this survey are still being analyzed. There was no evidence in the qualitative interviews that women students believed that they were being discriminated against or discouraged. The survey contains more systematic data on this question; and preliminary analysis has found no evidence even of perceived discrimination. The pre-med courses at Stony Brook are generally very large classes in which most students are unlikely to be known personally by their instructors. Thus, there is very little opportunity for a teacher to discriminate even if there was the inclination to do so. The pre-med advisor at Stony Brook is a woman, and male and female students alike report being encouraged in their decisions to pursue careers in medicine. If any discrimination against women pre-med students is being practiced at Stony Brook, it is not an overt type but would have to consist of very subtle cues by which women might be more likely than men to be discouraged from pursuing medicine as a career. If the women pre-med students report that they were not discouraged, as those we surveyed did, it is hard to see the mechanisms through which a type of "subtle" discrimination is actually working.

Our qualitative interviews suggest (but certainly do not prove) that we must look to differing social attitudes toward the occupational achievement of men and women in order to explain the pre-med persistence gap. There are several existing cultural theories which could be used to explain the persistence gap observed between male and female pre-med students. The "role deviance" theory assumes that because there are few women who are physicians, the absence of role models will cause young women to decide that pursuing a career in medicine would involve violating norms which define appropriate female behavior (Douvan 1976). Thus, the use of the term "deviance." Although it is true that there is currently not a large number of women physicians, if the perception by young women that pursuing a career in medicine would be deviant, how can we explain that men and women are equally likely to want to become physicians when they enter college? If this theory were correct, we would expect it to influence high school students as well as college students. Also, why would "role deviance" not influence women who got A's but influence those who got lower grades?

Another cultural theory could be called the "role conflict" theory (Angrist and Almquist 1975). This theory assumes that women will not pursue high achievement occupations because they perceive the demands of these careers as conflicting with the demands of their primary roles as wives and mothers. In recent years, social attitudes toward women working have changed significantly. In the survey of pre-med students which we are currently analyzing, we find no support for the hypothesis that women perceive the pursuit of a career in medicine to be in conflict with their ability to marry and have children.

A third cultural theory is attribution or the "psychological deficits" theory (Deaux 1976). This theory posits that because of differences in the ways in which men and women are socialized, women have lower levels of self-confidence, which cause them to react differently to "failures" and to reduce their aspirations. Our research suggests that it is true that women pre-med students have lower levels of self-confidence than their male counterparts. The differences, however, are quite small. It is possible that this theory does explain part of the persistence gap we have observed.

We have put forth a different cultural theory, which we call "normative alternatives." This theory hypothesizes that both initial career choice and persistence in the pursuit of high status careers will be influenced by the prevailing cultural attitudes toward women working in specific occupations and by the cultural attitudes toward the significance of occupational success for men and women. There has been very substantial cultural change over the last 15 years in regard to the first dimension: attitudes toward women entering specific occupations. Today, it is not only accepted but even approved that women should pursue careers in such high status occupations as medicine, law, and business (Fiorentine 1988). There is still, however, a substantial culturally defined difference between the sexes in the importance of occupational success. Whereas occupational success is virtually the only route to adult status open to men, women can attain adult status in an affiliative way through marriage and family. Because there are normative alternatives open to women which are not open to men, there is substantially more pressure on men to be occupationally successful. This pressure leads men, on average, to be more willing to overcome obstacles and persist in the face of difficulty in their goals to enter high prestige occupations such as medicine.

We have collected some historical trend data which add credibility to the cultural explanation of gender differences in occupational outcomes. First, utilizing large nationwide surveys conducted by the American Council on Education, we have examined the career aspirations of college freshmen from 1969 through 1984 (Fiorentine 1988). In the early years, men were substantially more likely than women to want to enter high prestige occupations such as medicine, law, academia, and business. Today, women represent nearly 50 percent of the freshmen aspirant pool to these high prestige occupations. We have also examined the value systems of men and women college students over the same time period. In 1969 there were relatively large differences between men and women in the importance placed on such status attainment values as earning money and having a position of authority. In 1969, for example, 32 percent of a large nationwide sample of college freshmen women said that to be well off financially was "very important" or "essential." By 1984 this proportion had more than doubled, rising to 67 percent. Among men, the proportion giving this response in 1969 was 54 percent; in 1984 it was 76 percent. We can view these data in another way. Specifically, if we compare the responses of men and women in the two time periods, we find a 22 percentage point gap between the sexes in 1969 and a 9 percentage point gap in 1984. A complementary set of findings is also relevant. Whereas in 1969 "nurturance" values, such as having a family, were more important for women than were status-attainment values, in 1984 status-attainment values are as important to women as nurturance values.

We believe that these changes in values (importance of money and "status attainment") can explain the sharp increases in the proportion of women entering high status occupations since 1970. We also believe that it is persisting differences in culturally defined gender values which continue to explain the still lower levels of occupational achievement of women when compared with men.

Although we do not have any direct evidence yet, we hypothesize that our "normative alternatives" theory is a potentially useful explanation of gender differences in achievement in a wide range of occupations including science. Male scientists, we think, publish more than female scientists because it is more important to men to be occupationally successful than it is to women. Men know that they will be heavily evaluated by others on the basis of their occupational success. Lack of occupational success for men in our society is unequivocal; it means failure. The definition of occupational success is, of course, relative to the position occupied. For some scientists at places like Harvard or Stanford, failure could be defined as not being elected to the National Academy of Sciences. (See Merton's [1957] analysis of the concept of "relative deprivation.")

How exactly does our culture produce lower levels of achievement for women scientists? The normative alternatives theory assumes that gender culture is a very strong variable which influences everyone brought up in the society.[22] It is virtually impossible to be a male in the United States and not "know" that one will be evaluated on the basis of occupational attainment and its material rewards. It is also virtually impossible to be a female in the United States and not to learn that occupational success is good but *not necessary* for social approval. This would be true even for the most highly motivated and "driven" women, for whom occupational achievement is the most important thing in their lives. Even these women "know" that it is possible for women to receive social approval without attaining the same level of occupational success as men. If this is true, it means that achievement differences between men and women in a given occupation cannot be empirically explained by controlling for *any* individual level variables.

Consider occupational success as a result of a large number of variables such as class of origin, IQ, type of education, achievement motivation. Assume that we take all these variables into account and any other variables hypothesized or known to influence occupational attainment. We hypothesize that there will still be a sex difference in occupational attainment (in any given occupation) after these individual level variables are accounted for. For even among men and women who come from the same class background, have identical IQs, identical educational preparation, and high levels of achievement motivation as such concepts are measured by surveys, there will still be a difference between men and women *on average*. The men will still be under greater cultural pressure than the women to be occupationally successful. It will be more socially acceptable for a woman to "settle" for a lower level of occupational attainment than for a man to do the same.

The normative alternatives theory therefore predicts that as a result of this cultural variable male scientists, on average, will be more driven to strive for higher levels of attainment. They will, on average, be more likely to work longer hours, forgo vacations, try to obtain more grants and graduate students, and do all the other activities which result in publishing many papers. A woman who is a professor at a major university is already an enormous success. On average, she should experience less cultural pressure to concentrate all energies on occupational attainment.

Note that in this theory we use the term "on average." *This theory is aimed at explaining group, not individual, differences.* Thus, our problem here is not to explain why one woman will be more successful than another but to explain the group difference in occupational attainment between men and women. The focus of this study is not the dependent variable. There may be many variables far more important than gender in influencing the success of scientists. Our aim is not to understand the causes of success but the social and cultural mechanisms through which one variable (gender) influences success. Clearly, there will be many women who will have a greater drive than men to achieve occupationally. Indeed, some women scientists do achieve more and publish more than 95 percent of all male scientists. But if this theory is correct, the drive of women to achieve occupationally will *on average* be lower than that of men.[23]

The traditional way to test a cultural theory such as normative alternatives would be to try to specify the ways in which culturally defined gender values made men and women act or think differently, which then in turn influenced their achievement. Indeed, we did this above, when we said that pressure to be occupationally successful might cause male scientists on average, for example, to work longer hours than female scientists. Presumably a whole list of ways in which the cultural values affect men and women scientists could be developed, data collected, and an attempt made to eliminate the sex difference by controlling for these other variables.

Unfortunately, as Lieberson (1985) has pointed out, these traditional strategies for attempting to untangle the causes of complex phenomenon have not met with a great deal of success in sociology. In addition to the false methodological assumptions which are frequently made in this type of research, sociologists have made the error of confusing analytic with empirical processes. It is analytically possible to make a list of the ways in which gender culture might influence the behavior of men and women scientists. Empirically, however, culture may be so strong that its influence cannot be demonstrated by controlling for any set of analytically discernible effects. If, for example, gender culture was to cause men to "care" more about success, we would not be confident in the ability of sociologists to measure this with attitudinal questions on surveys. We believe that gender culture has such a powerful impact on women and men that it may be impossible to eliminate the observed difference in achievement between men and women by controlling for any individual level variable(s). Controlling for individual differences in order to "wash out" the influence of a cultural variable on which there is no variance is conducting the analysis at the incorrect level (ibid., chap. 5)

What we are implying is that gender culture is the basic cause for the average difference in achievement observed between men and women. The proper unit of analysis for this hypothesis is a specific society at a specific time. The best way to "test" this theory is to find societies in which the gender culture differs. If this theory is correct, then if we can find a society in which there is more nearly equal pressure on men and women to achieve occupationally, we should find similar levels of achievement. If it turns out that normative alternatives is a cultural universal in industrialized societies, then we should concentrate on measuring the degree to which pressure on men and women to achieve occupational success differs over time in various societies and see if this is correlated with gender differences in achievement.

The importance of cultural explanations has been downplayed in recent sociological research. Once culture is recognized by sociologists as a powerful determinant of inequality, new research strategies will be needed in order to demonstrate systematically how culture works. But even if culture or normative alternatives do turn out to be a basic cause of gender differences in occupational attainment, this does *not* mean that women are not the subjects of discrimination. We should also continue to look for strategic research sites where we can examine the operation of the social selection system directly in order to find out where and how discrimination has influenced the attainment of women.

EVELYN FOX KELLER

10. The Wo/Man Scientist: Issues of Sex and Gender in the Pursuit of Science

MY PURPOSE IN THIS ESSAY is to argue for a shift in the focus of our deliberations—from the subject of "women and the pursuit of science" to that of men *and* women in the pursuit of science. That is to say, I wish to insert a third pole into our discussion. By its very insertion, this pole raises a question where before there was none: a question about those very aspects of the relation between men and science that are assumed to be so normal and natural. Before we can adequately address the question of equity between men and women in the world of scientific pursuits, we must first find a way of introducing parity into the questions we ask—and our starting assumptions—about men's and women's relations to the pursuit of science.

On the surface, there appears to be a simple way to do this. It consists of denying that there are significant (i.e. relevant) differences between men and women, making of them one genre, namely, human. Indeed, this assumption is almost basic to Western liberal ideology,[1] even if it has not been honored in Western liberal practice. With such a move, our discussion again becomes bipolar, but with a significant difference: the two terms are no longer "women and science," but "humans and science." Conceptual parity is effectively introduced by fiat: men's and women's relations to the pursuit of science ought to be equal because these relations are the same. Any differences that might in fact be observed are due not to the

Prepared for the series of workshops on Women in Science convened by the editors of this volume in 1986.

nature of men, women, or science, but merely to the persistence of residual prejudices (i.e. expectations born out of ignorance).

It is this apparently simple approach that has been adopted by recent generations of American women scientists in their efforts to attain equal status in science. I will argue, however, that such a formulation is inadequate not only in practice, having failed to secure equity for women scientists, but also in principle. Above all, it does not take into account the extent to which beliefs (call them prejudices, if you will) are internalized by actual men and women, and even by the practices of science. Accordingly, collapsing the two terms, men and women, into one (colloquially, "man") prevents us from perceiving that "universal man" has historically been modeled on a particular cultural experience of manhood; it also obscures the ways in which norms of masculinity have been covertly absorbed into science.

True parity in conceptualizing the relations between men and women and the pursuit of science requires a more complex taxonomy—one that not only preserves all three terms (men, women, and science) but simultaneously acknowledges the socially constituted character of each term. This last point must be underscored, precisely because it is so frequently hidden from view. It requires us to make two crucial distinctions: one between sex and gender and the other between nature and science. If sex is a biological category into which we are born as male or female infants, gender is a cultural category that shapes our development into adult men and women. In this sense, gender represents a cultural transformation of sex. In much the same way, science represents (or gives to us in representation) a cultural transformation of nature. Nature, in short, does not appear to us unmediated. Representations of nature take their shape from the instruments, theories, and values that particular scientists bring to the task of "revealing" nature. Just as the character (and values) of a culture are reflected in its socially agreed upon definitions of masculinity and femininity (i.e. its gender ideals), so too, the particular instruments, theories, and values that scientists employ in their attempt to represent nature are reflected in the picture of nature that emerges from their desks and laboratories. In other words, while sex and nature might perhaps be said to be givens, gender and science cannot. This intrusion of culture—between sex and gender on the one hand, between nature and science on the other— fatally undermines any confidence we might have had in the monolithic character of either of the categories, gender or science.

In an attempt to illustrate the political and conceptual inadequacy of the "liberal" approach to the question of women in science (i.e. the denial of substantive and pertinent differences between men and women), I will

begin with a brief and necessarily sketchy review of the problems encountered by American women scientists who have employed it over the past
century. I will then discuss the advantages of the more complex taxonomy
I have suggested. First, however, it seems necessary to indicate the historical context in which a "liberal" strategy seemed a giant step forward.[2]

Historical Background

During the late nineteenth century, the strategies used by women who
aspired to enter the world of science were often aimed more toward accom-
modation than toward equity. As such, they might be described as "pre-
liberal." Many women scientists resigned themselves to (or sometimes
actively sought) a secondary demarcation within the realm of science.
They accepted a feminine subsphere within the larger male sphere, no
doubt hoping that a kind of equality could be attained within that smaller
realm. Although this strategy may have created educational opportunities,
its ultimate inadequacy as a professional strategy was soon recognized. It
doomed those women who had managed to enter the world of science to
women's work—i.e., work that was poorly remunerated, low in status,
and generally regarded as intellectually inferior (see, e.g., Rossiter 1982).

As a consequence, strategies of separation gave way to strategies of
integration. Here, women scientists saw a natural ally in scientific claims
to objectivity and neutrality. If science is independent of its producers,
there should be no place for gender (or, for that matter, for race, religion,
or other social markers) in scientific discourse. And if scientific minds are
truly disembodied, it is irrelevant whether the body of a scientist is male
or female. Accordingly, women with scientific minds should be able to
claim access to science equal to that of men.

The question in many people's minds was of course whether women
actually did have scientific minds—a question invoking the specter of
difference in a frame that by definition undermined women scientists'
claims to equity. For the last one hundred years, this question has defined
the principal contest between feminists and nonfeminists in the struggle to
define the relationship between women and science. By the early part of
this century, scientific feminists—both in the sense of feminists who were
scientists and in the sense of feminists who grounded their political theory
in scientific principles—had come to see how readily claims of difference
are translated into conditions of inequality. With this recognition came the
conviction that it was necessary to stake any claims to equality on the
repudiation of the myth of a feminine mind. In an effort to discredit traditional views of innate, biological differences between the sexes (particu-

larly in the domain of cognitive capacities) and to demonstrate the importance
of environmental and institutional factors, they worked hard to gather
psychological, sociological, and anthropological data on men and women.
Their need to refute claims of natural difference seemed absolute—on their
success, they thought, would rest the entire future of women in science.
So confident were they in the correctness of their views on the one hand,
and in the objectivity of science on the other, that they felt it was sufficient
to rely on agreed upon standards of scientific rigor to demonstrate and
expose the flawed reasoning pervading popular belief.

In retrospect, we can see that the confidence these women scientists
had in the standards of scientific rigor was excessive. Feminist scientists
were not the only ones interested in the science of sex differences, and in
time, the subject was wrested out of their hands. Indeed, some of their
data, collected to demonstrate the social obstacles facing women scientists,
was later reinterpreted by other scientists to demonstrate the futility of
increased opportunity for women in science.[3]

In one of the saddest episodes of women's history in this country, the
tide soon turned against these early feminists, eroding many of the gains
they had earlier made. By the 1950s, the proportion of women scientists
had declined to roughly half of what it had been during the early part of the
century. In 1956, almost a century after admitting Ellen Swallow Richards
as its first female student, MIT convened a special committee to consider
whether or not to continue admitting women students; the recommendation
of the committee was the termination of coeducation at the school (see
Keller 1981).

During this dark period, virtually all traces of the earlier feminists in
science, proudly asserting themselves as women even while insisting on
their intellectual equality, had vanished. In their place arose a generation
of women who sought to sustain their struggle to be scientists by tacitly
agreeing to expunge from their professional identity the fact that they were
women. They too were committed to equity, but in ways that differed from
those of their predecessors; they sought safety in the absence of any distinguishing characteristics, interests, or mental attributes that might mark
them as women. Women scientists in the 1950s sought survival not only
in the promised gender neutrality of science but in the promise of their own
gender neutrality as well. "Making it" meant making it as a scientist indistinguishable from other scientists. However, since other scientists were
male, this meant eradicating any sign of difference between themselves
and the men in their profession. If a difference was to be marked, it was
more likely to be their difference from other women. It is perhaps not

surprising, then, that, *qua* women, they effectively disappeared from American science. Their numerical representation was no longer assiduously recorded by academic feminists; in many cases, it was not recorded at all (Rossiter, personal communication; see also Keller 1981). Often, by their own choice, their tell-tale first names were withheld from publications.

The overall impression is that these women no longer relied on science to prove their intellectual equality as a class. They maintained absolute confidence in the fairness and objectivity of science to prove, and reward, their performance as individual scientists—especially if they succeeded in eradicating the marks of their membership in the gender-class from which they seemed to have escaped. Unfortunately, hindsight permits us to see that this strategy failed to protect these women scientists from the effects of an increasingly exclusionary professional policy—it only helped obscure the effects of that policy. Not only the numbers but also the status of women scientists (as individuals and as a class) deteriorated steadily.

Even when the fortunes of women in science began to improve in the late 1960s and early 1970s—partly in response to Sputnik and the impetus it gave to increasing the number of young people entering science, and partly in response to the emerging women's movement—women scientists' commitment to the idea that gender was irrelevant to the practice of science remained strong. A personal anecdote may be relevant here. In 1974, I decided to examine the subject of women in science in one of a series of lectures at the University of Maryland. I recall, even now, the trepidations I felt. Merely to introduce the question of women in science within a professional setting seemed then like a bold and even dangerous move.

Supported by national efforts to expand our scientific workforce, however, real improvements gradually began to be made. Women started to count themselves again, to recognize one another, and even to counsel one another. The dramatic improvement in the status of women at MIT over the last twenty years provides clear testimony to the effectiveness of such collective efforts. But the reclamation of their group consciousness as women, at MIT and elsewhere, simultaneously rekindled their other need to transform (if not to efface) male consciousness of them as women. It revived the paradox that Nancy Cott (1987) describes as having plagued the entire history of modern feminism—a paradox that faced women scientists with particular urgency. Almost as a consequence of the reemergence of their group identity as women, their commitment to intellectual sameness—to the repudiation of difference—returned with renewed force.

Once again, they found themselves fighting against claims of innate sex differences in mental attributes, particularly those which now came from within the scientific community itself.

These claims have undergone a particularly vigorous revival in recent years, keeping pace, as several authors have pointed out, with the successes of modern feminism. Accordingly, the scientific rebuttal of sex-difference claims has resurfaced as a principal concern of contemporary feminist scientists. Over the last decade, a number of critical reviews devoted to this end have been published, some even suggesting that the very inquiry into a biological basis for male and female behavioral differences is an inherently sexist endeavor.

The need for these efforts is obvious, but so is their vulnerability and (perhaps inevitable) defensiveness. Each new claim of a hormonal or physiological correlate to suspected or known sex-linked behavioral attributes needs to be freshly examined, its fatal undermining flaw identified and exposed. Fueled at least in part by continuing reports of observed differences in behavior and performance between men and women scientists (or between men and women students), this effort is ongoing, and, as the scientific quality of such research improves, it has become increasingly demanding. Furthermore, the very terms of the debate about sex differences lend support to characteristic kinds of cultural myopia. But before we can correct our short-sightedness, we must shift the discussion from "women and science," to "men, women, and science," bearing in mind, as we make this shift, that none of these categories is itself monolithic.

Typically, the debate about sex differences focuses on the nature or nurture of women and implicitly assumes the nature or nurture of men as a norm. In such discourse, gender is not understood as a culturally specific norm that refers equally to both men and women; it tends to be read and heard primarily (if not exclusively) as a reference to a universal category of women. Accordingly, the possibility remains unexamined that normative conventions of white middle-class male socialization may have been incorporated into the very standards employed in assessing the behavior and performance both of women and of men growing up under different norms. I offer three examples to illustrate this point.

One particular claim of differences between male and female mental attributes has become a principal focus for feminist critics in science; it derives from differences observed in the performances of girls and boys on Standard Aptitude Tests (Benbow and Stanley 1980). Critics have pointed out that these test scores correlate poorly, if at all, with creative performance of actual mathematicians; but the question of what these test scores *do* describe has not been pursued. To what extent are they merely correlates of what some boys in our culture are encouraged to do? To what extent do our routinely accepted criteria of scientific intelligence faithfully measure actual ability to perform intellectual tasks, and to what extent are they inadvertent reflections of particular gender-biased cultural norms?

A second example involves a possible cause of continuing inequity that was considered at a conference on women in science convened by the Macy Foundation in 1981.[4] Concluding that most traditional forms of discrimination were no longer visible, it was suggested that one definitive and ongoing handicap *could* be identified: women were not trained to be sufficiently competitive to survive and prosper at the cutting edge of contemporary research. The recommendation that follows from this observation seemed obvious: retrain women to be more competitive. The question that was *not* raised was whether the norms of competition currently accepted by scientists are culturally accepted norms of male socialization or whether they are necessary to good research. To be blunt, is it necessary or even good for physics that, as one eminent physicist has said, "only blunt, bright bastards make it in this business" (quoted in Traweek 1984)?

A third example is more banal. It has to do with the common practice of measuring scientific productivity by counting the total number of papers of which one is an author. The number of papers on which the name of a principal investigator appears will, in general, depend directly on the size of the laboratory or group he or she heads. Suppose that (for whatever social or psychological reasons) most women scientists prefer to lead small groups. These scientists would obviously tend to have a lower "productivity." If such measures of productivity are used to infer quality of science, it would follow that large labs (or groups) produce better science than small ones—a proposition that no one would accept if it were stated so directly.

Caught on the horns of an impossible dilemma, women scientists of the twentieth century have tended to seek equity through the refutation of difference claims. Historical experience has taught us the vulnerability and ultimate inadequacy of that strategy, but it has also taught us that assertions of difference tend in practice to be self-defeating. Acknowledgment of gender-based difference has almost invariably been employed as a justification for exclusion. To the extent that scientific performance is measured by a single scale, to be seen as different is to be found wanting. In the face of such a universal standard, the hope of equity, indeed, the very concept of equity, appears to depend on the disavowal of difference.[5]

In hindsight, however, one can see the pitfalls of this strategy: If a universal standard equates difference with inequality, the same standard would translate equality into sameness, guaranteeing the exclusion of any experience, perception, or value that *is* other. As a consequence, "others" are eligible for inclusion only to the extent that they can excise those differences and eradicate all traces of that excision. Yet such operations not only fail to provide effective protection against whatever de facto discrimination continues to prevail; they often prove only partially successful, leaving in their wake residual scars that prevent those who do survive from becoming fully effective "competitors." Successful assimilation has thus tended to require not *equal* ability, but *extra* ability—the extra ability to compensate for the hidden costs incurred by the denial or suppression of a past history as "other."

As long as we accept a conception of science as a monolithic venture—defined by a single goal and a single standard of success—*neither* the assertion *nor* the denial of difference can procure equity for women in science. Indeed, the very assumption of a universal standard mitigates against equality for the carriers of any residual difference, whatever its source. This is *the* dilemma that has entrapped white women scientists throughout their entire history, much as it now entraps scientists (men and women) of color.

Fortunately, our understanding of the nature of scientific knowledge has moved a long way from such a univocal conception during the last few decades. Recent developments in the history and philosophy of science have led to a reevaluation in which it is acknowledged that the goals, methods, theories, and even the actual data of the natural science are *not* written *in* nature; *all* are subject to the inevitable play of social forces. Empirical and logical exigencies may constrain the representations of nature that emerge from the desks and laboratories of scientists, but they do not determine them. Social, psychological, and political norms are inescapable, and they too influence the questions we ask, the methods we choose, the explanations we find satisfying, and even the data we deem worthy of recording.

Such a shift in our conception of science provides a way out of the dilemma faced by earlier feminists in science. To the extent that we acknowledge a multiplicity of goals and standards in science, it becomes possible (at least in principle) to argue for the inclusion of difference—in experience, perceptions, and values—as intrinsically valuable to the production of science; hence, it becomes possible to envision equality without sameness. But a trap resides in this proposition as well—a trap that derives from the familiar and widespread temptation to map difference onto sex. Behind this temptation is the assumption that difference means duality and implies that women as a class will do a different (or "feminine") kind of science. However, such a proposition ignores the lessons we have learned from the history of women in science. It also fails to do justice to the enormous variability evident among actual women.

If recent work in the history and philosophy of science has sensitized us to the influence of social forces in the development of science, recent work in feminist scholarship has sensitized us to the role of social forces in shaping the development of men and women—i.e., in defining gender norms. It has also sensitized us to the historical importance of the complex psychosocial dynamics that have woven contemporary norms of gender and of science into an inextricable web. The proposition that, under conditions of equality, women would do a different kind of science, a more "feminine science," ignores the extent to which both women and "femininity" are socially constructed categories. More importantly, it ignores the extent to which our contemporary conceptions of femininity and science have been constructed in opposition to each other. If science has come to mean objectivity, reason, dispassion, and power, femininity has come to mean everything that science is not: subjectivity, feeling, passion, and impotence. Ignoring the force of cultural labels in the construction of these categories invites their acceptance as "natural"; for this reason, it militates against the possibility of reconstructing or reevaluating the labels themselves. Furthermore, if we do not attend to the force of cultural dynamics in the construction of the norms that have defined the words "masculine," "feminine," and "scientific," we remain oblivious to the uses made of these constructions.

I have argued elsewhere (Keller 1985) that the exclusion of the feminine from science has pertained to a particular definition of science: science as incontrovertibly objective, universal, impersonal—and also masculine. Such a definition both helps insure the invulnerability of science in the face of social criticism, *and* serves to demarcate male from female. It is a definition that sustains and is sustained by a division of emotional and intellectual labor—a division along the lines of sex.

In the past as in the present, this sexual division of labor has provided critical support for just those claims that science makes to a univocal and hence absolute epistemic authority. That same authority has, in turn, served to denigrate the entire excluded realm of those values that have been labeled "feminine." Because science itself plays a role in these complex cultural and historical dynamics, any discussion of men and women in science must take into account the presence of gender-laden cultural biases in the very definition of science. Failure to do so is to overlook an important channel through which particular cultural values are imported into the norms invoked for distinguishing "good" science from "bad" science. It is in just this process that de facto discrimination is often practiced against individuals or groups who happen to bring with them any of those values that, for reasons having nothing to do with scientific productivity, have been prejudged as undesirable.[6]

Finally, it is in the name of scientific productivity as well as equity that we must recognize both the value of difference *and* its extent. The enormous variability (both cultural and individual) that exists among actual men and women goes well beyond biological variability, either between or within the categories of male and female (i.e. sex). First, there is the cultural variability between different concepts of "masculinity" and "femininity" (i.e. gender). But in addition, there may be even greater variability in the degree to which individual men and women conform to or diverge from the gender norms (or stereotypes) of their particular cultural frames. To forget these last sources of difference is both to ignore the diversity of human culture and to do an injustice to those women scientists whose very existence as scientists has required their transcendance of the stereotypes of our own cultural heritage. If the denial of difference between men and women has proved to be neither functional for women scientists nor realistic in its application, the denial of differences among men and women can certainly fare no better.

III

WOMEN'S
CAREERS

The Obstacle Course

CYNTHIA FUCHS EPSTEIN

11. Constraints on Excellence: Structural and Cultural Barriers to the Recognition and Demonstration of Achievement

IT IS WIDELY RECOGNIZED that women have made few contributions to science, the professions, and the arts compared to those of men. Many have asked why this is so. The differences in performance clearly suggest to some that the answer is obviously related to women's psychological, cognitive, and analytic incapacities. These observers have paid scant attention to the social context that produces and inhibits productivity, relying instead on unproven suppositions regarding the nature of competence, talent, and genius, as well as the distribution of these attributes in men and women.

A cultural belief held widely in the United States and elsewhere proposes that great scientists and artists possess innate traits that generate achievements, that such individuals develop independent of external situations or influences, and that the products of their imagination or skill, if powerful enough, somehow will rise to the surface—that "genius will out." Those who do not add to knowledge or high culture are believed generally to lack the capacity to do so.

The assumptions which explain individual achievement or incapacity also tend to explain the collective behavior of groups. The greater or lesser productivity of one group at any moment in time tends to be linked to some basic quality or capacity of the individuals comprising the group. Germans and Swedes have enjoyed the label of industriousness. Now the sharp increase in Japanese productivity (and American insufficiency) is frequently attributed to a change in the American character rather than to such "external" institutional directives, such as government encouragement of certain industries or inattention to others.

The model holds that inborn traits and character development are important sources of creativity. This notion centers on a belief in the integrity of the individual independent of the group and, in part, on a belief in a just world, one with a physical and moral order in which intrinsic worth is revealed and ultimately rewarded.

To some extent, this reasoning follows observation of empirical reality. Some individuals and some groups do produce better science and art more regularly than do others; some groups produce very little. But sociological insight reveals that the empirical pattern of one situation may be repeated for different reasons. As Robert K. Merton (1957) has observed in his analysis of the dynamics of the self-fulfilling prophecy, when a statistical regularity is observed a belief develops that this *should* be the case, and furthermore, that there is good reason for the phenomenon. This reasoning leads to the repetition of the pattern. For example, the Soviet Union, having achieved excellence in the quality of its violinists and chess players, devotes considerable resources to training and supporting its youth in these endeavors, thereby continuing to produce world-class performers. The United States, where resources for such development are more haphazardly allocated, lags behind.

But encouragement and support are hardly the whole story. Deterrence is another possibility. Certain groups and individuals have encountered resistance to demonstrations of ability; they have been denied forums, or positions, or access to a network of colleagues. Certainly, women as a class and a group have systematically encountered resistance of this kind.

Because there are few well-known women in creative fields, it is assumed that women have rarely contributed to the creative process. We do not possess data to confirm or refute this. But it is common to assume that women are more free today to express whatever creativity they may possess because they are less intellectually and physically constrained by the restrictions imposed by their reproductive roles.

This seems logical, but is it? Science—i.e., "scientific" endeavors such as invention and discovery leading to generalizable principles—and art may not have been sex-segregated activities early in history. Perhaps women were the creators of the paintings in the Lascaux caves or the discoverers of fire. Invention and art were, after all, done at home. Limiting the scientific enterprise to special locales (away from the home and in the university laboratory) is new. It effectively limits the pool of scientific workers to those having access to such locales, as did the concentration of education within the Church. Restriction of location, while permitting concentration of tasks and resources, also restricts the pool of scientific personnel to the like-minded and similarly experienced. Science today rewards the specialist and devalues the contribution of generalists; its specialties have boundaries which block "outsiders," even from other scientific disciplines (Barber 1952).

Today, the level of sophistication and the theoretical abstraction of science has also institutionalized esoteric behavior in practice. As scientific endeavor has become increasingly complex, it has also become increasingly insulated and exclusionary. Further, the development of specialized knowledge has been accompanied by the development of an occupational community[1] long characterized by selective sponsorship and selective access (Ben-David 1971; Goode 1957). There are as many excuses to justify the boundaries of professional communities as there are defenses of geographic borders. The definition of science as "men's work" is one, although other traditional boundaries have included the restriction of members of certain religious or national groups as well. As Magali Sarfatti Larson (1977) has pointed out, both the medical and legal professions created "monopolies of competence" by such practices as restricting education to elite institutions, eliminating evening programs, and denying access to members of certain ethnic groups and the working class.

Professional communities also prescribe styles of work, sequence of preparation, and location. Expectations and evaluations of performance accompany these designations,[2] and justifications for the appropriateness of these designations abound. As a result, gatekeepers in the professional world have suggested that women do not have the capacity to think abstractly, are not objective, and do not have the necessary commitment to practice the professions or to do science (Lorber 1985; Rossiter 1982). Even today, the issue of women's capacity to think scientifically is far from being put to rest, and there remains considerable debate[3] as to whether women can think "as well" (i.e., as abstractly and objectively) as men. This debate emanates from a cultural disposition to see differences between the sexes in dichotomized ways (Epstein 1985) and not to take into account intra-gender variation. This view suggests social outcomes—such as the smaller proportion of women engaged in scientific work—are in fact caused by women's own attributes and choices.

A Structural-Cultural View

A body of social science research over the past several decades has examined and analyzed the social context in which women's choices are made and their capacities are developed. The view offered here will explain women's disproportionately low participation in science and in all the professions, based on this research. I propose that women's participation as producers of knowledge or other creative work in society is not primarily the result of women's independent choices or abilities. Rather, the low proportion of women participants in these activities reflects the priorities and choices made in the society as a whole and within each field, including science itself, and its interaction with other institutions, such as education and government. Similarly, the traditionally low proportion of women in certain occupations is a reflection of cultural attitudes and institutionalized practices directly affecting them. I further suggest that the perception of women's contributions and, indeed, the definition of what a contribution is, are also determined by cultural values and institutional frameworks.

For example, quilts, identified as women's artistic expression, were never considered to be "art" until recently. Their painterly qualities were recognized only in the early 1970s, when the Whitney Museum in New York hung them in a show called "Abstract Design in American Quilts" after some reassessment that was probably precipitated by activity of feminist art groups.[4] And in science, Rosalind Franklin's research that indicated the structure of DNA was regarded as "dull" and downplayed in comparison to the honors accorded Crick and Watson, no doubt because Franklin was a woman.

In the social sciences, topics associated with women—their work, social position, and so on—were (and are) considered less important than those regarded as gender neutral or those focused on men's activity. Thus, women contributing to research on women have their work labeled as "women's studies" and do not qualify as preferred candidates for posts in top-ranking university departments.

Because social scientists, like other people, tend to frame their analysis and discourse within a culture that attributes different qualities and aptitudes to men and women, it is not surprising that the discussion about women's achievements reflects presuppositions based upon attitudes toward men and women as well as attitudes toward excellence in general.

Americans prize individualism and believe in individuals' ability to control their destiny and exploit their natural talents. This notion is reflected in one strain of scientific thought through beliefs that the individual has control over his or her destiny, that excellence will out, that people can pull themselves up by their own bootstraps, and that the meritocratic ideal is translated into behavior. At the same time, and not entirely consistently, Americans believe that people are "born" and not "made," that what people become is a function of their family heritage and early experience. There is a strong tendency to ascribe talent, aptitude, and ability to individual biology. Great social scientists from Mill to Marx to Merton have recognized the ways in which people's location in the social strata structures their choices, at times narrowing or widening opportunity; yet, in society, this perspective is often at odds with the beliefs which focus on the individual. At the same time, there is a tradition that focuses on people's rootedness in groups or in social categories (class, educational status or occupation). The discourse of analysis wavers: the same people who believe in the autonomy of the individual also believe that individuals manifest the characteristics of their groups or class. Of course, these "characteristics" are often stereotyped views.

The most useful approach does of course incorporate both sets of beliefs. It is obvious to most social scientists that individuals can express preferences and make choices, including the choice of what they "will be," and that their choices are tempered and organized by social structure. However, social scientists have not sufficiently identified the processes by which structural factors affect choice. Most studies have focused on the statistical association of structural variables with behavior and attitudes but have not identified the process by which it occurs.

Although considerable attention has been paid to socialization, the acquisition of values through imitation, education, and other forms of social learning, we actually know little about the process of learning itself. Furthermore, the extant model assumes that it is a process most powerful early in life, creating personality and competency in certain areas. Early socialization is more presumed than proved, and it is assumed to be the "cause" of many gender differences in adult behavior. From my review of research (Epstein 1988), it seems more plausible to accept the model of socialization as an ongoing evolving process or as one marked by abrupt changes caused by events. Later socializing experiences can certainly be as powerful as earlier ones. (See Brim 1976, Brim and Kagan 1980; Epstein 1988.)

Furthermore, excessive theoretical reliance on socialization obscures the impact of social control. No society leaves behavior to the "choice" of its members. Social controls are regularized in institutions and fortified by ideological doctrines, structure discussion, attitude, and action. There is rarely a perfect fit between individual choices and cultural prescriptions, except under conditions of perfect coercion. For example, even in the past, some men and women chose work nontraditional for their sex.

The influence of social control must not be underestimated. Rules, laws, and the use of force have structured women's access to education and training; and so have convention and persuasion. Mendelsohn's sister may have been *persuaded* not to claim authorship of her compositions and to affix her brother's name to her own work (Marek 1972); and formal institutional rules have forbidden women to play in certain orchestras.

Women were denied admission to most professional schools until the turn of the century and then faced strong informal quotas. The same held for women seeking scientific training. They were simply not welcome in a number of the great university departments, and some schools maintained formal restrictions for decades after. Harvard, one of the most important elite law schools (because of its size and ability to place its students with influential firms and courts), did not accept women until 1950, nor were women admitted to the Harvard Business School until 1963. Other institutional restrictions which prevented women's development of intellectual skills were also common. Important symbolically as well as practically for women on the Harvard campus was the ban on women's use of some of the university's libraries until quite recently.[5]

Elisabeth Hansot and David Tyack (1988) have reminded us in their analysis of gender distinctions and similarities in the American educational system that social institutions (such as schools) vary considerably in the ways in which gender is made salient. They propose that an institutional approach to the study of gender distinctions in behavior, rather than an emphasis on self-selection or individual characteristics, better explains the clustering pattern of women and men in certain fields. The institutional approach reveals how institutions "allocate money, status and power; they have goals, means, resources, boundaries and systems of control." As they suggest, "institutions sometimes have explicit rules governing gender relationships, but they also have organizational cultures in which many gender practices are implicit, often the more powerful for being taken for granted." The added focus on what has been called "the new institutionalism" by James March and Johan Olsen (1984) identifies how institutions are "political actors in their own right" and are not simply reducible to exogenous forces or to the individuals who compose them. Equally important, "institutions have linkages with each other and mutual interactions that help to define the terrain in which they operate" (p. 735).

The emphasis on institutional controls and the reconceptualization that identifies an institution's "intent" is nowhere seen more clearly than in the legal profession. What holds for the law may well hold in science. In my studies of women in the legal profession, I sought to examine the problems women encounter in a male-dominated occupational sphere (Epstein 1968, 1981). Historically, legal institutions excluded women from professional practice. As a result, women found it difficult to get training or gain admission to law schools, and rules and traditions severely limited their opportunities to practice. These practices were justified by wider appeals to recognize the basic differences between the sexes. Both the skills and emotional makeup of women were viewed as inappropriate to the practice of law. Thus, stereotypes served the purpose of limiting women's entry into the law, as seen in the decision of Chief Justice C. J. Ryan of the Wisconsin State Supreme Court in 1875, when he argued against admitting Lavinia Goodell to the bar of that state:

Nature has tempered woman as little for the judicial conflicts of the courtroom as for the physical conflicts of the battlefield. Woman is modeled for gentler and better things. Our . . . profession has essentially and habitually to do with all that is selfish and extortionate, knavish and criminal, coarse and brutal, repulsive and obscene in human life. It would be revolting to all female sense of innocence and sanctity of their sex . . . and faith in woman on which hinge all the better affections and humanities of life. (Sachs and Wilson 1978: 96–97)

Ultimately, of course, women were admitted to the Wisconsin bar, as they were to the other bars in the country, but the views about women being unfit for courtroom strife and that their efforts should be devoted to "better things" were still prevalent in the 1960s and were seriously challenged only in the 1970s.[6]

In the past, stereotypes provided rationales for those who sought to exclude women from the legal profession, although some gatekeepers, like Dean Harlan Stone of the Columbia Law School, provided no rationale at all when he rejected a prospective applicant in the 1920s simply because she was a woman. When I interviewed her years later as part of my study of women lawyers, she told me that she asked him why, and the thundering reply from the soon-to-be chief justice of the United States, exercising the power of his university office,[7] was "We don't because we don't."

Restriction on women's participation had to affect their competence as professionals as well as their ability to meet the criteria set by the professions. I am not alluding here to what is commonly referred to as a male standard of behavior or style but rather to women's competence to engage in legal analysis and debate, to contribute to science, and to practice as competent diagnosticians and technically able practitioners in medicine. We note that almost all achievers in these realms have had to engage in active professional settings in which productivity was a valued goal, to be members of organizations with resources at their disposal, to have able assistants, and to be associated with acknowledged masters (cf. Zucker- man 1977, for example). I have elsewhere noted (Epstein 1970) that in the past, women mainly have excelled primarily in those occupations in which one can act alone.[8] Women have had outstanding careers as singers and writers, but seldom as conductors or directors. Only now, when women have more power and the authority that goes with it, can they produce music, art, and science that is dependent on equal participation in or leadership of teams of co-workers and for which they have been permitted to attain the necessary skills to become proficient and inspire the confi- dence of others.

It is important to note that not only have women been at a disadvantage in the professions responsible for the practice and the development of knowledge, but these institutions have also been important in creating the rules in society by which women have had to live. Thus, professions such as law, medicine, and science have also provided some of the standards by which women have been appraised and judged to be competent or incom- petent as human beings. For example, the law was divided for several decades over the issue of whether or not women could be considered "persons." Albie Sachs and Joan Hoff Wilson (1978) have documented the legal decisions both in Britain and in the United States which defined women as members of family units under the protection of a male head of household and made it impossible for them to be considered autonomous and independent persons. These made women dependent upon men eco- nomically and politically. Subject to men's wills, women did not have control over their time in order to study, write, or think. Only unmarried women or those whose husbands were particularly sympathetic to their interests could engage in scholarly or professional activity.

Medicine also helped to create definitions of women's nature, accord- ing to a series of models outlined in the work of Barbara Ehrenreich and Deirdre English. Ehrenreich and English (1978) pointed out that women had long been defined by a model of physical vulnerability which made them subject to numerous "experts" who have designed regimes of care and evaluation of their physical and mental life.[9] As a result, many women have been denied participation in the normal affairs of the public world because of their presumed propensity to mental instability and physical fragility. One ought also include the models offered by the psychoanalytic establishment which suggest that women's nature would cause them to turn inward, making exploration, adventure, and assertiveness implausi- ble—traits considered important to participation in the major professional and economic institutions. Scientists and social scientists also have been responsible for defining women's abilities as different from and lesser than men's and in denying them access to training and thus providing rationales for creating deficiencies in competence.

Ironically, one residue of the old models of women's nature is to be found in the work of the feminist psychologist Carol Gilligan, who accepts and advocates the notion that women possess different psyches than men, although she insinuates that their different orientation to the world is to be valued. Gilligan and the social scientists sympathetic to this view have not produced convincing evidence to support their contentions about differences except when reporting attitudes, as opposed to behaviors. Lilli Hornig, the executive director of Higher Education Resources Services, pointed out (personal communication, May 27, 1983) that women, like men, hold stereotypical views about themselves which show up in polls and which are interpreted as real differences (as opposed to biases in spheres as wide-ranging as political behavior and personality traits).

However, striking changes in society have opened opportunities to women and have produced a surge in their professional activity and contributions in recent years. During this time, we have seen that there is broad overlap in the demonstrated competence of men and women entering professional schools. For example, my work in law shows that women have performed as well as men on the same examinations (LSAT) and by other objectively determined criteria, such as performance on the job.[10] Judith Lorber's study of women physicians and her overview of research shows that women also do as well as men in medical training. Neither their scores on the National Board of Medical Examiners (NBME) nor their ratings by faculty in their junior year were significantly different (Lorber 1985).

The study of women in the professions has been useful not only in explaining quite revolutionary changes in many women's roles in our society but also in challenging explanatory modes of behavior for both sexes. The striking change in women's experiences substantially altered their views of themselves and their competence, and the evidence lies in their pursuit of careers in science, medicine, and law. These adult experiences resulting in personality change support a view of socialization as an ongoing process; opportunity structure is an extraordinarily important variable in creating motivation.

In law schools, women increasingly have achieved the highest recognition as editors of the top law reviews in the country—and have gone on to become clerks to judges in the federal courts, also a highly sought position, one that leads to placement later in the largest and most influential law firms, judicial posts, and law school professorships.

Furthermore, we have learned just how much the work itself creates change. For example, we have learned how people develop competence on the job and how competence increases with experience. My study of lawyers illustrates this well. Women lawyers who were originally motivated to take back seats as professionals because they felt it was inappropriate to demonstrate assertive behavior often found that when they were urged or required to become courtroom litigators they enjoyed this role and performed well. Brilliant oratory and thinking on one's feet require practice. Many academics experience heightened gratification as writers and teachers in their careers as professors or researchers because it is a requirement of the university that they teach and publish. Commitment to these pursuits is generated "on the job" and through the occupational experience—not because of early preferences.

The thrust of research has led us to beware of any notion of differences based on an innate sense or early socialization processes. Yet there are differences between men and women, stemming from social structures and the cultural views that interact with them, that position women in peripheral and marginal positions, whether by choice or not.

In the past, women were not only discouraged from professional activity, they were actively prevented from engaging in it. Quotas kept their representation in graduate and other professional schools low, and knowledge about prejudice in employment kept applications low as well. Although some schools lagged in admitting women to professional education—Harvard only admitted women to its medical school in 1945 and, as noted, to its law school in 1950 and business school in 1963—some were receptive at an earlier period. In some schools at the turn of the century, women were enrolled in percentages similar to those of today (42 percent at Tufts in 1900; 26 percent of the graduating class of Boston University in 1902). The evidence shows that they studied and graduated with excellence (Walsh 1977). At Boston University, for example, they accounted for 60 percent of the honors! Yet, finding work was difficult for women professionals. Many hospitals and law firms refused to hire them, and those that did made their life on the job difficult by directing them into dead-end positions and refusing to reward their efforts with the usual promotions and pay increases.

We will never know how many women might have become brilliant research scientists, engineers, or lawyers if they had had the same opportunities for access to training programs or the same on-the-job apprenticeships as men. We know, from the survivors who came to prominence in a different, more hospitable period, that the talent pool existed. Sandra Day O'Connor, the first woman Supreme Court justice, faced slammed doors at law firms in 1952 after graduating third in her class at Stanford Law

School—except for one firm, which offered her a job as a legal secretary. Ruth Bader Ginsburg, the first tenured woman professor of law at Columbia University (now a federal court judge in Washington, D.C.) graduated first in her class at Columbia Law School in 1957, but could not get a job, and took a fellowship year in Sweden instead.[11] Rosalyn Yalow, Gerty Cori, Maria Mayer, Barbara McClintock, all Nobel laureates, were offered professorships only belatedly, none as early as men of comparable achievement. At the turn of the century, laureate Marie Curie toiled in the basement of a research institute, working outside the French science establishment under inhospitable conditions.

Some women researchers simply had their contributions plundered by men who walked off with the prizes. In England, Rosalind Franklin's x-ray spectograph evidence of the double helix structure of DNA was pilfered by Crick and Watson, who went on to publish the proof of its structure and become Nobel laureates. Jocelyn Bell Burnell's discovery of pulsars was attributed to her senior co-workers, who carried off the Nobel Prize. Controversy about the importance of their contributions still surrounds the cases of Franklin and Burnell, but it is clear that neither woman found their work environment as hospitable as did their male colleagues.

Women have contributed to our fund of knowledge for generations (Epstein 1976), although they have suffered from inattention to or invisibility of their accomplishments. Part of this is due to culture and part to structure. Even today, women in academe proportionately hold more research, non-tenure-line positions than men. Estimates range from 13 to 20 percent, or about two to three times the proportion of men (Zuckerman 1987:133). In the past, women were not permitted to control their money or property; and, until very recently, they were denied control of their intellectual property. Some inventions credited to men were actually conceived by women, who did not have the legal right to own property or to control their earnings during the nineteenth century. Fred Amram, who has researched women inventors, has found evidence that the McCormick reaper, patented by Cyrus, was actually invented by his wife, a seamstress (*On Campus with Women*, Fall 1981). In intellectual life, mired in permanent jobs as research assistants, women often found their ideas incorporated into the work of men, sometimes being generally acknowledged but often not credited through footnotes and coauthorship. This was especially so for women who did not hold titled positions in organizations but who contributed to their husbands' work as research assistants and editors. In the past decade, it is curious to note how many men's work, previously published under their own names but with acknowledged assistance of their wives, is now being published under both names. Ariel Durant began to publish with her husband, Will Durant, for example, in their famous series, *The Story of Civilization,* in 1961, after he had published six volumes under his own name. Another example is the social scientist Gabriel Kolko, whose acknowledgment to his wife Joyce in *Wealth and Power in America* (1962) states, "This book is in every sense a joint enterprise and the first in a series of critical studies on which we're presently engaged" (p. xi). Joyce Kolko appeared later (1972) as the coauthor of a subsequent book.

It is difficult to document the actual contributions of women in research, writing, and artistic settings. Those who had personal ties to the men they served did so "voluntarily" and without a sense that they were part of a behavior pattern that characterized the lives of most talented women. Without a hope of forging an independent career, it seemed a good compromise to integrate intellectual work with family devotion. Choices were also organized by age and position. Most women associate with men who are also older and more established, and such preferences are backed by authority. A few accounts of renowned men in the arts whose wives also have elicited biographical notice hint at the social processes that diminish women's contributions and notice. In music, such women as Cosima Wagner (the wife of Richard Wagner and daughter of Franz Lizst) or Alma Mahler (the wife of Gustav Mahler, who also was married to or involved with the novelist Franz Werfel, the architect Walter Gropius and the artist Oskar Kokoschka) did not consider independent careers. A composer before her marriage, Alma Mahler did orchestration for her husband, but he forbade her to write music independently after their marriage (Marek 1972). Zelda Fitzgerald was warned by F. Scott not to refer to their common experiences in her writing because the material belonged to him as resources for his own fiction (Milford 1970). Zelda Fitzgerald deferred to this request with distress, but many other women benefited from a kind of "family wage" paid in prestige as well as money and the institutionalized role of reflected glory that was an expected mode of reward well into the 1960s. Few women rebelled. Many even accepted this as "just."

In his analysis of distributive justice, George Homans (1961) has noted that women cannot expect the same investment reward as men. According to his assessment, women have an unchanging value which they cannot increase with experience, as men can. Nor can they expect the same rewards. When their work is not acknowledged by husbands, fathers, or lovers, many have found comfort in expressions of personal affection as alternative rewards. Of course, until recently in professional life, family ties offered women an access which they might not have had if they were independent practitioners. Given a blocked opportunity structure, many professional women had to rely on particularistic relationships with men who could offer training and jobs, while men might follow a more universalistically determined path.[12]

In certain spheres, however, women did have more possibilities to demonstrate independent competence. Like black people, women in law and medicine found some opportunity in government service (although this was not so for women scientists according to Zuckerman and Cole, personal communication, 1987), where one could find employment through competitive examinations and decreased competition from white men, who preferred the greater prestige and economic opportunities of practice and academia. But even in this "haven of universalism," as Robert Merton has called it, women and other minorities were segregated and stratified in particular employment "ghettos." Few were elevated into the top 2 percent of the civil service, or headed laboratories, or were permitted to practice in the office of the attorney general until the 1970s. There is also considerable variation, of course, in the extent to which universalism or particularism marks the possibilities of mobility for men. Certain societies are more open than others in the extent to which traditional patterns, networks of association, and circumstances of birth "count."

Stratification and ghettoization are also characteristic of most professional domains. Institutions position women, and powerful individuals within these institutions do not commonly challenge tradition by crossing these lines by personally sponsoring women. Of course, the women who had sponsors were fortunate; their male mentors were among the few who were willing to resist convention. And because women did get some training and serve as assistants and invisible helpers, the pool of women's talent has not been entirely lost, unlike that of most black people, who do not have a chance to contribute at all. But it is in the nature of human motivation that when people are not appropriately rewarded for their efforts and contributions, they cease to aim high, or to exert the extra effort that characterizes those who push knowledge forward. Thus, today, we recognize as great only the hardiest, most autonomously driven women, who survived their punishing experiences. We also honor the handful of women who managed to collaborate with male mentors who were deviant in their time and institutions and were intelligent enough, secure enough, visionary enough, and sometimes just practical enough to encourage and promote the careers of able women.

Other patterns followed by women in professional life in the past show how institutions shaped women's careers. Trained in medicine or law, women practiced in a limited number of specialties: in medicine, psychiatry, dermatology, and pediatrics; not surgery or academic medicine (also closed to Jews until the end of World War II). In law, the areas of wills and estates, matrimonial cases, and the backroom work of research were considered women's work. Women were not chosen as law clerks or offered positions in academic law teaching. Women were not seen or heard in the courtroom except as unpaid litigators (in legal aid situations) for indigent clients. Many reasons were given for the specialties—again, the rationales focused on women's alleged incapacity to bear up in a specialty like surgery or to endure the courtroom conflict involved in litigation. The rationale was that the professional areas in which they clustered were those best suited to women's psyches and to family obligation. Yet, the sociologist's eye sees a pattern that typified women's specialties. The reality was that these areas were lower in prestige and offered lower remuneration (Epstein 1970). Women themselves were often heard to parrot the reasons men offered for this segregation within professions (Epstein 1981). In many cases, women were persuaded to become participants in their own exclusion.

It is important to note that until the 1970s, women who were admitted to professional schools were subjected to especially trying ordeals. In law schools, symbolic segregation and the undermining of women students was common. In some, professors held "ladies' days" when only women students were called upon. Dean Harriet Rabb of Columbia University, a top litigator in a number of EEOC cases, has recalled the practice from her days as a law student at Columbia. One professor asked, "Will all the little virgins please come to the front of the class?" while another obliged women to rise at the beginning of class on Valentine's Day and remain standing for the entire hour while he would ask them embarrassing and difficult-to-discuss problems. "Better go back to the kitchen," was one Harvard professor's retort to women students who stumbled in recitation in the days when the grueling "Paper Chase" mode of classroom interaction between students and professors was the norm (Epstein 1981). These are only a few examples of the gender-related ridicule women students had to endure.

Medical schools, of course, had their own brand of harassment, often in the form of practical jokes and black humor. Women students sometimes had to confront male cadavers whose genitals had been substituted with those belonging to a body of another race and denigrating questions on rounds. The fine line between humor and harassment was always difficult. If women objected to the treatment they received they were faulted for lacking a sense of humor, though they knew the humor was meant to humiliate them rather than initiate them into the inner group. In a recent M.A. thesis by Betsy Salkind, "Can't You Take a Joke?," women students at the Massachusetts Institute of Technology were found to be the main targets of peer harassment, some if it sexual (Project on the Status and Education of Women 1986;5).

Social Change and the Creation of Competence

Change on many fronts was accomplished by changes in the law and a growing confidence about achieving equal treatment. The passage of the Civil Rights Act of 1964 and the revolt against institutional controls of "establishments" in the 1960s, including the civil rights movement, the youth movement, and the women's movement, combined to initiate change in the system of rules and definitions. Young women activated by the civil rights movement, who had marched and "sat in" at lunch counters, started looking for institutionalized means and skills to combat injustice on a continuing basis. Young women were also looking for a way to create more meaningful careers than their mothers were able to have—those mothers whose situation had resulted in the problems outlined by Betty Friedan in *The Feminine Mystique,* the book that touched off the women's movement. Women began to look at the professional schools, where they could attain a defined status—lawyer, doctor, scientist—and avoid the amorphous situation faced by generations of college graduates whose studies in the liberal arts had opened the way only to becoming secretaries and assistants. Everywhere, college women were looking for a track that would lead toward career goals, and they entered medical and law schools in impressive numbers. There were also leaps in women's enrollment in schools of business, public administration, and engineering. Of particular interest to the sociologist and policy maker was the abrupt shift in career direction that girls, presumably socialized to become homemakers, demonstrated as they began to train for traditionally male fields. Women now constitute 40 percent of all law students, one third of all medical students, and one third of all students in graduate schools of business administration. Although not that high, the proportion of women in schools of engineering (10 percent) and in the hard sciences is also significantly increasing, as is their representation among degree recipients. The increasing numbers have gone beyond a threshold creating visibility. Professors at MIT and other technically oriented schools have expressed amazement at the change in numbers as women throng together with men.

Hiring practices in the fields formerly composed almost entirely of men have also changed dramatically for a number of reasons; legislation is one of the most important. Suits against the law schools were instituted in 1969. In 1970 the Women's Equity Action League (WEAL) filed a class action suit against every medical school in the country. It is now illegal for any employer to refuse to hire women applicants. Furthermore, with the professional schools producing so many women with such good records, leading law firms and hospitals must hire women as associates and residents if they still want the top 10 percent of professional school graduates. Now that the doors are open, women are more forceful in demanding equality. They are insisting on experience in a wide range of specialties before they choose one particular area, and they are resisting pressure to assume stereotyped roles. Nowadays, women are seeking to work in the litigation and corporate departments of large firms and are applying for, and getting, surgical residencies and grants to do research.[13]

It is particularly interesting that, as women today choose (and are chosen for) specialties they were once considered incompetent to handle, specialties for which women as well as men believed they had no talent or interest, they find these specializations quite congenial. Numbers of women enjoy the challenge of negotiating big business ventures, the high drama of the courtroom, and the exhilaration of scientific discovery.

It would, however, be giving an inaccurate picture not to emphasize that women in the professions still tend to gravitate toward "human services" and public interest work, and the work that tends to be less financially rewarding, less prestigious, and less honored than the specialties at the forefront of knowledge in their fields. As in the past, women lawyers are often involved in gender traditional spheres—joining public interest firms, taking cases of the poor for low pay, and doing matrimonial work. But much of the traditional work has a new twist. Women have developed practices that help other women in sex discrimination in employment; they have worked toward more equitable settlements in matrimonial law and child custody; they have worked with policy and social agencies to make the legal process more sensitive to women who are victims of rape and assault. Particularly important, women are now practicing these specialties at the appellate court level and "making law."

In medicine, women still disproportionately work in public health, but they have attempted to deliver a more human brand of medical care to women whose diseases have seemed less interesting and important to male medical specialists and who were regarded as "crocks" prone to psychosomatic disorders. Women physicians also are engaged in medical research, not necessarily on women's diseases.

These changes do not mean that women have achieved equality. Women still experience resistance from men who feel that the courtroom, the operating room, and the boardroom are male domains and who resent the new competition from competent women professionals. And some men concede equality in some spheres but not in others. Even such norm setters as the president of the United States and his colleagues maintain their memberships in clubs that exclude women and deride their abilities in private company. They continue to send the message to women that their presence in male domains creates discomfort for the whole community. It is well documented in the study of professions, one must be an "insider" to receive information and news about what is happening. Invisible colleges disseminate information to members first, often informally or at conferences attended by small groups of invited colleagues. Women today have more access to the invisible colleges, but in many fields they are still a small and marginal minority.

A recent survey of women in the Institute of Electrical and Electronics Engineers revealed that over half of the respondents reported unequal treatment, some exclusion from informal work-related social activities, and limited acceptance by colleagues of their professional decisions. While 61 percent of the women surveyed saw no sex bias in terms of assignments to interesting projects, 55 percent felt they were not receiving equivalent preparation for top-level careers (Project on the Status and Education of Women, 1986). This sort of early neglect is believed to account partially for discrepancies in salary between men and women who are otherwise equivalent in experience, education, and degree of professional responsibility.

There are other "insider" problems associated with access to informal relations. In my 1981 study of women lawyers (Epstein, 1981), I found that they were "outsiders within." The general process is one of ambiguity, as many women are placed in "no win" situations by men who resist their participation in the profession. As one woman put it, "You hardly ever meet a man practicing law who didn't regale you with the horrible experiences he'd had with ballsy, nasty, aggressive women, and how different you were," a view that survives in certain kinds of practices and courtrooms even today. Women are often placed in a double bind. Ironically, women, who a decade ago were regarded as not tough enough to engage in courtroom debate, are now considered *too* uncompromising when they engage in adversarial exchanges. Women who smile a lot in interaction with their male colleagues are regarded as insufficiently serious, but when they do not smile, men fault them for being "stiff." Many colleagues still interact with women on the basis of the stereotyped views altering their perceptions. The scientist Rosalind Franklin was characterized by James Watson (1968) as tough and cranky, a view quite contradictory to that of her friend and biographer, Anne Sayre (1975).

Today's emphasis on the culture of the workplace should alert us to the unstated and often unobserved impediments to women's ability to work at highest capacity and to have confidence in their capacities. Culture prescribes attitudes and styles of behavior which become fixed as norms defining social roles. Even though there may be a wide discrepancy between the norms and actual behavior, norms do not necessarily change to conform with behavior. In fact, many people subscribe to norms even while everyone they know deviates from them. Individuals lie but subscribe to the notion that they should tell the truth; they believe in monogamy but have affairs. Norms also prescribe the kinds of behavior appropriate to gender roles, roles that are accepted in principle even as they are violated in practice. For example, it is probably true that most women are competitive. They want to have brighter and more obedient children than their neighbors, or a nicer home than their relatives. They can be quicker and smarter than those around them. Yet I have found in interviews with lawyers and at conferences on women's issues that women will rarely admit that they are competitive and don't believe that their sex is so. Women hold stereotyped views about themselves that damage their confidence: they believe that they take second place to men.

There are both psychological and sociological reasons for such behavior. First, individuals want to obey their groups' norms and do not like to think of themselves as deviants, especially when the norms are extensions of values that define what a good person is. In the case of women, a good person is often a "feminine person." A woman wants others to have a good opinion of her, and to be like the others who constitute her reference group.

Second, people are aware that there is punishment for deviation. It may range from simple disapproval—the raised eyebrow—to absolute ostracism. Furthermore, punishments are ongoing. They are not usually one-time occurrences; a person is consistently reminded by others that he or she is not abiding by the norms.

Thus, women whose competitiveness, assertion, and intellectual challenge is expressed outside of socially approved domains are subject to punishments of disapproval and rejection, even when the competitiveness is taken to be appropriate to male occupants of the roles concerned. I found, for example, that women attorneys who are courtroom litigators and express competitiveness in an entirely appropriate setting often evoke the disapproval of their colleagues (not only their opponents), who think less of them for having expressed their competitiveness in an open and highly visible arena; as a result, they also are considered to be unpleasant associates. What is entirely acceptable and even prized in men is not acceptable for women in this setting. Some women worry that taking these jobs will make them "masculine" or "hard"; such fears undermine their forceful-

ness. Self-confidence is an important component of ability because it enhances risk taking; a lack of confidence may make a person withdraw from
a daring intellectual enterprise.

The culture generally sends messages that women are not as able as
men. This continues regardless of substantial evidence to the contrary. For
example, a 1986 study at Purdue University showed that there has been a
change in women's persistence in engineering programs. It was found that
women "persisters" tended to have higher SAT scores in both math and
verbal abilities and slightly higher grade point averages than men. Yet, the
women rated themselves less strong in these areas than the men, who rated
themselves higher in spite of their lower scores (*U.S. Women Engineering,*
Nov./Dec. 1985). We do not know how these women engineers' self-
evaluations will affect their future careers, but we do know that it is not a
good sign.

Women lawyers learn that men are more resentful when they lose to a
woman in court, and it is conceivable that male scientists also are more
resentful when a woman achieves priority in research on problems in which
they are all engaged. Judith Stiehm's work on the military (1981) shows
that men in basic training exceed the norms that were set in the company
of men when women are introduced into training. It is not unlikely that
men who compete with women do try to "overproduce" or institute practices that are disadvantageous to them, such as working longer hours,
sometimes by only "hanging around," when women's schedules are restricted. They may also try to undermine women's confidence, or shut
them out of the informal work group.

General cultural views are, of course, differentially manifested in institutional settings, and institutions do vary with respect to the customs by
which competition or cooperation is conducted. Women simply have a
better chance to succeed in some settings and very little chance to succeed
in others. Women, like men, find that when others honor their contributions, listen to their ideas, and acknowledge their authorship, they perform
at higher levels. This is often the case in domains or specialties where there
have been women achievers before. Women crystallographers and astronomers, for example, have achieved more, partially because of the legitimation achieved by women scientists who came before them. There are
also disproportionately more of them in the specialties where women see
precedence for their participation.

The legitimation women experience by being members of elite educational institutions and benefiting from the sponsorship of established people also serves to advance their careers and enhance their capacities.
However, women de-escalate ambition and reduce career commitment

when there is no hope of reward (Epstein 1976), or the rewards are inappropriately low (Epstein 1971, 1974 [1982]). Aspirations move in the direction of past performance (Cyert and March 1963).[14] Role models and peers can also influence one's sense of possibility (Epstein 1970; Tangri 1972).

The resources of institutions also affect receptivity to women. Where resources are lavish, women have a better opportunity to have a reasonable share of them, and also to have the good will of their colleagues. When resources are scarce, conflict often moves from objective determinations of merit to *ad hominem* and generalized deprecations of the person and the group to which she belongs (Coleman 1957).

We are moving into a period in which the number of women in the knowledge-producing professions is large enough to foster the kind of excellence we have seen demonstrated by men. It seems foolish to spend more analytic time wondering whether women have the capacity to be geniuses and creators of knowledge without permitting them the same encouragement and opportunity that men have had and ceasing to subject them to negative disclaimers about most women's presumed incapacities. If equality of encouragement and opportunity could be accomplished, a basis would be created for the legitimate review of women's performance. By that time, however, the question may no longer even be asked.

OWEN M. FISS

12. An Uncertain Inheritance

EQUALITY FOR WOMEN will not come through law alone. This is true in science as in any other domain. Law is but one social strategy for improving the status of women. It is, however, an especially important one, which has gained additional momentum and force from the struggles for racial equality. As a result of these struggles, the American legal system has now become fully committed to a principle that prohibits discrimination, and in recent years, women have turned this principle to their advantage. This essay tries to sketch the analytic structure of the antidiscrimination principle and to identify the dilemmas that courts have faced in giving this principle force and effect, first at the behest of the black civil rights movement and now of women.

I. The Roots of the Principle

The Civil War not only ended slavery but also reversed the basic stance of the Constitution toward blacks. A system which had tolerated slavery now promised equality. The war produced three amendments, the most important being the Fourteenth Amendment, which prohibits the states from denying to any person "equal protection of the laws."

The task of enforcing and interpreting this guarantee fell primarily to the judiciary, and the judiciary took the promise of equality to be essentially a protection against discrimination. The states were prohibited from discriminating on the basis of race. This development was foreshadowed by the Fifteenth Amendment, which uses the antidiscrimination principle to protect the right to vote.[1] As a matter of judicial doctrine, however, the formulation of the principle occurred in the late nineteenth century, in a case called *Strauder* v. *West Virginia*.[2] Ever since, there has been a close

I am especially grateful for the assistance of Tracy Fessenden, Matthew Haiken, and Nancy Marder.

and intricate bond between antidiscrimination and the constitutional ideal of equality. Indeed, over the past century, the rule against discrimination based on race has become the principal legal strategy for achieving racial equality.

The implementation of this strategy was long in coming and problematic in execution, but a turning point came in 1954, when the Supreme Court in *Brown* v. *Board of Education*[3] used the antidiscrimination principle to invalidate the system of separatism known as Jim Crow. The resistance to *Brown* was intense and widespread, but ultimately that decision was accepted by American society. The triumph of *Brown* resulted in the deepening of our egalitarian commitments and a tightening of the connection between the idea of racial equality and the rule against discrimination. The Jim Crow system tried to perpetuate the subordination of blacks through laws and other social practices that openly discriminated against them. Whites were sent to one school, blacks to another. *Brown* decreed an end to all this, and in so doing brought into being a period of American history that might properly be called the Second Reconstruction, a period that began in the early 1950s, reached a zenith in the mid-1960s, and came to an end in 1974, if not shortly before.

During the Second Reconstruction, two developments occurred in antidiscrimination law that would later decisively affect the women's movement. First, the rule against discrimination came to be embodied in various statutes. Although *Brown* v. *Board of Education* and the cases that followed in its wake had enormous sweep and power, they were nonetheless limited because the Fourteenth Amendment, on its face, addressed only the states and thus presumably did not reach various areas of human activity that were controlled primarily by private agencies. Employment is a good example; the equal protection clause controlled governmental, but not private, employment practices. Congress responded and filled the gap by enacting Title VII of the Civil Rights Act of 1964.[4] Here Congress sought to extend the Fourteenth Amendment from government to private employers, and it did so by writing into law, not the general equal protection guarantee, but the antidiscrimination principle. The same process of supplementation occurred with Title VI of the Civil Rights of 1964 (the forerunner to Title IX of the Education Amendments of 1972).[5] Title VI was not intended to overcome the state action requirement of the Fourteenth Amendment, as was Title VII, but rather to make certain that the federal government did not become involved in, or a party to, unconstitutional practices. Once again, the chosen instrument for achieving that end was a rule against discrimination. Title VI prohibits the federal government from making grants to or otherwise aiding or supporting an activity that discriminates on the basis of race.

The second development involved the broadening of the rule against discrimination so that it would protect disadvantaged groups defined in terms other than race. As a purely historical matter, the Fourteenth Amendment was intended to prótect the newly freed slaves; but its language was not confined to protecting only that group. The equal protection clause protects "any person." As a result, antidiscrimination, the principle governing the application of that clause, was thought to bar discrimination based on any "invidious" criterion. It was therefore natural, though not by any means inevitable, that during the Second Reconstruction, when the Fourteenth Amendment was used more and more forcefully to protect blacks, the antidiscrimination principle was also used to protect other groups, including women. Similarly, when Title VII was passed, it barred not only racial discrimination but also discrimination based on sex. Because Title VI of the Civil Rights Act of 1964 did not cover discrimination based on sex, Title IX of the Education Amendments of 1972 Act was enacted to fill that gap.

Through this historical process the rule against discrimination became for women what it had been for blacks: the principal legal instrument for achieving equality. Sometimes this rule is more limited in its scope for women than it is for blacks. Exceptions are made now and then for the all-male club, specialized types of jobs, and athletics. For the most part, however, these exceptions are narrowly defined, and if there is any long-term historical trend, it is to give the rule against discrimination based on sex as much sweep and scope as it has had with race. In fact, in the struggle for gender equality, the antidiscrimination principle has already assumed the stature it achieved in the racial area, namely, that of a moral axiom. It has constitutional and statutory roots, it is found in both federal and state law, and it is applicable to virtually all forms of activity, public and private.

II. Two Theories of Antidiscrimination Law

As an inherited strategy, antidiscrimination came not only with much of the moral fervor of the black civil rights movement but also with many of the same intellectual dilemmas. These dilemmas primarily stem from the fact that the antidiscrimination principle lends itself to two quite different interpretations, one emphasizing process and the other substance. Although the law now emphatically prohibits various employers and educational agencies from discriminating on the basis of sex, it is still unclear what precisely the law means by "discrimination."

Imagine a graduate program in astrophysics that excludes women. The admissions officer uses sex as a criterion of admission and denies admission to all applicants who are women, even when their paper credentials, such as scores on the GRE, are better than those of some of the men admitted. There is no doubt that this practice would violate the antidiscrimination principle, but there is an ambiguity about what aspect of this situation constitutes the violation. The procedural interpretation of antidiscrimination focuses on the use of a forbidden criterion of selection, in this case, sex. The claim is that the law prohibits the use of various invidious criteria, such as sex, race, and religion, in the process of making choices or selections. It is argued that the very use of such a criterion corrupts the admissions process and that this procedural flaw alone constitutes the wrong to be remedied.

The substantive theory, on the other hand, focuses not on the use of an invidious criterion but rather on the effect or consequences of using such a criterion—the fact that women are excluded from the graduate program. This exclusion is condemned because it deprives women, as a group, of valuable opportunities for study and employment, and thus perpetuates the subordinate position of women in society.

Process theorists do not ignore the exclusionary effect altogether but rather use it in a different way than do substantive theorists. They might, for example, use the exclusionary effect in an evidentiary fashion. Suppose that the admissions officer denies that he (the use of this pronoun is deliberate) uses sex as a criterion of selection. Let's also assume that the number of applications is distributed equally between men and women, yet the result of the selection process is to admit only men.

If the admissions officer used only a single "objective" criterion—GRE scores—then it would be quite easy to determine whether he was being true to his word. All the applications could be ranked in terms of their GRE scores, and if any woman scored higher than any of the men admitted, then the inference would be irresistible that the admissions officer surreptitiously used sex as a criterion of admission. But with a more complex, and more realistic, admission process—one that employs personal interviews and letters of recommendation, as well as GRE scores—such an "objective" ranking of candidates for admission is not possible. The process theorist might, however, turn to the result (no women admitted) to support an inference of discrimination. One and only one thing could explain that outcome, and that is the use of the forbidden criterion. The admissions officer will be given a chance to explain, or to rebut the inference, but failing a good explanation it will be assumed that an improper criterion was used, and for that reason—not because of the effect—the admissions process will be deemed "discriminatory" and thus illegal under the process theory. The substantive theorist, however, would see the demographic pattern—all men and no women—not as the basis of an inference of wrongdoing, but as itself the wrong proscribed by the law.

Process theorists also might use effect to justify the rule against discrimination itself. If asked to explain why gender is a forbidden criterion, a process theorist might elaborate on the consequences that would be likely to follow from its use. Given the current distribution of power within the profession and the fact that the power of selection is by and large in the hands of men, allowing decisionmakers to use sex as a criterion of selection will inevitably lead to the exclusion of women and the perpetuation of existing patterns of subordination. Men prefer men, at least when it comes to positions of power and prestige. In this effort at justification, those who subscribe to the process theory make reference to effect and thus adopt the central premise underlying the substantive theory, namely, the desirability of ending the subordination of women as a group.

A difference nonetheless persists between the substantive and process theories, because this is not the only justification the process theorists could offer for the rule against discrimination. Nor is it the typical one. Process theorists usually justify the rule on grounds that are independent of effect: They believe that using gender in the selection process is a form of unfair treatment—judging individuals on grounds unrelated to merit—and that the function of the antidiscrimination law is to assure fairness in the selection process. Vigorous enforcement of antidiscrimination law might also improve the status of a disadvantaged group, but that would, on this justificatory theory, merely be an incidental, though agreeable, by-product of a rule justified on other grounds.

III. The Hard Cases

In the situation I began with, in which the decision maker openly uses sex as a criterion of selection to exclude women, the process and substantive theories of antidiscrimination overlap. They produce an identical legal judgment, in the sense that both theories render the selection process unlawful. In situations of this type, therefore, the difference between the two theories is largely academic, important for a proper theoretical understanding of the law but not of much practical significance. There are, however, two other situations, which have become increasingly prevalent in recent years, where the difference between the two theories has great practical import. One is where a criterion that appears innocent on its face is used as the basis of selection and produces or perpetuates a pattern of exclusion. The other is preferential treatment or affirmative action, where sex is used as a criterion of selection but only to increase the number of women in the program. In both these kinds of cases, which might be seen as belonging to a second generation in the evolution of antidiscrimination law, the substantive and process theories of antidiscrimination diverge and produce different legal judgments.

A.
THE EVALUATION OF FACIALLY INNOCENT CRITERIA

In speaking of facially innocent criteria, I have in mind something like performance on a standardized test. One should be careful, however, not to assume that the use of such criteria necessarily involves what I have called a second generation problem. Often the seemingly innocent criterion is a sham; the decisionmaker purports to use the results from the standardized test as a selection criterion but actually manipulates or falsifies the scores of various candidates in order to give the place or position to a man. When that occurs, the "real" (as opposed to the "stated") criterion of selection is sex, and once again, the two theories of discrimination will produce the same result: the selection process will be deemed unlawful.

There are also situations when the facially innocent criterion is honestly applied but has no apparent connection or relevance to the demands of the job or program. Imagine, for example, that this standardized test purports to measure the candidate's ability to understand spatial relations and that, for some reason, women do less well on this test than men. Also assume that the test is used for admission to a graduate program in psychiatry. Then it would seem that the test is unrelated to the task and in that sense nonfunctional, and if it could be further hypothesized that the test had been chosen for the purpose of excluding women, or with gross indifference to its adverse impact on women applicants, the process and substantive theories would once again converge and produce the same legal judgment. It could easily be said that the "real" criterion of selection is sex.

A more troublesome problem arises, however, when the spatial relations test is honestly administered and honestly chosen (e.g., the spatial relations test is used for admissions to a graduate program in physics). In that case, the complaint against the test stems entirely from the fact that the test has a disproportionate adverse impact on women—it has the effect of excluding a greater proportion of women applicants. The judicial task is to decide whether the use of the facially innocent criterion violates the norm against discrimination when it is assumed that the criterion—the standardized test—is (1) honestly used; (2) adopted to serve legitimate interests and, in fact, furthers those interests; and (3) has the effect of excluding a greater proportion of female applicants.

If the emphasis is exclusively on process, then the use of the test, under the conditions I specified, would be lawful. The result of using the test might be to exclude a disproportionate number of women, but, strictly speaking, the criterion of selection is not sex. Moreover, the aims of antidiscrimination law, as understood by those theorists who emphasize procedural fairness, would be fulfilled, for it could not be said that the process of selection is unrelated to merit or otherwise corrupt. Substantive theorists would, of course, take a different view of the matter. They would argue that, whatever might be said on behalf of the process, the fact remains that women are being excluded from the program and thus denied valuable educational opportunities and the positions of power and prestige tied to those educational opportunities. As with race, an exception might be allowed if the facially innocent criterion served a *compelling* rather than just a *legitimate* interest, but short of that, the substantive theorist would insist that the use of the innocent criterion should be barred because of its disparate or disproportionate adverse impact on women.

In the context of the Fourteenth Amendment, in which antidiscrimination appears only as a judicially constructed gloss on equal protection, the actual language of the antidiscrimination norm as articulated by the courts— "do not discriminate on the basis of sex"—does not pose an insurmountable barrier to the substantive theory and the invalidation of the criterion. The courts created the norm and can easily adjust it to fulfill what might be deemed its overriding or fundamental purpose (ending the subordination of disadvantaged groups). With statutes, for example, Title VII or Title IX, overcoming the limitation imposed by the language of the antidiscrimination norm is more problematic. Courts are not free to rewrite statutes. Three interpretive strategies might nevertheless be developed to overcome these difficulties.

One is to make a distinction between the "letter" and the "spirit" of the law and to insist that the underlying purpose of the statute is to end the subordination of women and that the words of the statute should be construed accordingly. A second is to emphasize the original connection between the statute and the equal protection clause. The statute was meant to codify the principle that governed the application of the Fourteenth Amendment; if the amendment is deemed to prohibit a practice, so should the statute. Third, a court might expand the notion of "functional equivalence" in order to meet the requirements of the statutory language. That notion was implicitly used in the situation where the facially innocent criterion was scrupulously followed but chosen in the first place as a method of selection for impermissible reasons (i.e. to exclude women). In that case, the purpose or motivation for choosing the criterion became the basis for characterizing the criterion (a spatial relations test for psychiatry) as the functional equivalent of, or proxy for, gender. Here the test is being administered fairly and some legitimate interest is being served by the use of the test, but its disproportionately negative impact on women candidates renders the use of the test the functional equivalent of using sex as a criterion of selection.

Aside from these problems posed by the statutory language, the substantive attack on the facially innocent criterion might run into difficulty because disproportionate impact, the key for invalidation, is a phenomenon that can be understood only in terms of groups. According to the substantive theory, the facially innocent criterion is unlawful because it excludes more of one group (women) than another (men).[6] Such a group orientation seems to be at odds with the view, affirmed by the Supreme Court on occasion,[7] that the rights of the Fourteenth Amendment belong to individuals and are in that sense personal in nature. Substantive theorists see, however, a crucial link between individual and group well-being. They believe that the social status of any particular woman is, in important ways, determined by the status of women as a group, and that in order to secure a measure of equality for women as individuals, the law must eradicate those practices that perpetuate the subordinate position of the group. Substantive theorists also point to legal doctrines both within and outside the antidiscrimination context that display a concern for the welfare of groups (e.g. religious minorities, workers, and consumers). A group orientation might be at odds with the individualistic tenets of classical liberalism, but it is, so they insist, thoroughly compatible with American legal principles.

Another difficulty for the substantive theory arises from the fact that it creates certain remedial costs since, by hypothesis, the contested criterion serves some legitimate interests. If the criterion were disallowed, as the substantive theory requires, those interests would be compromised or sacrificed. Society would have to suffer the risk that scientific progress might be impeded. Those costs might be avoided if some alternative selection procedure could be found that did not have an exclusionary effect but that nevertheless furthered the institution's and society's legitimate interests. Methods other than a standardized test, for example, might be found by the physics department for measuring ability in spatial relations. But some account would have to be taken of the costs of developing new selection criteria and the chance that this search for an alternative might come to naught.

Admittedly, the costs of disallowing the facially innocent criterion are not overwhelming, since, also by hypothesis, the criterion serves only a legitimate, and not a compelling, interest. It may well be that the interests furthered, though legitimate, are actually remote and rather tangential. But a sacrifice of a legitimate interest is still a sacrifice, and must be acknowledged. What moves the substantive theorist is not the view that the legal intrusion is cost-free, either to society or the particular institution, but rather that the intrusion and its attendant costs are justified by a higher social good—improving the status of a disadvantaged group. Such a trade-off among competing interests is possible, indeed common in the law, but it entails a judgment of some moment and is avoided by process theorists altogether. They read the law as doing no more than prohibiting the use of criterion (sex) that is presumably unrelated to merit.

For these reasons, the resistance to the use of the substantive theory of antidiscrimination to evaluate facially innocent criteria has been considerable, even in the racial context.[8] Some headway was made before the collapse of the Second Reconstruction, but even then, what emerged in the Supreme Court doctrine was not a pure substantive theory but one that incorporates both procedural and substantive elements. This hybrid focuses on past discrimination. It is predicated on the view that antidiscrimination law prohibits not only present discriminatory practices but also those practices that perpetuate past discrimination. A particularly striking application of this theory occurred in 1969, when a literacy test was disallowed as a qualification for voting because it perpetuated the discrimination that had occurred in educational systems operated in years past under Jim Crow.[9]

The procedural element in this hybrid theory consists of the fact that the past discrimination (which is being perpetuated by present practices) is defined in terms of process—the explicit use of race as a selection criterion (the Jim Crow system). The present practice perpetuates a past unfair treatment and is itself unfair. On the other hand, substantive elements are also present. For example, the victims of the past discrimination and present beneficiaries of the remedial order are defined in group terms. Moreover, in fashioning relief under this hybrid, courts appear willing to assume an identity between past victims and present beneficiaries and to impose remedial burdens on agencies or institutions that did not themselves perpetrate the earlier discrimination. And like a substantive theory, the theory of past discrimination might disallow the use of a selection criterion even though it presumably furthers legitimate interests. This sacrifice of legitimate interests can be justified as a way of avoiding the perpetuation of a previous discrimination, but the very fact that a sacrifice is required differentiates the theory of past discrimination from a pure process theory.

In sex discrimination cases, the theory of past discrimination is also likely to be invoked and acknowledged as a hybrid. Certain facially innocent criteria will be attacked not simply because of their disproportionately

adverse impact on women but also because they perpetuate past discrimination. The hybrid theory relieves the courts of having to stretch the theory
of functional equivalence in order to transform a facially innocent criterion
(e.g. the test) into the forbidden criterion (sex). All that would be required
is proof linking the criterion's disproportionate impact to past discrimination. In the case of blacks, whose past was largely defined by one century
of slavery and another of Jim Crow, the burden of this proof was virtually
nonexistent. All differences in performance between blacks and whites
could be assumed to be a product of those social systems that were, so
obviously, massive instances of racial discrimination. In the case of women,
however, the burden might be more considerable.

There are, of course, the easy cases, like the one that figured so prom-
inently in our discussions—a requirement that a job applicant have a graduate degree from the Princeton physics department, which, prior to 1969,
apparently openly excluded women. Such requirements could easily be
attacked under the theory of past discrimination. But how should one treat
performance on a spatial relations test? One of the participants referred to
research that indicated that differential performance of men and women is
genetically based (hormonal development in teenage girls, etc.). If that is
the case, which seems most unlikely, then the theory of past discrimination
would not preclude the use of the criterion. A more plausible position is to
view the difference in performance on the spatial relations test as the
product of the socialization process (e.g., the games kids are given to
play), but even then, the applicability of the past discrimination theory
remains in doubt. These socialization processes are deeply imbedded in
varying expectations for men and women and in many respects invidious,
but compared to the processes that account for the differential performance
of blacks and whites on literacy tests, it is harder—though obviously not
impossible—to conceptualize them as a crude form of process-discrimi-
nation.

B.
THE LEGALITY OF PREFERENTIAL TREATMENT

If, in the case of women, the burden of proof entailed in the theory of
past discrimination cannot be surmounted, facially innocent criteria will
have to be judged by the original versions of the substantive and process
theories. In that case, the difference that I described earlier will emerge—
the process theory will allow those criteria, while the substantive theory
will disallow them when they have a differential impact and do not serve a
compelling interest. A difference between the two theories will also arise
in evaluating another second generation practice, preferential treatment or

affirmative action, in which gender is used as a criterion in the selection process but now, supposedly, to advantage women.

Affirmative action programs seek to increase the number of women in science and do so in two different ways: by vigorously recruiting more women to enter the field and by establishing preferences for women in a competitive selection process. In the latter case, which is the subject of my concern and the more common meaning of the phrase "affirmative action" today, women are given an edge in the competitive process on the basis of their gender. This can occur either when all other factors are equal, that is, when sex is used as a tie-breaker, or when the women who are selected rank below men in terms of the standard, meritocratic criteria. The principal question for antidiscrimination law in this context is whether such preferential treatment is permissible.

As a social practice, preferential treatment based on gender might be justified on a number of grounds. It can, for example, be seen as a way of accelerating the process by which existing patterns of subordination are eradicated by bringing more women into especially prestigious and powerful professions in our society. In that sense, preferential treatment may be seen as an impatient expression of the substantive theory of antidiscrimination (in fact, some substantive theorists, viewing inaction as a form of action, argue for preferential treatment as a legal duty). But the justification for preferential treatment can also be based on process-type considerations. For example, some might view it as an auxiliary policing device, to help official law enforcement agencies detect violations of law. The idea is that as long as men are in control of all selection processes, there is a genuine risk that they will continue to use their power in surreptitious ways to perpetuate their dominant positions. When that power is shared by women, however, these unlawful practices are less likely and might be more easily discovered. Preferential treatment might also be based on considerations only remotely related to discrimination. It might, for example, be seen as an adjunct of the counseling function. By increasing the number of women in science, we teach younger women that careers in science are possible and available for them. We encourage their ambitions. We also provide them with role models and persons to whom they are more likely to turn for advice and support.

Some acknowledge these social or moral justifications for preferential treatment but nonetheless insist that it is unlawful. No matter how desirable these programs might be as a matter of social policy, they are, so it is often argued, barred by the laws prohibiting discrimination. Those who take this position tend to focus on situations of scarcity, in which an opportunity given to one necessarily cannot be given to another. The rule against sex discrimination is then invoked on behalf of the rejected male applicant (i.e. the one who would have been admitted or hired if not for the preference). As part of this argument, some weight will be placed, of course, on the precise words of the antidiscrimination norm, especially in its statutory form: What the law prohibits, so we are reminded, is discrimination based on sex, either sex. But a more fundamental objection can be raised to the preferential treatment, one that goes to the fairness of the selection process. The claim is that the rejected male applicant is being treated unfairly because he is being judged on the basis of a criterion (his sex) that presumably has nothing to do with his merits. In fact, the claim is that he is being subjected to the same kind of unfair treatment that a rejected female applicant would be subjected to in a first generation situation—where she is rejected on the basis of her sex. The discrimination suffered now might be reversed, but it is still discrimination.

Obviously this argument has its greatest appeal to process theorists. Those who subscribe to the substantive theory of antidiscrimination generally acknowledge the claim of individual unfairness, but because they do not view antidiscrimination law as a device to ensure individual fairness, they are not prepared to use this harm to create an absolute bar to affirmative action. The infliction of individual unfairness is not viewed by them as a violation of the law. Rather, it is treated as part of the overall cost of achieving equality, analogous to the costs entailed in disallowing the bona fide use of facially innocent criteria—just one more sacrifice that must be suffered in order to eradicate gender hierarchy. The language of the law is interpreted accordingly. It would be a cruel irony, substantive theorists insist, to use these laws to bar programs that try to accomplish what these laws were intended to accomplish—improving the social status of women.[10]

Although substantive theorists refuse to treat the infliction of individual unfairness as an absolute bar to preferential treatment, a sensitivity to this harm is reflected in the kinds of affirmative action programs that they are prepared to accept. This sensitivity is manifest in a Supreme Court decision upholding an affirmative action program of a local transportation agency.[11] In a competition for a promotion to road dispatcher, a woman was chosen in part on the basis of her sex. The male applicant who ranked (slightly) higher than she did under the standard, meritocratic criteria but who did not get the job brought suit under Title VII, claiming that he was discriminated against on the basis of sex. The Court rejected this claim and approved the affirmative action plan, but only after it found, first, that the plan tried to correct a "manifest imbalance in traditionally sex segregated job categories" and, second, that the plan had not "unnecessarily trammeled the rights of male employees or created an absolute bar to their advancement."

In making the latter judgment, the majority opinion stressed a number of factors: the plan did not set aside jobs for women but simply allowed sex to be considered a plus factor; no persons were excluded from consideration; both the woman awarded the job and the man who was rejected were eligible for promotion; the difference between the candidates under the standard criteria appeared trivial; the promotion did not unsettle any firmly rooted expectation of the rejected applicant; and, since this was only a promotion case, the rejected applicant retained his employment with the agency, at the same salary and with the same seniority, and remained eligible for other promotions—all factors that minimized the unfairness to the rejected male. The Court upheld the affirmative action plan but only after it stressed that the plan "visits minimal intrusion on the legitimate expectations of other employees."

Wholly apart from the concerns voiced on behalf of the rejected male applicant, and perhaps of greater concern to substantive theorists and to those who favor affirmative action programs, is the risk that these programs might actually become counterproductive—that they might not enhance but actually impair the social status of women. Although the number of women in various scientific programs and institutions might increase, the value of their achievements might be demeaned (especially when gender is used as something more than a tie-breaker). Because those who are admitted to an educational program or employed on the sex-based preference are, by definition, persons who would not have been admitted or hired according to the traditional, meritocratic standards, the achievement represented by their selection might be taken to be less than what it first appears. And this element of doubt would not be confined to the women who actually are the direct beneficiaries of the preferential program, if for no other reason than because they cannot be identified with any certainty. It will extend to all women. Once preferential treatment is legitimized and becomes a pervasive social method, the accomplishments of all women are in danger of being devalued.

Some members of the black community—often those who have already made it into positions of power and prestige—voice a similar concern about race-based preferential programs.[12] They claim that their achievements, and the achievements of the entire black community, have been devalued by affirmative action programs. I have no doubt that, in time, these concerns will surface in the women's movement and have an equally divisive effect. There is not much the substantive theorist can do in response, other than plead for caution and make certain that preferential programs are used in a way that minimizes the danger of counterproductivity. The premise is that, on balance, the net effect of these programs will be to improve the status of women. It is important to recognize, however, that for process theorists something more than caution is needed. They view preferential treatment programs not as a risky way of trying to further a worthy social objective but as a bald violation of antidiscrimination law. They seek not a modulation and calibration of these programs, but a complete bar.

IV. The Limits of Antidiscrimination Law?

We can thus see in the context of preferential treatment, as we did in the evaluation of facially innocent criteria, that the substantive and process theories arrive at decidedly different results, in terms of what is legal or illegal. These differences are important and will become increasingly so as the second generation problems come to dominate the legal landscape. But attention to these differences *within* antidiscrimination law should not obscure the fact that antidiscrimination law is a limited strategy for achieving equality for women. Antidiscrimination law seeks to govern what I have called the selection process, that is, the process by which scarce opportunities (jobs, positions in educational programs, etc.) are awarded among a pool of applicants. It does not govern the life processes and methods by which the pool of applicants is constituted.

Antidiscrimination law would bar a sign that read "Only men need apply" and would allow (as opposed to require) vigorous recruitment measures, either as a way of welcoming all applicants (process version) or as a first step toward increasing the number of women applicants (substantive version). Moreover, by assuring that no discrimination will take place in the selection process, women may be encouraged to apply, or to embark on educational careers that will ultimately put them in a position to apply. However, antidiscrimination law has little control over the diffuse norms, expectations, and general conditions of social life that often determine the careers of individuals or entire groups of people.

It is difficult to imagine, for example, any theory of antidiscrimination law that would enable it to become an effective instrument for separating the child-bearing and child-rearing responsibilities, so as to dispel the idea that a woman's place is in the home or that women are especially charged with the responsibility of caring for children. As long as this expectation endures, it will be impossible for women to enter the job market on an equal footing with men, and it is unlikely that all but a few (the childless, or the "super moms") will contemplate professions as demanding in terms of time and commitment as science. Something must be done, perhaps even by the state, but it is hard to see antidiscrimination law as *the* remedy, precisely because its focus is on the selection process. It is also impossible to imagine any theory of antidiscrimination law that would give Judge

Ginsburg her wish, uttered with great eloquence at the very end of our deliberations, when asked to identify the single most effective thing that could be done to improve the status of women in science. What she sought was child-care facilities on a scale and at a level that would enable women to pursue their careers, in science, or any other domain, with the intensity and ambitiousness that success in American society usually requires.

In race, increasing attention has come to be focused on the so-called black underclass,[13] and the dynamics that brought that phenomenon into being, and we have come to understand more clearly the limited jurisdiction of antidiscrimination law. No one seriously looks to Title VII as the "solution" to the problem of black teenage pregnancy. A proper recognition of the limits of antidiscrimination law does not, however, trivialize the contribution this particular legal strategy can make toward the realization of true equality. For one thing, the status of minorities and women in society are in important ways determined by the process by which scarce educational and employment opportunities are awarded. Antidiscrimination law, especially under a substantive theory, makes an important contribution toward reforming those processes and turning them to the advantage of traditionally disadvantaged groups.

There is, moreover, no inconsistency between the vigorous enforcement of antidiscrimination law and the development of broader and more varied strategies to end hierarchical relationships, whether between the races or between genders. The civil rights movement did not end with the Second Reconstruction but naturally and dramatically led to the Great Society and the War on Poverty. Similarly, it should be emphasized that a full and robust commitment to antidiscrimination is neither inconsistent with, nor rendered redundant by, the search for wider and more varied social strategies for improving the social status of women. By deepening our commitment to distributive justice and revealing all that it requires, antidiscrimination law acts not as a bar to that endeavor but as an inspiration.

IV

A THEORETICAL EXPLANATION

JONATHAN R. COLE / BURTON SINGER

13. A Theory of Limited Differences: Explaining the Productivity Puzzle in Science

THIS ESSAY USES a new general theory of limited differences to propose an explanation for a long established, but poorly understood, pattern of scientific productivity.[1] The theory attempts to explain the empirical fact that male scientists, on average, publish about twice as many scientific papers as their female counterparts, and this disparity increases over the course of careers.[2] Our aim here is to illustrate how a fine-grained explanatory theory of limited differences can account for this. We have chosen the productivity puzzle in science as a strategic research site,[3] but the general theory of limited differences should apply to many other societal patterns of inequality and social stratification—from racial differences in income and occupational prestige over careers to differences in occupational choice among racial and gender groups.

The first section describes the phenomenon to be explained. Section two presents the elements of the theory of limited differences and indicates how the structure of a kick-reaction system can explain the publication process in science. In section three we review prior attempts to explain the gender-differentiated productivity patterns in science. The fourth section formalizes scientific development and the productivity of men and women

J. R. Cole was supported by grants from the National Science Foundation (SES-84-11152), the Josiah Macy, Jr. Foundation, and the Russell Sage Foundation. B. Singer was supported by grant ROI-HD19226 from the National Institute of Child Health and Human Development. Useful comments on earlier drafts were made by Salome Waelsch, Jeremiah Ostriker, Gerald Holton, Stephen Cole, John Bruer, Harriet Zuckerman, Joel Cohen, Margaret Marini and participants at a Macy conference on women in science.

in terms of the theory of limited differences; and in section five we illustrate that process with detailed career constructions from a micro-simulation implementation of the theory. Section six explains how competition among scientists is the driving force affecting the dynamic features of the theory. We conclude with a discussion of testable features of the limited differences theory and outline a research agenda, focusing on measurement problems that must be resolved if the proposed theory is to be further validated or refuted.

The Skewed Distribution of Scientific
Productivity and the Productivity Puzzle

Science is a highly stratified institution. A small proportion of scientists hold the lion's share of powerful and prestigious positions as well as honorific awards,[4] and this inequality in rewards is paralleled by equally skewed rates of scientific productivity. That is, the numeric count of published scientific articles and books. Most scientists publish a very limited number of papers; a small percentage publish a great number. Between 10 and 15 percent of all scientists publish about half of all the science produced.[5] This pattern is as true for women scientists as it is for men. The theory of limited differences pertains to all Ph.D. scientists, but since most do not produce more than three or four papers in a career, we intend to concentrate on the elite group of scientists who are the major producers. Therefore, it is this population of primary producers of science, that is, the upper tier of the stratification system, and the factors that influence their rate and amount of production that is the principal focus of this paper.

A second pattern of scientific productivity remains poorly understood. Male scientists publish more than females. This sex-related pattern has been demonstrated in more than 50 studies (see Cole and Zuckerman 1984). It is as true today as it was in the 1920s, 1930s, and through the 1960s.[6] To cite but one example, a recent summary of scientific productivity patterns for 526 "matched" men and women who received their Ph.D.s in 1969–70 showed that for the 12 years following their degrees the female mean to male mean productivity ratio was .57 for published papers, .42 for median number of papers.[7] For each type of comparison, the gender difference increases over time (Cole and Zuckerman 1984).[8] Using mean numbers of papers, the ratio of publications of women to men in the first 7 years of the career (i.e., the tenure-relevant years) was .63; for years 8 through 12 it was .51. The ratios of the medians change from .51 for the earlier years to .30 for the later years. This "fanning out" process of sex differences in publications for the 526 scientists and for four cohorts of matched men and women Ph.D.s dating from 1932 to 1957 is illustrated in Figure 13.1.[9] The picture of the 1970 cohort shows that almost all of the fanning action occurs among the top 25 percent of producers. The theory of limited differences attempts to explain these sex differences in scientific productivity.

We will also try to explain similarities that have emerged from studies of men and women scientists over the past two decades.[10] For example, there is virtually no association between sex status and: (a) admission to graduate schools of varying prestige or assessed quality; (b) receipt of post-doctoral fellowships; (c) acceptance or rejection of manuscripts submitted for publication; (d) success rates for grant applications; and (e) number of early career honorific awards received. A priori, one might expect that there would be important disparities by sex in early career experiences that would reveal productivity differences within a few years of Ph.D. completion. That this does not occur is what brings forth the notion of a productivity puzzle.

Disparities only emerge gradually. They reveal themselves in *cumulative* numbers of publications, *total* citations to published work, promotion to tenured positions at the most prestigious science departments, and receipt of top honorific recognition, such as Nobel Prizes, Fields Medals, and Lasker Awards. Observations of small fragments in time of the careers of men and women scientists whose initial conditions at the start of graduate school are roughly the same reveal virtually no distinctions in productivity by sex. It is the cumulative, *long-term* nature of the development of productivity and, in turn, reward differentials that represents the challenge for an explanatory theory.

An obvious possible explanation for the gross disparities in productivity exhibited in Figure 13.1 is simply sex discrimination. Although this has undoubtedly played some role—particularly in earlier cohorts—recent focused interviews of men and women scientists[11] revealed that most women indicated that they had not personally experienced discrimination. Nevertheless, most had heard of "other cases" of sex discrimination in science. The theory proposed herein views sex discrimination as only one of many causes of the cumulative productivity differential between men and women scientists. There is partial empirical support currently available for a "limited differences" explanation. Full validation and further refinement of this theory is an important challenge for the future.

13.1 Publication histories for five cohorts of men and women scientists.

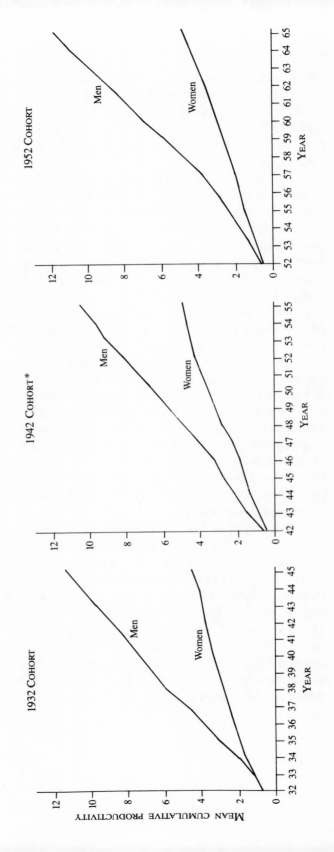

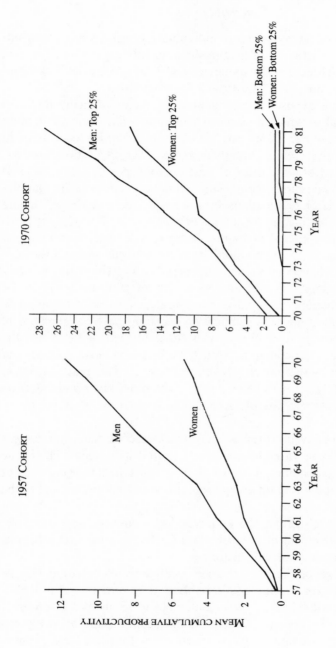

1957 Cohort

1970 Cohort

MEAN CUMULATIVE PRODUCTIVITY

YEAR

Men

Women

Men: Top 25%

Women: Top 25%

Men: Bottom 25%

Women: Bottom 25%

*Source: J. R. Cole and H. Zuckerman, "The Productivity Puzzle: Persistence and Change in Patterns of Publication on Men and Women Scientists," in P. Maehr and M. W. Steinkamp, eds., *Advances in Motivation and Achievement*, pp. 217–256. Greenwich, Conn.: JAI Press, 1984.

A Theory of Limited Differences:
General Outline

We focus on the dynamics of individuals, each of whom is embedded in a network or networks of relationships constituting the social system of scientific specialties. System dynamics and details of the network structure of science are not part of the present formulation.

Individuals are exposed to a sequence of events of many different types, some or all of which may occur more than once, depending upon the substantive context. There may also be a priori order restrictions in time on some of the events; "acceptance into Ph.D. institution," for example, must occur before "offer of post-doctoral position." Associated with each individual is an outcome variable(s)—e.g., manuscript completions and manuscript publications in a science career setting,[12] annual wages and/or annual income in studies of black-white earnings differentials, scores on age-graded mathematics tests as in studies of U.S. vs. Japanese schools. A "kick-reaction" pair corresponds to each event.[13] Examples, in the context of science careers, of kicks (which may be positive, neutral, or negative) are: acceptance or rejection by a top Ph.D. institution; positive and/or negative funding decisions on grant applications; positive and/or negative publication decisions based on manuscript submissions; marriage to a spouse who either hinders or facilitates the scientist's career. Associated with each kick is a positive, neutral, or negative reaction by the person who experiences it. This reaction acts— almost immediately, or with some delay—with other kicks and reactions to influence the outcome variable(s). Kicks and reactions thus have "memories."

The evolution of events, their associated kick-reaction pairs, and changes in levels of outcome variables are characterized as a vector stochastic process where the conditional probabilities of current state occupancy or changes in state are based on an individual's prior history, subject to the following general constraints:

i) with high probability, the kick-reaction sequences with memory, delays and establishment of competencies for future kicks and reactions determine the outcome variable histories;

ii) with few exceptions, the influence over a short interval of any single kick-reaction pair on an outcome variable will be small (or "limited"). Two individuals with similar, even identical histories up to a given time, but who experience opposite kinds of kick-reaction pairs to any given event—e.g., (negative kick, negative reaction) vs. (positive kick, positive reaction)—will *not* exhibit dramatically different outcome variable dynamics over a short time interval in the immediate future;

iii) all or nearly all kick-reaction pairs influence durations until future events by small amounts. Recurrent events have gently changing duration distributions regulating inter-event intervals. Major changes in waiting-time distributions between events and/or changes in levels of outcome variables will occur over long times or between pairs of events separated by "many" intervening events;

iv) there are a *few* special events for which, in distinguished subpopulations (call them A and B), the probability of a negative reaction to a negative kick for a member of group A exceeds the corresponding probability for a member of group B. Correlatively, the probability of an "improvement" in the outcome variable for an individual with a negative reaction to a negative kick on a special event is less than the corresponding probability for an individual with a positive reaction to a negative kick on the same special event, all other features of the past histories being the same;

v) conditional probabilities of specific changes in outcome variables at a given time are insensitive to the order of occurrence of a subset of the possible events that may have occurred in the past. The specific form of the kick-reaction pairs associated with these events will influence current outcome variable changes; however their order of occurrence will not matter.[14]

When applied to science careers, this general framework implies that cumulative productivity differentials between men and women scientists—identified as group A and B, respectively in conditions *(iv)*—are the result of small—or limited—differences in their reactions to a limited set of kicks. It is the cumulative effect of these small differences that produce, analogous to a "multiplier effect,"[15] major productivity differentials between men and women over a career. Small within-sex differences in a few kick and/or reaction intensities also lead to large career productivity differentials in the population of women scientists alone and among men scientists alone. It is the specific substantive content of the kick-reaction pairs which influences productivity in science and which, more generally, gives specific content to the theory as it is applied to different phenomena.

Consider an example of a career event which is hypothesized to increase slightly the probability that women scientists will be less productive than men. A set of graduate students in physics must select Ph.D. sponsors. All of the faculty in the physics department are men who are willing to sponsor the work of male graduate students; but 10–15 percent of them are unwilling to sponsor female graduate students. This is not an over-whelming disadvantage, but it is a disadvantage nonetheless. It will not affect most women scientists, who would not have wanted to work with that 10–15 percent of reluctant professors in the first place. Most women will not have any experience of a negative kick in the choice of sponsors. However, the entire set of women has a slightly smaller pool of eligible sponsors from which to choose. Some women would have selected these men, but for their refusal to sponsor women. If among these sponsors there are some excellent and powerful scientists, then a small proportion of women students will experience a slight negative kick resulting from not studying with them. The reactions to this kick will vary among the women who experience it. Some will fight even harder against such discrimination; but a subset may have their aspirations dampened and motivation reduced slightly by the experience. Such a difference is small but has clear impli-cations for future events in the careers of the affected scientists and to a limited degree is sex dependent.

Take another illustration. A slightly higher proportion of women than men of roughly equal quality have grant applications rejected for research support of the same size. Consequently, they have fewer resources for their research, less travel money for giving talks, papers, or attending conferences—in short for becoming "visible" to their peers. The sex dif-ference may be limited indeed, but the bias, where it occurs, may decrease slightly these women's productivity potential. In conjunction with prior disadvantages, the somewhat poorer probability of funding reduces the probability of quickly completing manuscripts and publishing papers.

These small negative kicks adversely influence productivity potential in the early phase of the career. They then influence a more significant event, the tenure decision. If tenure is denied, then further negative kicks and reactions can follow, exacerbating still further the difference in pro-ductivity potential.

In the case of scientific productivity, the major driving forces behind the manuscript production and publication processes are two interrelated goals: priority for scientific discovery and accompanying peer recognition (Merton 1957); and success in the competition for resources to pursue research at a high level. These primary goals are mediated by the action (kick)-reaction pairs experienced by individual scientists. Positive kicks and reactions are associated with increased incentive and hence with in-creased manuscript completion and publication. Negative kicks and reac-tions such as grant rejections and lack of peer recognition—for example, very low use of work by peers—serve as a major disincentive, often lead-ing scientists simply to abandon the race for major scientific discoveries.

Prior Explanations for the Productivity Patterns

Of course, there have been a number of theoretical and empirical efforts to explain the skewed productivity among scientists and the puzzling sex differential. In addition, there are other classes of explanations that have not been applied to this problem but that are plainly theoretically relevant. Most efforts within the sociology of science have focused on three classes of explanations: theories of initial conditions, theories of evolving social processes, and structural constraints.[16] We briefly review these explanations and indicate their relationship to the theory of limited differences.

THEORIES OF INITIAL CONDITIONS: THE SACRED SPARK

Productive scientists, this orientation holds, are those with "a sacred spark," those with the aptitude to tackle and solve difficult problems. While few would question that variations in ability play a formidable role in distinguishing between creative and uncreative science, this position ignores the role of culture, social structure, and personality in contributing to the productivity process. It is problematic primarily because there is simply no evidence to support the claim that gender differences are related either to the initial physiological or to biological conditions that are claimed to result in the productivity differences we are attempting to explain.[17]

THEORIES OF INITIAL CONDITIONS: PSYCHOLOGICAL TRAITS AND SOCIALIZATION PATTERNS

Motivational intensity required for high levels of productivity is assumed a priori in this orientation to be dampened in women because of the formidable early cultural and structural barriers that women face and must hurdle before reaching the starting line for a predominantly "masculine occupation" (Berryman 1983; Kahle and Matyas 1985; Marini and Brinton 1984; Bielby 1991). Socialization processes lead young women to be less confident about their scientific ability, less assertive in advancing their ideas and opinions, less apt to pursue their goals aggressively, while simultaneously being more ambivalent than men about their work and family roles. In due course, women and men come to the starting line for scientific careers carrying baggage of substantially different weights. Differences in scientific production follow naturally from these differences in background and current attitudes and traits. But no mechanism is proposed to explain how these "initial conditions" facilitate or impede subsequent events which unfold during a career—in short, how they are linked to actual productiv-

ity. The theory of limited differences as specialized herein to scientific
careers does deal with this latter process. It is incorporated in the differential probabilities of positive reactions to negative kicks on special events
such as NIH grant decisions.

THEORIES OF EVOLVING SOCIAL PROCESSES: REINFORCEMENT
AND SOCIAL LEARNING

Theories of "reinforcement" and "social learning" are based upon observations of the full process of scientific production over a span of years.
Reinforcement theory assumes that high levels of initial productivity receive positive reinforcement (through conference invitations, citations, job
offers, awards, etc.), which increases the probability of subsequent high
scientific productivity (Cole and Cole 1973). Conversely, poor early performance, going unrecognized, is negatively reinforced and leads to lower
future production. Social learning theorists hold that individuals' reactions
to events, or stimuli, will be influenced both by their past experience with
the stimuli, with cognitive processes that influence the perception and
retention of the event, and with anticipated future effects of a particular
response (Bandura 1986). As Bandura suggests:

In the social learning theory view, people are neither driven by inner forces nor
buffeted by environmental stimuli. Rather, psychological functioning is explained
in terms of a continuous reciprocal interaction of personal and environmental
determinants. Within this approach, symbolic, vicarious, and self-regulating processes assume a prominent role. (1977: 12–13)

Reinforcement processes and social learning undoubtedly operate to
influence manuscript production, but as they have been formulated, they
do not specify the emergent structural and cultural properties in social
systems that operate to influence manuscript production, specifically sex
differentials and the fanning out process. Furthermore, they emphasize the
internalized psychological components of action rather than the dynamic
structural bases for actions and reactions.

THEORIES OF EVOLVING SOCIAL PROCESS:
CUMULATION OF ADVANTAGE

Processes of "cumulative advantage," first articulated by Robert K.
Merton (1968), attempt to explain time-bounded patterns of skewed productivity (and recognition) in terms of increased opportunities for scientific
resources, both capital and human, that accrue to those who are productive
early on—and especially to the productive located at prestigious scientific
institutions. Changing distributions of resources become the basis for creating

even greater productivity distance between the "haves" and the "have-nots."[18] The cumulative advantage literature has focused on the increasing inequality of scientific publications and citations but has failed to establish the crucial step-by-step linkage between the changing distribution of resources and productivity inequality. The growing inequality has been *assumed* to be the result of cumulative advantages (see Allison, Long and Kraus 1982).

The theory of limited differences, while incorporating the idea of cumulative advantage, is much more fine-grained and specific about the mechanisms which generate productivity differentials. The kick-reaction sequences are the primitive ingredients in the theory of limited differences; there is no comparable *explicitly formulated* mechanism for the evolution of individual careers in the extant literature on cumulative advantage. In addition, the fact that reactions to particular events are allowed to depend upon prior events, past kick-reaction pairs, and initial conditions is what allows us to integrate socialization processes, psychological theory, and cultural value systems explicitly into the evolutionary dynamics of the limited differences formulation.

Furthermore, in the case of the scientific productivity differential between men and women, the theory of limited differences, through analysis of the sequencing of kick-reaction pairs, suggests a method for assessing whether or not one group rather than another will ultimately monopolize resources, rewards, and the productivity process. The theory allows us to determine the extent to which one group or another "dominates" the distribution of specific positive or negative kicks, and analysis of reaction systems enables us to identify differentiated responses by men and women to similar positive and negative career events.

STRUCTURAL CONSTRAINTS

Sex discrimination has been used to explain the sex differences in publications, and undoubtedly it has been a source of structural constraint for women scientists. Differential opportunities based on sex, or on other individual attributes that are unrelated to performance, can be translated into competitive advantages in the acquisition of resources and facilities necessary for high productivity.[19] In fact, discrimination, whether based on sex, religion, national origin, age, or race, enters the theory as a significant substantive element in determining limited differences.

But discrimination is viewed here as one among many sources of structural constraints affecting publication probabilities. For example, women's domestic responsibilities associated with child-bearing and child-raising could account for the lower rate of productivity of women scientists. This hypothesis has been studied in some detail. It turns out that women with children are as scientifically prolific, on average, as those without them (Cole 1979; Cole and Zuckerman 1987). But a small subset of women, distinctly limited in number but greater than the number of men, are adversely affected by building families, and this represents a limited difference that will influence productivity for that small subgroup. Thus, marital and fertility histories are a central feature of the application of limited differences dynamics to science careers.

Since scientific production is almost invariably carried out within social organizations, these organizational contexts can influence the form and substance of productivity. Some environments may be conducive to research; others hostile to it—and this holds for all scientists. But there may also be organizational structures that limit the productivity of women more than men. These range from barriers to participation to subtle exclusions from informal interaction within laboratory settings (Bielby and Baron 1984; Fox 1981a; Long 1978; Pfeffer 1982; Reskin 1978a,b). Only a rough beginning at empirical research aimed at measuring the actual effects of organizational structures on scientific productivity has been carried out (see Long and McGinnis 1981). These studies have been unable as yet to specify adequately how dynamic interactive processes between the "environment" and the individual influence productivity and, more specifically, what features of organizational environments adversely affect women's productivity relative to similarly situated men.

There are, then, a set of existing social and social psychological theories that purport to explain gender differences in scientific productivity and the fanning out process. Each of these theories has useful elements, but individually they take us only a limited way toward explaining the productivity patterns in question.[20] Aspects of each are incorporated into the theory of limited differences. We turn now to a description of basic elements in the theory and to its application to solving the productivity puzzle.[21]

Formalization of Science Career Development

We specialize and interpret the general outline of the theory given above in the context of science careers.

A PRIMARY EVENT LIST AND DELINEATION OF INITIAL CONDITIONS

A set of career events, hypothesized to be the basis of an explanation of the productivity puzzle in science, is listed in Table 13.1. This is by no means an exhaustive list; however, it does contain the items which both empirical studies to date (see, among many others, Astin 1969; Astin and Bayer 1972; Bayer and Astin 1972; Centra 1974; Clemente, 1973; Cohen 1980; Crane 1969; Cole and Cole 1973; Gaston 1973; Hagstrom 1971; Hargens, McCann and Reskin 1978; Helmreich et al. 1980; Spence, Helmreich and Stapp 1975; Spence and Helmreich 1978, 1979; Zuckerman 1977, 1989; Cole 1979; Allison and Stewart 1974; Long and McGinnis 1981; Cole, Rubin and Cole 1978; Cole, Cole and COSPUP 1981; Reskin, 1977, 1978a, 1978b; Zuckerman and Merton 1971a, 1971b; Zuckerman and Cole 1975; Over 1982; Over and Moore 1980) and prior theoretical proposals suggest should be the most important events.

TABLE 13.1 *Events Influencing Scientists' Productivity Histories*

E_1 = Decision on Ph.D. institution
E_2 = Decision on Ph.D. sponsorship
E_3 = First post-doctoral job or post-doctoral fellowship
E_4 = Publication decision: acceptance or rejection of paper
E_5 = Marriage or cohabitation
E_6 = Birth of child
E_7 = Perceived quality of research: critical reception of publications
E_8 = Funding decision
E_9 = Marital disruption or cessation of cohabitation
E_{10} = Tenure decision
E_{11} = Moderate honorific recognition (e.g., Guggenheim, Sloan fellowships)
E_{12} = Major honorific award (e.g. Lasker, NAS membership, Fields Medal, Nobel Prize)
E_{13} = Laboratory directorship
E_{14} = Job offer from outstanding department
E_{15} = Critical reception of paper prior to publication

Note: For each event E_i, there will be nine logical combinations of kick-reaction pairs.

There are some a priori order restrictions to be imposed on these events which indicate that some of them must occur in time prior to others. Introducing the relation $<$ to mean "before," we require:

(i) $E_1 < E_2 < E_3 < E_{10}$

(i.e., acceptance into Ph.D. institution must occur *before* acceptance of Ph.D. sponsorship, which, in turn, must occur before a tenure decision.)

(ii) $E_5 < E_9$
(iii) $E_5 < E_6^{(1)} < \ldots\ldots < E_6^{(m)}$ ($E_i^{(j)}$ means j^{th} occurrence of ith event)
(iv) $E_2 < E_{11}^{(1)} < \ldots < E_{11}^{(v)}$
(v) $E_{11}^{(1)} < E_{12}^{(1)} < E_{12}^{(2)} \ldots.$
(vi) $E_{15} < E_4 < E_7$

With the exception of these constraints, any ordering of events is possible in principle. Science careers will be assumed to start when an individual applies to a graduate program in some scientific field or specialty. The details of the process of self-selection which leads some individuals to this choice, as opposed to other career options, is an important topic which lies outside of the scope of the present formulation. Thus gender differences in early socialization and a variety of attitudes and expectations about what is or is not achievable in a scientific career will be assumed to be the primary source of variation across individuals when the career process initiates. Persons clearly differ in basic ability and motivation even when initially self-selecting to begin a science career; however, there are, as yet, no good measures of early ability and motivation which distinguish men from women at this stage. There are also no effective *early* screening measures which will indicate who among persons in the same discipline, prestige level of graduate school, and with comparable undergraduate record are likely to be the major producers of science in their cohort.

OUTCOME PROCESSES

Two interrelated outcome variables will be central to the present specification of science careers: manuscript completions and publications. These variables are related in a publication process as delineated in Figure 13.2.

When drafts of manuscripts are generated, they are frequently circulated among close peers for comment and criticism—this gives rise to the event, E_{15}. Following this event, manuscripts are completed and, with high probability, are submitted to journals for review, thus leading to a publication decision—the event E_4. A favorable decision is followed by a manuscript publication. However, an unfavorable decision puts manuscripts into a feedback loop which may lead to a revised manuscript completion and a subsequent publication or may lead to the scientist simply giving up on the paper. The process exhibited in Figure 13.2 has separate compartments for "writing following an outright rejection for publication" and "writing following an editor's request for some revisions" because of the very different attitudes that scientists will have while in each of these regimes. This distinction then leads to different probabilities of manuscript completions and resubmissions, an important feature of a science career.

The probabilities associated with transitions along various paths in this set of events vary dramatically by field, specialty, and even research area. For example, the probability of publication given an initial submission in virology is approximately .9 or roughly .75–.9 in physics, while the same conditional probability is roughly .2–.3 in some subspecialties of sociology and economics.[22] Such differentials are apt to lead to different expectations by scientists in varying fields and hence to different intensities of reactions to rejected manuscripts.[23] Thus the (submission → publication) link is primarily subject-matter determined while the (peer evaluation → completion), (completion → submission), (revision → submission), and (rejection → revision → submission) paths are much more heavily influenced by individual drive, motivation, and career aspirations.

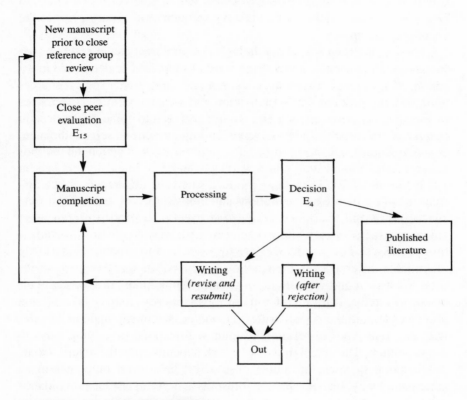

13.2   The process of scientific publication.

KICK-REACTION PAIRS

With each event from the list in Table 13.1, as it occurs in an individual's evolving career, there is associated a kick-reaction pair. Kicks and reactions can each be of three kinds: positive, neutral, and negative.[24] We will denote these alternatives by the system of symbols:

KICKS	REACTIONS
k^+ = positive kick	r^+ = positive reaction
$k\cdot$ = neutral kick	$r\cdot$ = neutral reaction
k^- = negative kick	r^- = negative reaction

Thus for each event there are nine possible kick-reaction pairs. The probabilities of occurrence of pairs such as (k^+, r^+) and (k^-, r^-) will be substantially larger than the probabilities associated with, for example, (k^+, r^-), (k^-, r^+), and $(k^\cdot, r^-)$. A given kick-reaction pair will influence the outcome variables through conditional probability distributions whose structure is described in the next section. In addition, the probability distribution on kick-reaction pairs associated with a given event will depend on an individual's kick-reaction history and outcome history prior to the event in question.

There is a strong social psychological component to the reactions associated with particular kicks in the form of increased or decreased motivation. Motivational differences result not only from psychological sources but also from processes of socialization and social structure. Differences in socialization between men and women can lead to some differences in individual traits such as aggressiveness, competitiveness, self-confidence, degree of confidence, and comfort in an environment in which the individual represents a minority (Duncan and Duncan 1978; Maccoby and Jacklin 1975; Marini 1987). These may produce small sex differences in the distribution of expectations, aspirations and motivation and somewhat different tolerance and resistance to negative events. In short, differences in reaction systems of men and women scientists may result from socialization processes. These processes are hypothesized to produce empirically identifiable differences in the reaction systems of men and women scientists. Most men and women scientists may "look alike" in terms of their reaction systems, if for no other reason than that self- and social selection processes lead to these similarities. However, scientists, male or female, with different types of reaction systems will respond in varying ways to the same kick. The actual impact of a kick depends upon the reaction to it.

Reaction systems, of course, also affect behavioral outcomes in an anticipatory way: fear of rejection forestalls action and produces avoidance behavior. Reaction systems affect and are affected by the cognitive styles of scientists. Some scientists will be risk-adverse, fearing negative kicks. Others opt for tackling risky problems and take chances in their efforts to be published in the top journals, or to be optimally funded for their research.

There are also structural constraints on flexible reactions to kicks that have little to do with psychological traits. Clearly, scientists in different social structural locations have differential opportunities to react positively or negatively to kicks (Fox 1983). Some are in situations where they can "do something about" a negative kick, others are not. Institutional structures not only affect reactions to kicks but influence the sequencing of future kicks and the duration of time between manuscript completions.

Social and cultural customs and mores, such as marital patterns, also can constrain types of reactions to kicks, as is the case when the geographic mobility of a woman is restricted by her spouse's job (Marwell, Rosenberg and Spilerman 1979). Ceterus paribus, women scientists are more apt than men to be structurally constrained in their choices. For both men and women the sequence of reactions to the same or similar events will change with successive kicks.[25] In particular, the resiliency of positive or neutral reactions will diminish with a succession of negative kicks.

PROBABILITY SPECIFICATIONS AND MEMORY EFFECTS

We represent science careers in terms of: (a) a sequence of early events—in particular, those which occur up to the first post-Ph.D. position—where there is an accumulation of kicks and reactions which strongly influence subsequent mid-career development; (b) the period from first job beyond the Ph.D. to first major award (this is where the basic publication record is established); and (c) the post-initial major award period, where substantial publication is reinforced, accelerated because of growing resources, or dampened because of increased obligations outside of the research role. Many scientists, even among prolific producers, will never move to phase c, but large proportions of those traveling in this fast lane will receive substantial honorific recognition.[26]

The Early Events Module

We consider the events E_1, E_2, and E_3 which are, of course, constrained by the order relation $E_1 < E_2 < E_3$. In addition, the events E_4, E_5, E_6, E_8, and E_9 may be interdigitated with $E_1 - E_3$ subject also to the order restriction listed on p. 289. Early event histories will consist of sequences of three or more events from the above list, and E_1, E_2, and E_3 must occur in each sequence. We denote by $|E_1$, $|E_2$,,.... the possible *sequences* made up of at most the above eight distinct events subject to order restrictions and allowing some events, such as birth of a child (E_6), to occur more than once prior to E_3.

For example, we may set

$|E_1 = \{E_1, E_2, E_3\}$
$|E_2 = \{E_5, E_1, E_6^{(1)}, E_2, E_3\}$
$|E_3 = \{E_1, E_2, E_5, E_3\}$
.
.
.
.

Then within sequence $|E_i$ we denote the kick-reaction pairs by $(K_{j_1}^{(|E_i)}$, $R_{j_1}^{(|E_i)})$, $(K_{j_2}^{(|E_i)}, R_{j_2}^{(|E_i)}),\ldots$ where j_1 is identified with the subscript of the first event in $|E_i$, j_2 is identified with the subscript of the second event in $|E_i\ldots$, etc. For example, in sequence $|E_2$, $j_1 = 5$, $j_2 = 1$, $j_3 = 6$, $j_4 = 2$, and $j_5 = 3$.

With this notation at hand we represent the joint distribution of events and kick-reaction pairs as the product of conditional probabilities

$$\text{Prob} \left(|E_i;\quad (K_{j_1}^{(|E_i)}, R_{j_1}^{(|E_i)}),\ldots, (K_{j_\mu}^{(|E_i)}, R_{j_\mu}^{(|E_i)})\right) \tag{1}$$

$$= \prod_{l=0}^{\mu-1} \text{Prob} \left((K_{j_{\mu-l}}^{(|E_i)}, R_{j_{\mu-l}}^{(|E_i)}) \left| \begin{array}{l} \text{kick-reaction pairs} \\ \text{prior to event } j_{\mu-l} \end{array} \right.; \quad |E_i\right) \text{Prob} (|E_i)$$

where $\mu = $ number of events in $|E_i$

Each kick-reaction pair can assume any one of the nine possible values (k^+,r^+), $(k^{\cdot},r^+)$, (k^-,r^+), $(k^+,r^{\cdot})$, $(k^{\cdot},r^{\cdot})$, $(k^-,r^{\cdot})$, (k^+,r^-), $(k^{\cdot},r^-)$, and (k^-,r^-). Numerical specification of $\text{Prob}(|E_i)$ is guided by empirical frequencies in existing surveys of scientists. The general conditional probabilities in equation (1) must be further restricted to conform to particular proposals about the influence of memory on current perceptions of kicks and associated reactions. Two specifications which are relevant for science careers are:

(A) For a given sequence, $|E_i$, past kicks and reactions prior to the 1th event influence the probability of the 1th kick-reaction pair only through the sums

$$\sum_{m=1}^{l-1} W_{j_m}^{(|E_i)} \text{sgn} (K_{j_m}^{(|E_i)})$$

and   (2)

$$\sum_{m=1}^{l-1} V_{j_m}^{(|E_i)} \text{sgn} (R_{j_m}^{(|E_i)})$$

where

$$\text{sgn} (K_{j_m}^{(|E_i)}) = \begin{cases} +1 & \text{if} \quad K_{j_m}^{(|E_i)} = k^+ \\ 0 & \text{if} \quad K_{j_m}^{(|E_i)} = k^{\cdot} \\ -1 & \text{if} \quad K_{j_m}^{(|E_i)} = k^- \end{cases}$$

and

$$\text{sgn } (R_{j_m}^{(|E_i)}) = \begin{cases} +1 & \text{if} \quad R_{j_m}^{(|E_i)} = r^+ \\ \quad 0 & \text{if} \quad R_{j_m}^{(|E_i)} = r^\cdot \\ -1 & \text{'if} \quad R_{j_m}^{(|E_i)} = r^- \end{cases}$$

The weights $\{W_{j_m}^{(|E_i)}\}$ and $\{V_{j_m}^{(|E_i)}\}$ indicate the relative influences of past events on the probability of a current kick-reaction pair. The weight sequences are associated with specific orderings of events—namely, $|E_i$— and need not be invariant under permutations of them. The parameterization (2) implies that the full past influences current probabilities—if all weights are non-zero—and that the longer the sequence the less influence any single kick-reaction pair in the past will have.

(B) If an early event and its kick-reaction pair only influence a specific future event, this effect is captured in the specification

Prob $((K_l^{(|E_i)}, R_l^{(|E_i)}) \mid$ kick-reaction history prior to l^{th} event)

$= \text{Prob } ((K_l^{(|E_i)}, R_l^{(|E_i)}) \mid (K_j^{(|E_i)}, R_j^{(|E_i)}))$ \hfill (3)

for a distinguished event—the jth event—occurring at an earlier time, $j < l$. An example of this is where the jth event is "marriage by a woman scientist in undergraduate school" but to a man whose career imposes rigid geographical immobility for the couple. In terms of the event sequence formalism, this is a history for which $E_5 < E_1$. Now we define the lth event to be $E_3 = $ (offer of first post-doctoral position) and assume that the position is at an outstanding institution—thereby giving rise to a positive kick—but that it is located outside the geographical range which would preserve the husband's job. Thus a negative reaction is associated with the positive kick—i.e., $(K_1^{(|E_i)}, R_1^{(|E_i)}) = (k^+, r^-)$. The marriage itself, at the time of its occurrence, is associated with $(K_j^{(|E_i)}, R_j^{(|E_i)}) = (k^+, r^+)$. The dependency restriction (3) implies that all events other than the marriage have no influence on the current kick-reaction pair. The idiosyncratic detail of geographic immobility of a spouse is not formally incorporated in the probability specification; however, non-zero conditional probabilities for the sequence of kick-reaction pairs

$(K_j^{(|E_i)}, R_j^{(|E_i)}) = (k^+, r^+) \rightarrow (K_l^{(|E_i)}), R_l^{(|E_i)}) = (k^+, r^-)$ \hfill (4)

are interpreted to mean that some major obstacle associated with the marriage gave rise to the negative reaction on the lth event.

Mid-Career Dynamics

Development of manuscripts for publication usually begins prior to Ph.D. completion in the sciences and, in some fields, even in undergraduate colleges. We assume that once the manuscript completion process begins, new manuscripts are produced at independent but not identically distributed intervals until the start of a first post-doctoral position. Kicks and reactions in the early events module are not assumed to influence manuscript completions prior to receipt of the Ph.D. degree. However, kicks and reactions in the feedback loop of the publication process—Figure 13.2—will slightly increase the manuscript completion rate when positive reactions occur and slightly decrease it when negative reactions occur.

Once the first post-Ph.D. position is attained, then the waiting times between successive manuscript completions have means and variances which are functions of the cumulating kick-reaction experience to the full range of events listed in Table 13.1. These means and variances decrease slightly with each positive reaction and increase with negative reactions. Thus, the intermanuscript completion intervals are decomposed into episodes separated by occurrences of events outside the publication module, and, condition on the kick-reaction pairs associated with these events, the conditional mean and variance of the waiting time distribution for manuscript completions is adaptively altered.

The cumulative number of manuscript completions and publications as well as their rate of occurrence in particular time intervals influences the probability of kick-reaction pairs on special events such as grant decisions and major and minor awards. Indeed in the post-Ph.D. regime, events occur in a continuously evolving stream where the inter-event time intervals and the character of the associated kick-reaction pairs is governed by the prior kick-reaction history and the productivity record. Qualitatively, r^+ reactions and increasing manuscript completions increase the probability of (k^+, r^+) on future events and the probability of r^+ when k^- occurs. Thus past success generates resilience to future negative kicks, such as grant rejections. Waiting times until occurrence of both minor and major awards[27] also depend on productivity and citation ranking of the individual scientist among peers in his (her) subspecialty. For minor awards, the higher the ranking on at least one of these variables, the shorter the expected waiting time until reception of awards and the shorter the expected duration between successive awards.

Major awards in most scientific fields are dominated by the most prolific and visible scientists—perhaps the top 10 percent. Major awards, such as Nobel Prizes, have a ratchet effect. Upon receipt of one, the influence of past history on the durations between manuscript completions is reset to a "post-award" level and no longer depends significantly on the earlier kick-reaction history.

Beyond the First Major Award

There is considerable variation in reactions by scientists to the receipt of major awards. Some continue research at an increased pace; others leave the laboratory altogether; still others have temporary reductions in scientific productivity followed by reestablishment of a prolific rate of publication.[28] Those who shift into administrative roles have dramatically reduced manuscript completion rates; their publication probabilities are assumed to be unrelated to past kick-reaction histories. For those continuing research as their primary activity, the previous reaction history no longer really influences manuscript completion rates. After receiving major awards, the primary influences in manuscript completion rates are assumed, a priori, to be kicks associated with grant rejections. Eminent scientists are not immune to negative peer reviews, lower than expected priority scores, and rejections of grant applications. While they tend to submit more proposals than their less distinguished colleagues, they generally have larger laboratories to sustain. Even the occasional rejection of a large budget proposal can represent a significant negative kick for the productivity of their labs. Indeed, the investment in large blocks of time to "keep the lab going" leads some of these eminent scientists to modify their future research aspirations and overall career goals. Finally, after receiving a major award, some scientists change specialties or fields of inquiry.[29] When this happens, we view the manuscript completion rate for these transfer scientists as roughly equivalent to a new Ph.D. and with the same influence of negative reactions—if they occur—on their productivity.[30]

LIMITED DIFFERENCES: SOURCES OF DISPARITY BETWEEN GROUPS

The formulation of the evolutionary dynamics of science careers in the previous section makes no distinction, in principle, between different subpopulations—e.g., men vs. women scientists. Indeed, within each of these groups, the full range of qualitative principles listed as generic for the generation of productivity and kick-reaction histories is operative. Disparities between men and women are introduced as small (or limited) differences in probabilities associated with kick-reaction pairs for a small subset of the events in Table 13.1. In particular we assume that:

(i) For funding decision—event E_8—
$$\text{Prob}_{[\text{women}]} ((k^-, r^-) \text{ on } E_8 \,|\, \text{past history})$$
$$> \text{Prob}_{[\text{men}]} ((k^-, r^-) \text{ on } E_8 \,|\, \text{past history}) \tag{5}$$

Thus, given identical histories,[31] women tend to have negative reactions to grant rejections more often than men. Correlatively

$$\text{Prob}_{[\text{women}]}\,((k^-,r^+)\text{ on }E_8\,|\,\text{past history}) \tag{6}$$
$$< \text{Prob}_{[\text{men}]}\,((k^-,r^+)\text{ on }E_8\,|\,\text{past history})$$

(ii) $\text{Prob}_{[\text{women}]}\,(k^-\text{ on }E_2\,|\,\text{past history in early events module})$
   $> \text{Prob}_{[\text{men}]}\,(k^-\text{ on }E_2\,|\,\text{past history in early events module})$ (7)

This inequality is motivated by the fact that a small proportion of the outstanding scientists refuse to accept women as their students, as noted above, thereby limiting—by a small amount—advantageous post-doctoral positions and subsequent support groups recommending them for both minor and major awards.

(iii) $\text{Prob}_{[\text{women}]}\,((k^-,r^-)\text{ on }E_6 = \text{birth of a child}\,|\,\text{past history})$
    $> \text{Prob}_{[\text{men}]}\,((k^-,r^-)\text{ on }E_6\,|\,\text{past history})$ (8)

(iv) $\text{Prob}_{[\text{women}]}\,((k^-,r^-)\text{ on }E_{10} = \text{tenure decision}\,|\,\text{past history})$
    $> \text{Prob}_{[\text{men}]}\,((k^-,r^-)\text{ on }E_{10}\,|\,\text{past history})$ (9)

(v) $\text{Prob}_{[\text{women}]}\,((k^-,r^-)\text{ on }E_{15} = \text{critical reception of paper prior to}$
   publication $|\,\text{past history})$
   $> \text{Prob}_{[\text{men}]}\,((k^-,r^-)\text{ on }E_{15}\,|\,\text{past history})$ (10)

Correlatively

$$\text{Prob}_{[\text{women}]}\,((k^-,r^+)\text{ on }E_{15}\,|\,\text{past history})$$
$$< \text{Prob}_{[\text{men}]}\,((k^-,r^+)\text{ on }E_{15}\,|\,\text{past history}) \tag{11}$$

Inequalities *(i)* and *(v)* imply that women tend to get more discouraged by negative decisions on grant applications and critical commentary about their work than men.[32] Although this is not universally the case, the consequence of the negative reactions is to slow down the manuscript completion rate by a small amount. Over a period of 7–10 years this can result in major disparities in productivity between otherwise indistinguishable men and women scientists. Thus the full set of inequalities, *(i)–(v)*, coupled to the conditional probability specifications on p. 294 constitute the basic formalism of the theory of limited differences, as applied to science careers. Quantitative implementation of this formalism with a range of functional forms for the conditional probabilities based on past histories requires a *family* of microsimulation models, which will be reported on in detail in a later publication. The point of embedding this general evolutionary the-

ory of science careers in a *family* of models is that the manuscript completion and publication histories are relatively insensitive to a diversity of
perturbations in kick-reaction histories. This is a form of structural stability
of science careers; that is, most small variations in the details of the kick-
reaction histories do not lead to qualitatively different career paths.

Examples of Individual Histories and Their Interpretation

In order to clarify the character of microsimulation implementations of
the theory of limited differences, we construct three hypothetical examples
of science careers: one for a prolific and eminent male scientist; a second
for a woman who is less prolific but eminent; and a third for a less productive and noneminent woman scientist who might have been more prolific
but for her action-reaction experiences. These hypotheticals represent only
three of a myriad of possible careers and are intended to clarify the three
interrelated sequences of kicks and reactions, completed manuscripts, and
publications which develop over a career. They are portrayed schemati-
cally as shown in Figure 13.3. The cumulative effect of the early kick-
reaction pairs heavily influences the early and mid-career manuscript completion rate, based on the events E_1 (acceptance into Ph.D. institution), E_2
(acceptance of Ph.D. sponsorship), E_3 (first post-Ph.D. job), and, if they
occur prior to E_3, decision on first manuscript submitted for publication
(E_4), E_5 (entry into first marriage or cohabitation), or E_6 (birth of a child).

The case history for the eminent male scientist begins by noting that
his personal background and academic record prior to the Ph.D. produced
a sense of great self-confidence in his scientific ability. His reaction system
was geared toward success; he had high expectations for achievement.
And indeed, his first three events are all positive, experiencing (k^+, r^+)
pairs in terms of admission to the top Ph.D. department of his choice,
acceptance by a first-class sponsor, and receipt of a distinguished job upon
completion of his degree. These kick-reaction pairs serve as a major cumulative influence on his rate of manuscript completion. This produces a
strong incentive to succeed in competition with other scientists for important discoveries. The cumulated positive reaction intensities in the "early
events module" determine the initial manuscript completion rate immediately following the first post-Ph.D. job. This rate can, of course, be modi-
fied by later events. These early positive reactions then interact with the
positive outcome and reaction of the scientist to having his first grant
application ($E_8^{(1)}$) funded. This further increases the probability of high
rates of manuscript completion and submission for publication.

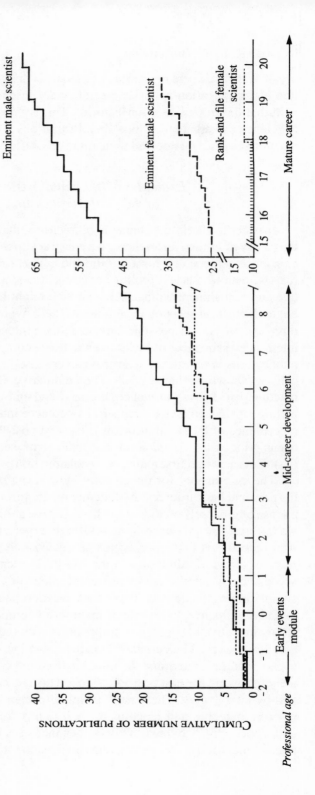

KEY

(,) kick-reaction pair–e.g, (+,-) means positive kick, negative reaction

$E_i^{(a)}$ means j^{th} occurrence of ith event

$\underset{m}{\triangle}\ \underset{n}{\bar{\triangle}}$ submitted manuscript m accepted for publication; manuscript n rejected

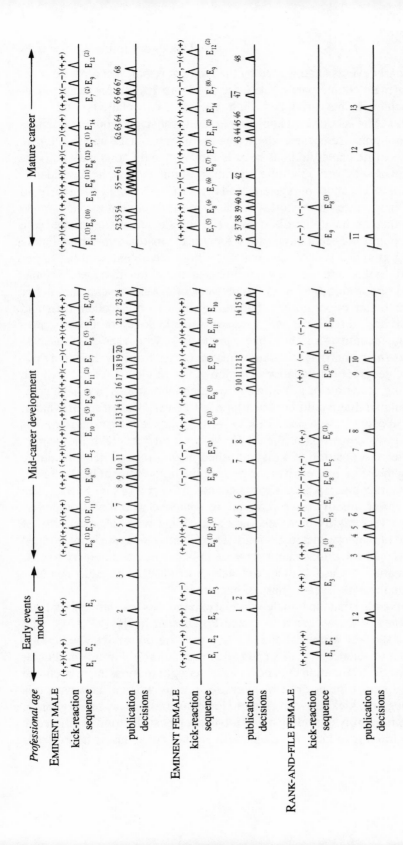

13.3 Simulation of manuscript completion and kick-reaction histories.

The early events experienced by the eminent female scientist are similar to her male counterpart's (see Case 2, Figure 13.3). Her early educational achievements produced high personal expectations and lofty aspirations. She is confident about her aptitude for science but experiences some cross-pressure because she wants to mix a marriage and family life with a scientific career and has been led to believe that this may be risky for a woman with lofty scientific aspirations. Nonetheless, her confidence abounds and her reaction system leads her to be strongly motivated to succeed in science. She works in the same field as the eminent male scientist that we have just discussed. The female scientist experiences a (k^+, r^+) pair for acceptance into a top Ph.D. institution. Immediately following graduate school, she marries a highly eminent scientist in her own field. In this case marriage represents an initial positive kick. She has increased opportunities to enter the network of leading researchers in her field—far better opportunities than those open to most other men and women of her professional age. These positive kick-reaction pairs represent strong incentives for her to begin publishing, which she does successfully. But she is married to a man whose career is firmly rooted in a very restricted geographical location. His job is not in close proximity to the outstanding academic or government research laboratories that have positions available that would best facilitate her career. The effect of what at the time appears to be a positive kick—i.e., entry into first marriage—is delayed until completion of the Ph.D., when it interacts with the woman being offered a position at a distinguished institution and her reluctantly declining the offer as a result of the geographic immobility of her husband. This discouragement, which is associated with the pair (k^+, r^-) for E_3, can serve to lower initially the aspirations of someone who might otherwise have been very highly motivated and skilled and with full capability of being one of the very best in her field. This negative reaction is interpreted as setting a lower initial manuscript completion rate than exists for the male scientist in Case 1, who did not experience a dramatic negative reaction in the early events phase of his career.

Observe also that the female scientist experiences a negative kick from being turned down by her first-choice Ph.D. sponsor $[E_2(k^-, r^+)]$, who refuses to sponsor women, believing that they are poor risks who are apt to drop out of science to get married and raise children. For this particular woman the discrimination engenders a further fight to show the first-choice potential sponsor the error of his ways. For many women, however, this kind of negative kick can lead either to lower aspirations or, subsequently, to lower probability of finding a first job in a top post-doctoral position due to poorer training or lack of national influence of her sponsor. It also leads to a slight advantage for male scientists in the early career stage and later in regulating the probabilities of receipt of both major and minor awards.[33]

Returning to Figure 13.3, observe that following E_3 there is very little difference in the manuscript completion and publication histories for male and female scientists, except that the completion rate for the male (driven by the stronger initial cumulated positive reactions) dominates that of the female scientist. In addition, negative kicks from a few grant rejections do not lead to negative reactions by the male scientist, whereas they do lead to such reactions, with a slight accompanying reduction in the manuscript completion rate, for the female scientist. Recall that we are treating (k^-, r^-) pairs on grant decisions as proportionately slightly more frequent among women than men, assuming a priori that a slightly higher proportion of women are somewhat more vulnerable than men to intense negative reactions from grant rejections, leading thereby to a slight reduction in their relative productivity.[34]

For the woman, the early large negative reaction of not accepting an optimal first job (E_3) and mild discouragement from the grant rejection (E_8) leads to a slower rate of accumulation of publications by the female relative to the male scientist working in the same field. The birth of a child (E_6) before the tenure decision does not produce a negative kick for the woman's productivity, but it does take her away from her department and colleagues and contributes to a delay in the decision on her tenure.[35] The woman scientist receives tenure (E_{10}), but somewhat later than her male counterpart—after roughly 15 completed manuscripts, compared with 11 for the man. Thus, the intensity and set of consequences of the positive reaction to the tenure decision is less than it was for the man. In terms of publication histories, this also reduces for some time her ability to attract the best graduate and post-doctoral students and to build the size of her laboratory. By the time the male receives his first major award $(E_{12}^{(1)})$ some 15 years after the Ph.D., the ratio of female to male publications is 30 / 52—or roughly .58. The accumulating publication disparity exhibited in these caricatures is a common but not universal feature of male–female differences among very eminent scientists. Each negative reaction for the woman on a few events contributes a small amount toward slowing down the manuscript completion rate—as it would for a man as well. Over a major portion of a career, say 20 years or so, this gives rise to a substantial disparity in lifetime productivity as measured by publication counts.[36]

For analytic purposes, the third phase of the career development process is assumed to begin after a scientist receives at least one major award. At this point a branching takes place. For the vast majority who continue to do scientific research, many begin to manage larger, quality laboratories. They obtain greater resources, their production of manuscripts increases dramatically, although their own relations to production often change. Thus, becoming the director of an excellent lab at a distinguished institution relatively early in a career often leads to substantial increases in output. Note that in our hypothetical examples, the eminent male scientist is made director of such a laboratory almost immediately after being honored with a major award; the woman does not receive this positive kick at all—although she might well have expected it. We assume that men are more apt than women to assume such directorships—and comparatively early on in the career.[37] This is viewed, in the present formulation, as a consequence of the slightly higher probability of outstanding male scientists having Ph.D. or post-doctoral sponsors who are particularly influential and who facilitate the visibility of their intellectual progeny and sponsor them for minor and major awards at early ages. Thus, a virtually undetectable "limited difference" in the early events module can have major consequences in the later career stage.

The second woman (see Case 3, Figure 13.3) was also labeled as an exceptionally able youngster, although she retained a sense that her success was more a result of hard work than ability. Working in astrophysics, she attends a distinguished graduate school, holds a major pre-doctoral fellowship, and receives excellent training from her first-choice sponsor (E_2). She publishes papers with her sponsor before receiving the Ph.D., accepts a first job offer at the most distinguished department in the country (E_3), and finds herself among the brightest and most dedicated young scientists she has ever encountered. Competition is fierce in the fast lane in which she is traveling.

As an assistant professor, she is the sole author of two papers in prestigious journals and receives two years of support from the National Science Foundation (NSF) for her research. Her sense of competence increases, but a paper she thought offered a particularly novel solution to a long-established problem is received poorly by some of her distinguished colleagues who are important reference individuals for her (E_{15}) and is rejected by a major journal (E_4). She begins to question her originality compared to the other bright and seemingly indefatigable assistant professors in her department. This slight loss of self-confidence is exacerbated when a research proposal of hers is rejected ($E_8^{(2)}$). Her motivation to complete several manuscripts and submit manuscripts and grants for peer review is dampened, leading to delays in her submissions.

Her marriage to a nonscientist (E_5) and the subsequent birth of two children ($E_6^{(1,2)}$) does not result in loss of time in the laboratory, but it does mark the end of all her "discretionary" time. However, the termination of her grant and the rejection of a second manuscript reduces her motivation and her career aspirations and leads her to question whether she can maintain the pace of research required by her department. The final blow to her aspirations and motivations is her denial of tenure (E_{10}) by her distinguished department. Not satisfied with being simply run-of-the-mill, she cuts back significantly on the pace of her research—reducing still further the probabilities that she will continue to be a prolific scientist.

This scientist was headed for membership in the productive elite but experienced a set of slight negative kicks, which accumulated over time and interacted with her self-doubt about her ability to compete with the best young minds in her field. The several paper and grant rejections sting her; the denial of tenure is an intense kick. Together these events and the concomitant reactions lead to a longer time between completion and submission of manuscripts. She slowly moves out of the fast lane and never receives a major award. It is important to emphasize that precisely the same event history could be constructed for men scientists. The inequalities detailed above imply that the probability of the sequence of negative kicks and reactions are more apt to be part of the career histories of women than of men scientists.

Competition

I was competitive beyond the run of younger mathematicians, and I knew equally that this was not a very pretty attitude. However, it was not an attitude which I was free to assume or reject. I was quite aware that I was an out among ins and I would get no shred of recognition that I did not force. (Wiener 1956: 87)

The formulations in this chapter may be viewed in many respects as a theory about the response of the community of scientists to an unstated driving force: competition. The social system of science is driven by competition in at least two forms. There is competition for ideas and hence priority in discovery and competition for the funds which are, in many instances, essential for the pursuit of particular lines of inquiry.[38] In the era of "little science," competition for ideas was the dominant form of this phenomenon. However, with the very large economic costs of resources for doing such things as high energy physics via particle accelerators, astrophysics via satellite observations, or climatology via deep sea sediment cores, competition for ideas is now augmented by and thoroughly intertwined with competition for funds.[39] This two-sided competition is particularly acute among the stratum of prolific, highly productive research scientists located at the major scientific institutions—those few who account for such a large proportion of all scientific discoveries.

Biographical reports and sociological studies ranging from large-scale surveys to studies of individual laboratories testify to the centrality of competition in the lives of scientists.[40] There are strong interactions between the action (kick)-reaction system as delineated in this chapter and competition processes, particularly due to the scarcity of resources for pursuing many types of scientific inquiry. Scientists' perceptions of the peer review systems of the National Institute of Health (NIH) and the National Science Foundation (NSF) bring to life the interrelationship between the action (kick)-reaction system, competition, and scarcity.[41] Scientists' productivity is linked directly to keeping the laboratory operating at a high pitch, and it is becoming increasingly difficult and time consuming to obtain the necessary support.

Competition for ideas and priority in discovery exists for some problems, especially in the upper echelons of any scientific discipline, with the competing parties having nearly complete knowledge of what their competitors are doing. The quintessential example of this is, of course, the race for determination of the structure of DNA by the Watson–Crick and Pauling labs.[42] The intense transfer of information via frequent conferences, private laboratory visits, and even telephone conversations between competitors and / or their close collaborators plays a major role in structuring research agendas and in regulating the duration of time between experiment completion, manuscript submission, and publication.

The increasing awareness of the centrality of competition processes as a driving force in science has, unfortunately, not been accompanied by the extensive empirical research which is required to document the fine-grained relationships between competition and the events presented in Table 13.1. Empirical research to date lacks specificity on the focus of competition, its types and intensities; it also lacks detail on the role played by reference groups and social networks in producing and maintaining competition. Furthermore, the fragmentary evidence currently at hand indicates that there is substantial heterogeneity across subspecialties in forms of competition. Because of the sketchy nature of the available evidence about competition in science and because of the complexity of the phenomenon itself, we have not attempted to formalize competition processes or their precise interrelationship to the action (kick)-reaction system already described. We view clarification of the details of competition processes in science as a topic of major importance for future elaboration of the theory of limited differences. In the present discussion, competition remains implicit in the action (kick)-reaction formulation.

Conclusions and Discussion

The theory of limited differences proposes an explanation for social patterns of group differences. It offers a theoretical explanation for dynamic patterns of increased differentiation, increased attenuation, or social stability in the relative standing of the groups over time. At a micro level of analyzing individual histories, it examines dynamic interactions in which small, limited differences in reactions lead to large changes in individual career histories over extended periods of time.

The theory avoids reliance on "causal" models that emphasize the action of one or two variables as determining agents, or on a battery of correlates where no interrelationships have been either theoretically described or empirically demonstrated. The theory allows us to specify precisely the interrelationships between concrete events in the histories of individuals, a set of reactions to these experiences, and the short- and longer-term consequences on processes of differentiation in scientific productivity.

To test the theory, a program of research focusing on conceptual and methodological problems is required. Included in the portfolio of problems are: determining the relationship between actions and reactions (and the adequate fine-grained measurement of the sequence of events); determining the relative intensities of a variety of actions and reactions that influence outcome variables; understanding the influence of the "anticipation" of events on the selectivity process; examining empirically the nature of time dependencies between events and their consequences; examining how action (kick)-reaction pairs are influenced by organizational and network structures; determining the precise relationship between micro-level outcomes and macro-level, system outcomes; and determining precisely the relationships between structural analysis, social psychology, and culture.

The simple aggregate trends, exhibited in Figure 13.1, indicating the increasing disparity ("fanning out") between cumulative publication counts of men and women scientists, can be modeled by exceedingly simple mathematical representations. Polynomial growth curves (Foulkes and Davis 1981; Ware and Wu 1981) with gender-specific parameters and Polya urn schemes (Feller 1968) with gender-specific selection probabilities are two of the most obvious possibilities. Unfortunately, simplistic models of this kind do not incorporate the fine-grained behavioral assumptions necessary to provide an *explanation* of the patterns in Figure 13.1 in terms of more primitive psychological and sociological constructs. The theory of limited differences is one proposed explanation. It is highly *non*-parsimonious in terms of models which can account for these patterns, but, on the other hand, it is rooted in fundamental behavioral processes. Finally, it suggests that criteria in addition to an ability to reproduce the patterns in Figure 13.1 should form the basis for assessments of whether or not empirical data can support the theory.

A minimum restriction is that we should require data on kick-reaction pairs associated with E_2, E_6, E_8, E_{10}, and E_{15} to support the inequalities (5)–(11) characterizing the sources of disparity between men and women scientists. It is important to emphasize that while gender differences in publication counts are not detectable over short time intervals, gender differences in the frequency of occurrence of kick-reaction pairs—i.e., for events E_2, E_6, E_8, E_{10}, and E_{15}—conditional on full or partial past histories should be ascertainable.[43] In addition, there should be *no* discernable gender differences in the frequency of occurrence of kick-reaction pairs for events other than those indicated above.

Having imposed this set of requirements on empirical evidence needed to support the limited differences theory, it is essential to address some basic—and as yet unanswered—questions about measurement processes. If we try to recover scientists' career histories from longitudinal surveys, then we need a defensible basis for structuring questions that will yield trustworthy responses for the nine types of kick-reaction pairs delineated herein. For events such as those in Table 13.1, we must know how far back in time retrospective questions can be posed in a formal survey so that kick-reaction pairs can be defensibly recovered.

More basic than the above questions is the issue of just what one means by an accurate report of a reaction—i.e., whether r^+, $r^.$, r^-. There is no independent way to assess, for a given person, the accuracy of a reaction report *and* a statement of its impact on motivation to complete manuscripts. While we can observe the consequences for manuscript completions of the kicks, k^-, k^+, which are often readily ascertained regardless of the elicited reaction, defensible and relatively objective assessments of reactions is probably not achievable by standard survey instruments. The closest that one is likely to get to a "gold standard" for reactions is participant observation studies in which sociologists are members of a laboratory—as at Rockefeller University, a Hughes Institute, Fermi Lab, or the Stanford Linear Accelerator Center—where it is possible to observe (unobtrusively) in detail, and continuously over time, the behavior of scientists following receipt of kicks. The observed behaviors would then lead to characterizations and designations of r^+, $r^.$, r^- by the observer; and this would represent the standard for comparison against scientist-elicited responses. There are already some participant observation studies of this kind (see, among others, Latour and Woolgar 1979; Knorr-Cetina 1981; Gilbert and Mulkay

1984) not conducted with an eye toward kick-reaction measurement, but certainly allowing for classification of behaviors and assignment of reaction types.[44]

Many more participant observation studies must be carried out if there is to be deep understanding of the psychological and social processes that are the basis for science careers. Furthermore, there is no substitute for this kind of study if there is to be a clear understanding of the competition processes which drive science careers. An unobtrusive observer, witnessing laboratory discussions of what competitors are doing and listening to the debate and rationale for problem choices, is a central feature of the measurement processes which can either support or refute the limited differences theory. An additional strategy for ascertaining reactions would be to have temporally specific interviews with both the scientist whose reactions are being measured and the fellow scientists who are themselves "witnesses" of the reactions. Through intensive questioning of role partners, it may be possible to increase the reliability of the participant observer's judgment of reactions to specific kicks.

Fine-grained nuances must be ascertained if kick-reaction designations are to be trusted; and it may be that standardized questionnaire surveys will be of limited value relative to within-laboratory participant observation studies. In particular, we expect that in the course of developing tests of the limited differences explanation of science careers, it will be necessary to develop further and elaborate on the structuring and analysis of vignettes (Rossi 1979). We envision the vignettes being prepared by the on-site observers in laboratories.

Shifting from measurement issues back to limited differences theory per se, there is another aspect of choice behavior by scientists that is not reflected in the theoretical formulation presented in this chapter but that deserves precise formalization as part of a research agenda for the future. The missing ingredient is the notion of a scientist's anticipation of future kicks of either positive or negative type and the influence of such perceptions on current motivation, hence on his (her) manuscript completion rate. Evidence from focused interviews (Cole and Zuckerman 1987) suggests that the perceptions about future events which influence productivity are unions of events and their associated kicks, rather than the precisely timed single events and kick-reaction pairs which govern the career history constructions described in this chapter. Whether a scientist perceives future positive or negative kicks on an event such as a grant decision, or candidacy for awards such as Guggenheim fellowships, or election to a professional society, depends on *both* past personal kick-reaction history *and* a consideration of what the competition is doing scientifically and receiving in the way of rewards. It will also be governed by his (her) perceptions of the composition of the judges who will act on his or her proposal or application (Cole 1987). Assessments of anticipation of future events and the influence of these perceptions on a scientist's productivity will almost certainly require participant observation studies of the kind mentioned above. A full delineation of anticipatory processes and their interaction with the kick-reaction paradigm and limited differences explanation for productivity differentials between men and women is, in our opinion, a major task for future theoretical and empirical development. The present essay should be viewed as a first step in an extensive program aimed at a much deeper understanding of the characteristics of scientific careers.

Notes

Introduction

1. Indeed, it is only in recent years that the contributions of women scientists are being actively studied by historians of science. See Rossiter (1982) and the growing number of biographies and autobiographies of women scientists. For two prime examples, see Keller (1983) and Sayre (1975). See also Levi-Montalcini (1988) for a riveting autobiography and the small library on Marie Curie.

2. On the invention of the word "scientist" and its genderless character, see Merton (1988). Rossiter (1982: 25) also notes the oddity that women were included in the directory *American Men of Science, sans* any sign that the descriptive shoe hardly fit.

3. In part, Curie's popular appeal seemed based on the exotic character of her research material; radium with its strange heat and distinctive glow seemed a mysterious substance. And, in part, it was the woman, Marie Curie, her achievements and her fierce determination, that excited admiration, not just in Europe but in the United States as well. She became so popular a figure that in 1921 a public subscription was launched to buy a gram of radium for her research. Upon its successful conclusion, Curie collected her gram of radium from no less than the president of the United States, who happened to be Warren Harding—or more precisely, she collected the gold key which would open the case containing the precious substance. See Weill (1971) for a brief bibliography of the large literature on Curie.

4. Curie was considered a difficult person after Pierre Curie's death, according to David Wilson's biography of Rutherford, and was handled by her scientific peers with "flattery rather than frank and open discussion." Quoted in Pycior (1987: 204) and Wilson (1983).

5. See Kramer (1973: 477–480) and Koblitz (1987).

6. See Kohlstedt (1987, quoted on p. 130). Mitchell's student at Vassar and later a pioneer in applying chemistry to nutrition, Ellen Swallow Richards was the first woman admitted to study at the Massachusetts Institute of Technology, but only as a "special student." See James (1980) on Richards.

7. On Sabin, see Breiger (1980).

8. The question of the number of women scientists in America during the nineteenth century is far from settled. Data for the early twentieth century are better but also far from satisfactory. See Rossiter (1982, chap. 2). See also Kohlstedt (1987) for an account of Maria Mitchell's effort, in 1876, to identify the American women known to be engaged in scientific activity as teachers, physicians, writers, or practicing researchers. As head of the Association for the Advancement of Women, she wanted to compile a roster of women engaged in scientific research. She mailed questionnaires to women scientists she knew asking them to identify others, in a fashion loosely resembling a "snowball sample." Kohlstedt does not indicate how many different women were named in Mitchell's survey but does report 79 replies. Mitchell was evidently well aware of the shortcomings of the procedure as a means of estimating the number of women scientists at the time.

9. These data cover the physical sciences, engineering, and the biological sciences. The summary figures for these fields conceal marked differences among them; women have always been more often represented in the biological than physical sciences, and this pattern has continued into the 1980s. So, too, women have consistently earned a larger share

of degrees in the social sciences than in the physical and biological sciences. In 1920, for example, 22 percent of doctoral degrees in the social sciences went to women, although this proportion declined from the 1930s to the 1960s (Harmon and Soldz 1963, calculated from table 26, p. 50).

10. Various explanations have been proposed for the marked reduction in the proportion of women taking degrees in the sciences. These include greater restrictions on job opportunities for women during the Depression years and the marked rise in births following World War II, which may have led to the declining proportion of women taking advanced degrees in other fields as well.

11. See Harmon and Soldz (1963: 52) for the data for 1960 and Harmon (1978, app. A) for the period 1960–1974.

12. Committee on the Education and Employment of Women in Science and Engineering (1983, table 2.1) for 1980, and Vetter and Babco (1989) for 1988 information. To preserve comparability, the data for the social sciences have been excluded from the overall percentage.

1. The Careers of Men and Women Scientists

1. On insiders and outsiders, see Merton 1972; on social pioneers, see Zuckerman 1987.

2. Demographic data on scientists and engineers are routinely reported by the National Science Foundation in its *Women and Minorities in Science and Engineering,* by the National Science Board in its biannual *Science Indicators* volumes, and by agencies such as the National Research Council and the National Institutes of Health.

3. Since I have not located any data on the age distributions of men and women scientists and engineers currently at work, I rely on inferences of this sort.

4. Rossiter (1978) proposes that such differences are responses to the differing job markets in various sciences. Women, she suggests, are more apt to find jobs in new and rapidly growing fields, where shortages make the hiring of women "tolerable." This hypothesis obviously needs careful examination over fairly long periods.

5. While mainly sociologists, these researchers also include psychologists and economists. A cadre of policy researchers associated with the Committee on the Education and Employment of Women in Science and Engineering of the National Research Council and from federal agencies has also made important contributions.

6. Moreover, the proportions of women doctorates holding faculty positions at all ranks has increased more sharply in the top 50 institutions in the 1970s than in all other institutions (CEEWISE 1979: 76).

7. In industry, the distributions are similar: 29 percent of men and 16 percent of women work mainly in managerial jobs (CEEWISE 1983: 5.3).

8. Not all assistant professors are newly minted Ph.D.s, but the great majority are.

9. Comparable data available on scientists and engineers in industry are sketchy but largely consistent with those for academics. The large differences in rank and managerial responsibility for the aggregate of men and women scientists and engineers working in industry suggest that such differences are concentrated among older workers (CEEWISE 1983: 5.3). See Perrucci (1970) for an earlier analysis.

10. The 29 percent difference is absolutely large, but women in the sciences and engineering in fact do somewhat better than women college graduates generally, who earn 33 percent less than men with the same educational attainments (NSF 1986: 7; data from the U.S. Department of Labor).

11. Again, the data available for industrial scientists are far less detailed than those for academics. These are quickly summarized: women scientists who work in industry have lower median salaries than men; the salary differential increases with years of experience, but, unlike academics, such differences turn up among newly hired men and women as well as among older ones. It is not known whether such differences between men and women are explained by their different field distributions (CEEWISE 1983: 5.5).

12. These differences stretch into the retirement years. As more than one older woman scientist has observed, having earned small salaries and being promoted late has the further consequence of producing small pensions on retirement. Since retirement pay is pegged to the amount of past salary and the number of years contributions to retirement plans were made, women's annuity incomes tend to be smaller than men's.

13. These findings are consistent with those reported for the NIH between 1966 and 1972

(Douglass and James 1973) and for NSF's program in political science (Sigelman and Scioli 1986).

14. Cole and I (1984) have also suggested that women may respond somewhat differently than men to positive and negative reinforcement of their work (as gauged by citation in the literature). This hypothesis also deserves further examination.

15. As Long (1987) noted, this record of equal numbers of citation per paper is surprising, given the fact that women, in general, have poorer appointments than men (and thus poorer research facilities) and also less active research programs.

16. See Cole and Singer, Chapter 13 in this volume, on the theory of limited differences, which brings together processes of self-selection, social selection, and cumulative advantage and disadvantage to help explain the "productivity puzzle."

17. Human capital economists emphasize that women's lower educational attainments, intermittent work histories, and part-time employment account in large measure for gender differences in attainments in the work force at large. Indeed, they may account, in part, for the gross cross-sectional differences observed here. They cannot be the whole explanation, however, since gender differences in career attainments appear in groups with the same human capital investments.

18. On the importance of specified ignorance in science and scholarship, see Merton 1987.

19. There is reason to believe that gender disparities in career attainments of lawyers and managers also grow as they age (Epstein 1987 [private communication]; Gallese 1985; White 1967). Comparative studies of other occupations are needed to establish the generality of this pattern.

2. Citation Classics

1. This sample included persons who in a national survey of academics in 1980 had indicated that over their career span they had published 21 or more articles or that they had published 5 or more articles during the two years prior to the 1980 survey. The sample of 543 includes approximately equal numbers of women and men. We selected all highly productive women and a matched random sample of men from each institution and field. The study covers all disciplines represented among academics. In order to compensate for age as a variable in the lifetime publication record, we identified the highly productive younger cohort by looking at their more recent publication record.

2. *Current Contents* is a resource publication produced by the Institute for Scientific Information. It lists the topics covered in over 1,000 scholarly journals. It reproduces the table of contents of these journals so that readers can scan and get a quick overview of the research studies and topics covered by the various journals in their field.

3. While all authors of the publications identified as *citation classics* are asked to prepare an essay, less than one third return these essays. Foreign scholars have a much higher rate of return.

4. While citation classic might have had multiple authorship, often the essay was written by one of the authors, usually the senior author.

5. It is important to recognize that 25 percent of all published papers are never cited even once and that the average annual citation count for papers that are cited is only 1.7 (Garfield 1979).

6. We used four citation indices: total citation count; total number of pieces cited; citation count of the most important piece; and citations of the three most-cited pieces.

7. Sex Differences in Careers

1. Segregation indices were computed from National Center for Education Statistics reported in Fox (1984). The index of dissimilarity measures the proportion of women (or men) who would have to change fields in order to have the same sex ratio in each field. For Ph.D. recipients, the index increased from 31 to 34 between 1971 and 1981. For students receiving the baccalaureate, the index declined from 46 to 36 over the same period.

8. Gender, Environmental Milieu, and Productivity in Science

1. Andrews (1976) measured creative ability with the Remote Associates Test, developed by S. A. Mednick. The test requires that the respondent think of a fourth word which can be colloquially associated with three other words not commonly linked.
2. The publication of both men and women is strongly skewed, with the majority publishing little and a minority publishing a lot (Cole 1979; Cole and Zuckerman 1984). But smaller proportions of women compared to men are among the prolific (Astin 1978; Bayer 1973; Cole and Zuckerman 1984).
3. In one quest of those data, I have conducted a national survey of Research Productivity Among Social Scientists: The Environmental Link, supported by the National Science Foundation.
4. Of the 532 persons interviewed, 128 were academic administrators, 127 department chairs, and 225 faculty members (Bowen and Schuster 1985:289).
5. These well-documented biases against women in academic rank and promotion (Ahern and Scott 1981; Cole 1979), compared to national and international honors and awards (Cole 1979:59), suggest that it may be at the institutional—department, college, or university—level that gender-based assessments are especially likely to operate. This remains to be verified.
6. Loring did not survey male faculty. It is possible that the men, too, believe that an "old boys' network" exists. However, as males, men have more probable access to the network than do women.
7. These data represent the incidence of collab-

oration as indicated in abstracting services, which include abstracts of monographs, texts, and patents, as well as papers. Depending upon the field and nationality of the journal, papers in core journals may contain higher proportions of collaborative work than do abstracting sources (Beaver and Rose, 1978; Zuckerman 1965, chap. 3). This is because core journals have higher proportions of eminent authors than do the abstracting sources, and the eminent, in turn, are more likely to collaborate.

Relatedly and interestingly, Nobel laureates have been "trend setters" in collaboration, shifting to joint research earlier and in greater proportions than scientists at large (Zuckerman 1977:177). Thus, as early as 1901–25, 41 percent of all the Nobel prize–winning research was collaborative, at a time when only 25 percent of scientific papers were coauthored. Later, as in the 1951–72 period, the (79 percent) proportion of prize winners with collaborative research was about the same as the (71 percent) proportion of joint authored papers at large.
8. Because of the possibility of identifying individuals in such a group of 29, the particular name of the unit is not disclosed. The unit is comprised mainly of natural and social scientists.
9. In 1977 and 1981, doctoral-level women in science and engineering were two to three times as likely as men to hold "off-ladder" positions as instructors, lecturers, or other adjuncts (National Research Council 1979, figure 4.1; National Research Council 1983, table 4.4).

9. Discrimination Against Women in Science

1. For the proportion of women in the various scientific disciplines, see National Science Board, Science Indicators: 1982.
2. A significant critique of this widely cited review was made by Block (1976).
3. For prior discussions of the definition of discrimination in this way, see Cole (1979) and Cole (1986). Clearly, how one defines discrimination determines how it will be studied and perhaps whether or not its presence is found. If discrimination is defined as unequal outcomes rather than a process, then discrimination exists wherever there is an unequal outcome; the problem we are interested in

researching has been eliminated by definition. Although this may make sense for some political purposes, for social scientists to call inequality "discrimination" does not add to our knowledge of how that inequality was created and will not help us reduce that inequality if that is our goal. For a relevant discussion of legal interpretations of discrimination, see Chapter 12, by Owen Fiss, in this volume.
4. Even if there were a correlation between a demographic characteristic and the ability of people in a particular group to perform a task, the use of that characteristic in social selec-

tion would represent a type of "statistical discrimination" as it would incorrectly assume that *all* members of the group would be less able to perform the given task. For a discussion of how the Supreme Court has dealt with statistical discrimination, see Cole (1979).

5. In practice it is sometimes difficult to distinguish between what is functionally relevant and irrelevant. See Cole (1979).

6. The distinction between cultural and structural explanations is itself an analytic distinction. Empirically, of course, the two are intertwined and influence each other. Thus, social structure may influence the change in culture; and a change in culture may, in turn, bring about changes in the social structure.

7. There are, of course, a few social scientists who recognize that we must study both cultural and structural processes. See the excellent review of "demand" and "supply" factors by Marini and Brinton (1984).

8. Some recent work by sociologists studying inequality in Poland illustrates the widespread tendency to interpret outcome differences as evidence for structural causes such as "discrimination" or "lack of equal opportunity." In a collection of articles on social stratification in Poland, the editors, Slomczynski and Krauze (1986), cite a study by Pohoski showing that the children of the "intelligentsia" are substantially more likely than the children of the "working class" or of farmers to obtain high levels of education. The editors conclude that the data show that "inequality of educational opportunity is still substantial" (p. 14). There is no indication that Pohoski had any evidence on educational opportunity. Indeed, the higher educational attainment of the children of the intelligentsia *could* be a result of superior opportunities. It could, also, however, result from both ability and value differences rather than differences in opportunity. In fact, Misztal (1986) shows that there are significant differences in the emphasis put on the value of education in the various occupational groups in Poland.

9. As evidence of how politicized the analysis of discrimination has become, we should note that because of this testimony Rosenberg has been strongly attacked in both public and private forums (see the account in *Ms.* July 1986). Rosenberg's testimony was found offensive not only because she saw the achievement of women as being influenced by the values they have internalized ("blaming the victim") but also because some feel it is wrong for an academic to give testimony (even if this testimony is true) which might harm "the cause."

10. For an extremely important discussion of the importance of selectivity and how it invalidates the "control" approach to multivariate analysis, see Lieberson (1985).

11. These arguments generally fail to control for role performance; but even if they did we would argue that they are unjustified because they would not take into account the possible self-selectivity which could account for any observed differences. If the argument presented in this paper is correct, it would have implications for many sex discrimination suits. The general logic employed by the plaintiffs in these suits is to show an inequality which cannot be "washed out" when other variables are controlled. These controls frequently include no measure of role performance and to our knowledge never consider the problem of selectivity. An important policy implication of this paper is that in most cases only evidence on the process of social selection enables us to determine whether outcome differences are the result of discrimination. This does *not* mean that organizations should not be legally required to institute affirmative action programs in which they attempt to increase female applicants for particular positions. It does mean that discrimination suits need to provide evidence on the *process* of structural discrimination, difficult as this often is.

12. Such a study could not adequately be conducted with cross-sectional data. Since the process of attainment can influence values, it would be necessary to have the measure of values prior to the measure of attainment. See also Lieberson (1985) on the necessity of using longitudinal data.

13. Even in this case selectivity could play some role, as it is possible that men might be more likely than women to request that they be considered for promotion or to actively seek outside offers which force their own department to consider them for promotion at an earlier time than they normally would. Some data on this process would help us determine the extent to which selectivity may influence academic rank.

14. This would assume that there are no differences in quality of role performance or other functionally significant criteria of evaluation between men and women scientists. If there are significant differences in role performance, then a situation in which there is no difference in outcome (no inequality) might mask discrimination against the group with the higher level of role performance. For example, if female scientists publish twice as much as male scientists but male scientists receive the same rewards as female, it is pos-

sible that discrimination might be operating. Direct evidence would be needed to test this hypothesis.

15. Qualitative data cannot be used to "prove" an hypothesis because we do not know the extent to which the qualitative examples are representative of the population. It is possible, of course, to sample *systematically,* then collect qualitative data and transpose such data into quantitative analysis which would be representative of the population. Even here, however, there would be significant problems if the qualitative data were not collected in a standardized way.

16. Data on medical students show that there are not significant differences in the proportion of male and female medical students who complete their studies. The overwhelming majority of students admitted graduate (Medical Education 1981).

17. In fact, there were a few years for which we were unable to find data (especially during World War II); but for most years we have data on all applicants.

18. We also had data enabling us to compare the qualifications of accepted and rejected male and female applicants as determined by MCAT scores and found no significant differences (Cole 1986). We also looked at the behavior of individual schools and found no evidence that individual schools had practiced any significant sex discrimination.

19. Skeptics might be amused by the fact that in this case we also accept the null hypothesis. In the control variable approach, however, it is for all practical purposes frequently impossible to prove discrimination; we can only demonstrate the *possibility* of discrimination. Our approach makes it possible to *prove* discrimination. For example, suppose we had data on applications of Jews and Christians to medical schools or Ivy League colleges before World War II. If there were indeed quotas on Jews, we could prove discrimina-

tion. Such a study, in which we could reject the null hypothesis, would be an important next step in demonstrating the power of the approach suggested here.

20. The conclusion that lower grades of women in science courses is a result of biologically determined ability differences between men and women would be another example of confusing outcome with process. If women do get lower grades than men, there are many possible explanations for this. It is *possible,* although unlikely in our opinion, that this could be a result of biologically linked ability differences. It is also possible that it is a result of structural factors such as women not being given the same type of preparation in science and math prior to entrance to college. It could also be a result of cultural factors as science and math may be culturally defined as areas more important for men than women. Currently we do not have the ability to examine the influence of biological variables on achievement. This problem is in Lieberson's (1985) term "undoable."

21. Later research in progress verified that this was an excellent indicator of the aspiration of freshmen to become doctors.

22. In Lieberson's (1985: 185–195) terms, this would be a "basic" as opposed to a "superficial" cause.

23. Also, if our interest is in explaining group differences, we can rule out from consideration any variable which is not correlated with gender. Thus, for example, social class, not being correlated with gender, would be by definition a variable which would not be of concern in trying to explain sex differences in occupational achievement. If we did look at the occupational achievement of men and women separated by their class, we might find some interesting differences or "interactions" in how class influenced the achievement of the sexes, but we could not explain the overall gap in achievement between them.

10. The Wo / Man Scientist

1. Here differences between men and women are denied a priori political significance.

2. In this, I lean heavily on Rossiter's excellent review, *Women Scientists in America,* on my research on the history of women at MIT (Keller 1981), and also, where pertinent, on my own personal experiences as a woman scientist coming of age in the late 1950s.

3. See, e.g., Rossiter (1982, chap. 5), for further elaboration of this argument.

4. Personal communication, Alice Huang, 1982.

5. In practice, of course, working scientists are well aware of the value of individual differences in talent and style. Only when such differences are claimed to be generic does this consciousness comes into open conflict with the belief that the value and quality of scientific performance can be assessed by a single all-encompassing measure: how "good" is he or she?

6. The story of Barbara McClintock provides a case in point (see Keller 1983). Proud of her iconoclastic individualism, determined to transcend all stereotypes of her sex, she succeeded in fashioning a vision of science that stands in stark contrast to the prevailing vision around her. Her "difference" from her colleagues derived neither from her sex, nor from her female socialization, but precisely from her position as iconoclast and "out-sider." As such, it serves to put into relief the particularistic values underlying the norms of conventional science. These values, I have argued, are not, as we have been taught, universal, but rather a heritage of the cultural equation between "scientific" and "masculine" that has helped shape the history of modern science. (For this last argument, see Keller 1985.)

11. Constraints on Excellence

1. See Joseph Ben-David, *The Scientist's Role in Society: A Comparative Study* (Englewood Cliffs, N.J.: Prentice-Hall, 1971); and William J. Goode, "Community Within a Community: The Professions," *American Sociological Review* 22 (1957): 194–200.

2. Some time ago, Harriet Zuckerman brought to my attention one example of the different and lower standards used in evaluating the eminence of women as compared with men in American society. Because Americans of achievement listed in *Who's Who*, a directory of prominent people in the United States, tend to be predominantly men, a separate volume, *Who's Who of American Women*, was introduced in 1958. In establishing criteria for inclusion, the editors noted in their preface to the first edition that they were "scaling down" the *Who's Who* standards because (as they said in their letter to potential listees) for women, "national or international prominence . . . is not a requisite." (See Preface to *Who's Who of American Women*, 1958–59, and a form letter, dated 1968.)

3. Particularly with regard to the discussion over mathematical skills and spatial relations. Today some academic feminist theorists assert that women bring another "way of knowing" to scientific inquiry.

4. The show was held in July 1971 and marked the first time a major museum hung quilts as art rather than as needlework (Harris 1987). Furthermore, the market in quilts jumped from prices in the $50–100 range to the thousands (with top specimens bringing in prices of $42,000 in one case, and $176,000 in another) (ibid.).

5. The Widener Library opened its doors to women in 1950. Before that, there was a separate reading room for women, with female graduate students being allowed limited access to the library. In 1966, all Harvard libraries were finally opened to women.

6. Ironically, ideas such as these are once more finding expression in popular culture and in the work of some feminist writers such as Carol Gilligan (1982).

7. As March and Olsen (1984: 739) remind us, "Holding office provides participation rights [to the incumbents] and alters the distribution of power and access."

8. Only in group activities such as film and theater, where women were needed as actors, because men could not substitute, did they have opportunity to excel.

9. Rosenberg (1982) also documented the debates over women's frailty generated by a member of the Harvard Board of Overseers, who maintained that higher education posed grave danger for the female sex. Women's unique physiology, he explained, limited their educational capacity and threatened their reproductive capacity. See also Cole's (1979) account of scientists' and social scientists' assessments of the capacities and potentials of women from the late 1800s until the present (Chapter 6).

10. The data do not reveal clear-cut differences using other measures. Standardized tests such as the SAT given to high school students do reveal that, on average, boys do better than girls. But girls get better grades than boys in science in high school. Yet, in a study of 2,200 Rhode Island twelfth-graders who were academically prepared for science careers with course work including calculus and physics, only a small number of girls indicated a career interest in engineering, science, or technology. The proportion of girls who do well represents a considerable pool of able candidates for further training. Indeed, the proportion of women earning Ph.D.s in the last decade rose from 23.3 percent to 34.3 percent (National Research Council 1987). In a study funded by the Rockefeller Foundation, Sue E. Berryman found that there is little difference between the sexes in mathematics achievement in the ninth grade. However,

girls' choices of fewer mathematics electives translates into lower twelfth-grade mathematics achievement, fewer women choosing college science majors, and fewer women science Ph.D.s (Berryman 1983). See also Eagly (1987), who reports that meta-analysis of studies on sex differences in cognitive abilities show virtually no differences. For a critical review of the sex differences literature, see Hyde and Linn (1986) and especially Linn and Petersen (1986), which indicates that there is no apparent basis for inferring causal connections to explain gender differences across the areas of spatial ability, aptitude in mathematics, and science achievement. Hyde's (1990) further meta-analysis review of gender differences finds insignificant differences in cognitive abilities with no differences in verbal ability; meaningful differences in only one type of spatial ability, mental rotation, and a moderate difference in mathematical performance.

11. Problems in finding employment suited to educational background was, of course, not limited to the field of law. The first members of the Department of Home Economics at Berkeley were actually women chemists who were unable to find jobs in chemistry. See "Gender Stratification in Higher Education," *Women's Studies International Forum* 10(2), 1987: 157–164.

12. I discuss this within the framework of a theory of role integration (1976). The theory asserts that women have been relegated to "helping roles" in occupations as well as in the home. Therefore, their contributions are absorbed but not accredited.

13. In medicine, women's representation in surgery increased from 1.1 percent in 1966 (Epstein 1970: 163) to 5 percent in 1985 (AMA 1986).

14. Kanter (1977: 256) notes this process in understanding commitment of women in business. She observes, "[Commitment] seems clearly tied to increasing rewards and chance for growth implied in high opportunity." Because commitment is closely tied to the existence of an opportunity structure, the expectations for future rewards such as job mobility, growth, or increased status are essential for commitment to exist (see also Bielby 1982).

12. An Uncertain Inheritance

1. "The right of citizens of the United States shall not be denied or abridged by the United States or by any state on account of race, color, or previous condition of servitude." U.S. Const. amend XV, sec. 1.

2. 100 U.S. 303 (1879).

3. 347 U.S. 483 (1954).

4. Civil Rights Act of 1964, Title VII, 42 U.S.C. § 2000(e) et seq (1982).

5. Civil Rights Act of 1964, Title VI, 42 U.S.C. § 2000(d) et seq (1982); Title IX of the Education Amendments of 1972, 20 U.S.C. § 1681 (1982).

6. Owen Fiss, "Groups and the Equal Protection Clause," *Philosophy and Public Affairs* (1976) 5: 107.

7. "The rights created by the first section of the Fourteenth Amendment are, by its terms, guaranteed to the individual. The rights established are personal rights." *Shelley* v. *Kraemer*, 334 U.S. 1, 22 (1948).

8. *Washington* v. *Davis*, 426 U.S. 229 (1976); see also *Personnel Administrator of Massa-*chusetts v. *Fenney*, 442 U.S. 256 (1979).

9. *Gaston County* v. *United States*, 395 U.S. 285 (1969); see also *Griggs* v. *Duke Power Co.*, 401 U.S. 424 (1971).

10. Cf. the opinions of Justice Marshall and Blackmun in *Regents of the University of California* v. *Bakke*, 438 U.S. 265, 387, 402 (1978).

11. *Johnson* v. *Transportation Agency, Santa Clara County*, 480 U.S. 616 (1987).

12. Glenn Loury, "Beyond Civil Rights," *New Republic* (October 7, 1985): 22.

13. Hochschild, Jennifer, "Race, Class Power and the American Welfare State," in Amy Gutmann, Democracy and the Welfare State, Princeton University Press, 1988; "Equal Opportunity and the Estranged Poor," Annals of the American Academy of Political and Social Science, No. 501, January 1989. See also Lemann, Nicholas, "The Origins of the Underclass, Part I," *The Atlantic* (June 1986): 31; Part II (July 1986): 54.

13. A Theory of Limited Differences

1. A formal presentation of the general theory is currently in preparation.
2. The male–female difference exists within every productivity stratum, for example, when career publication totals are divided into quartiles or total publications in 12–15 years following the Ph.D., or longer periods of time (see Figure 13.1).
3. On strategic research sites, see Merton 1987.
4. Throughout this paper "scientific productivity" will refer to the number of scientific articles that are published within specific units of time. Whether we discuss total counts or papers per year, we refer to the number of papers published. There is a large literature on problems in measuring scientific productivity and its relationship to both the quality of scientific work and its impact. See among many others Cole and Cole (1973), Cole (1979), Gaston (1973, 1978), Allison and Stewart (1974), Long (1978), Long and McGinnis (1981), Reskin (1977, 1978a, 1979), Andrews (1979), Allison (1980), and for recent groups of Ph.D.s, Cole and Zuckerman (1984). Suffice to say, publication counts are strongly correlated with impact as measured by peer appraisals and by citations, as well as with the prestige of honorific awards.
5. See Price (1963). Subsequent studies demonstrated that this pattern obtains in every scientific discipline studied, and for the United States and all other nations whose scientific output have been examined. While we have charted these patterns well, there have been no successful attempts to explain them.
6. Cole (1979) reports data on the relationship between sex status and publications for matched samples of male and female Ph.D.s, who received their degrees in the same year and from the same science department in 1922, 1932, 1942, 1952, 1957–58. The association is illustrated for each of these distinct cohorts in Figure 13.1.
7. Scientists were drawn from six fields: astronomy, biochemistry, chemistry, earth sciences, mathematics, and physics. Pairs of scientists were matched in the sense that they were selected from the same departments in the same years. Analysis was performed both on the aggregates of 263 pairs and on individual pairs. The results were much the same regardless of the type of comparison.
8. While patterns of citations to published science by men and women look much the same as the productivity patterns, evidence suggests that women scientists publish articles that receive just as many citations per article as do men. Thus, the differential in citations appears to result from the greater total output of the men.
9. For each of the five pictures shown in Figure 13.1, the random samples of scientists were matched by year of Ph.D., field, and department of Ph.D. Where possible, men and women were matched by specialty at the time of receiving their degree. Publication data were obtained from abstracts. For a complete description of these samples see J. R. Cole (1979).
10. For descriptions of the diverse set of samples that we have collected data on, see Cole and Cole (1973, 1976, 1985); Cole, Rubin and Cole (1978); Cole, Cole and Simon (1981); Cole, Cole, and the Committee on Science and Public Policy (1981); Cole (1975); Cole (1979); Cole, Cole and Dietrich (1978); Zuckerman and Cole (1975); Zuckerman and Merton (1971a, 1971b); Zuckerman (1970, 1977).
11. This observation was made repeatedly in recent extended interviews with 123 men and women of science conducted by Harriet Zuckerman and J. R. Cole.
12. Plainly, there are many other outcome variables in science, such as appointments to various positions and peer recognition. In this chapter we focus exclusively on the research role and on those who are the major contributors (operationalized by producers of x or more papers within y years of receiving the Ph.D.) to the development of science through publication. This is a small percentage of Ph.D. recipients in science and is nearly invariant across Ph.D. cohorts from 1920 to the present. There are, at present, no obvious early screening criteria to ascertain (at Ph.D. completion) who will fall into this group.
13. The term "kick" is drawn from the physical science literature and is used here in a completely neutral way since it refers to a positive or negative event or perturbation. Kicks are experienced by men and women, and no invidious comparison is intended by the use of this term.
14. Condition (V) implies that every detail of past history will not influence future events in a unique and idiosyncratic fashion. The same outcome history can arise in a multiplicity of ways.
15. The theory of limited differences calls forth a series of interesting metaphors drawn from the biological and physical sciences. We need

only look to Darwin's *The Origin of Species* (1859) for a clear articulation of the effects of small cumulating differences, which may not even be distinguishable for substantial periods of time, becoming the basis for highly notable variations in species. In his chapter on natural selection, or the survival of the fittest, Darwin has many references to the influence of small differences over time. Consider only one:

> during the modification of the descendants of any one species, and during the incessant struggle of all species to increase in numbers, the more diversified the descendants become, the better will be their chance of success in the battle for life. Thus the small differences distinguishing varieties of the same species, steadily tend to increase, till they equal the greater differences between species of the same genus, or even of distinct genera.

E. O. Wilson's concept of "multiplier effects" also recognizes how small differences can interact with the environment to produce larger effects:

> A small evolutionary change in the behavior pattern of individuals can be amplified into a major social effect by the expanding upward distribution of the effect into multiple facets of social life. . . . Multiplier effects can speed social evolution still more when an individual's behavior is strongly influenced by the particularities of its social experience. (Wilson 1975: 11–13)

Finally, we find parallels to the concept of an action-reaction system in the experimental embryologist C. H. Waddington's concept of "competence" developed almost 50 years ago. In his discussion of competence, Waddington (1940) notes:

> In the first place, it is a state of instability, since it involves a readiness either to react to an organizer and follow a certain developmental path, or not to react and to develop in some other way . . . one can compare a piece of developing tissue to a ball running down a system of valleys which branches downwards, like a delta. . . . The tissue, like the ball . . . must move downhill, but at some points there are two downhill paths open to it. At such branching points, it may sometimes require a definite external stimulus, such as evocator substance, to push the tissue into one of the developmental paths; in such a case, competences which occur later along this path will only be developed if

the evocator has acted. In other cases, a certain path may be followed merely because an evocator has failed to be present, and then the subsequent competences may appear to develop autonomously." (p. 45)

The theory of limited differences calls forth a series of additional metaphors drawn from other scientific disciplines. The image of the controlled chain reaction is one such metaphor. There an initial action can of course balloon quickly into a large difference if many "kicks" for one group line up positively and all of the kicks for the other group are negative. In such a limiting case, enormous differences between men and women would occur as their careers unfold. But the qualitative data suggest that men *and* women experience both positive *and* negative kicks that we hypothesize affect scientific productivity and career advancements. In fact, the chain reaction which might lead from small initial differences to enormous disparities is modulated by a set of competing and conflicting positive and negative forces. These are metaphorically the barium rods which slow down or even halt the initial chain reaction. Consider another metaphor. We place a big stone on top of a hill; we let it begin to roll down. Depending on tiny impulses it gets, it moves one way or another and will end up in a very different place at the bottom, depending on the smallest chance variations. Each small perturbation changes its trajectory for the future. And the longer the hill, the larger the possibility of spreading apart from the initial path. Still another metaphor is drawn from the kinetic theory of gases or fluids model. Here each scientist is viewed as a molecule. External events move the molecule in one direction or another. The path varies according to the number and types of pushes.

While these concepts help convey the image of actions and reactions as well as the concept of the long-term larger effects of initially small differences, in important respects each fails to capture a critical feature of the theory of limited differences. The fundamental distinction lies of course in "consciousness," that is, the ability of scientists to react to events in nonmechanistic ways that are not akin to reactions by either particles, molecules, or the biological systems described by Waddington or Wilson. Thus, these concepts drawn from biology are at best weak analogies to the distinctly socially structured action-reaction system developed here. Salome Waelsch brought Waddington's work to our attention and helped make us aware of the centrality of the reaction system for the theory of limited differences.

16. Here a truncated description is required. An in-depth critical appraisal of these earlier orientations will be published elsewhere.

17. There have been no agreed upon measures that predict scientific talent, imagination, or aptitude. IQ scores, at best a weak measure of scientific ability, have been found, first, to be uncorrelated with sex, as well as with publication counts and citations. Bayer and Folger (1966) found a correlation of .05 between IQ scores and citations to scientists' work (see also, Harmon 1963, 1965); Cole (1979) found a correlation of −.03 between publication counts and IQ for the first 13 years of the careers of men and women scientists receiving their Ph.D.s in 1957–58. Although men tend to have higher scores than women on the mathematical portion of the SAT and GRE, the explanation for this difference remains unclear. There is no evidence that after the groups are socially and self-selected into Ph.D. programs that this difference is reflected in subsequent performance.

18. For elaborations upon Merton's work, see, among others, Cole and Cole 1973: 237–247; Allison and Stewart 1974; Allison, Long and Kraus 1982; Zuckerman 1977; Cole 1979; Mittermeir and Knorr 1979; Zuckerman 1989.

19. In the first half of this century, the application of nepotism rules, of quotas on having members of certain religious groups, and of an unwillingness to have women working in certain laboratories represented discrimination that significantly influenced the scientific productivity and career histories of women and men who were adversely affected by these discriminatory practices.

20. Although these theoretical concepts have been used to explain sex differences in productivity, they are applied, almost invariably, as a fortiori or post factum interpretations of observed patterns. There has rarely been an attempt to test precisely these theoretical interpretations. Either data do not exist for direct tests of the theory, or the tests have been carried out with imprecise and often questionable "proxies" for key variables.

21. It is important to assert at the outset that the stochastic process formulation with strong memory effects and dependence among multiple variables developed herein represents a mathematical formalization of a very specific theory. Much of the modeling activity in the contemporary sociology literature is *not* of this character and is of an exploratory data analytic type where the goal is to assess which combination(s) of an a priori list of variables

are the important influences on a given outcome variable(s). With this particular goal, standard regression modes with interaction terms including dynamic autoregressive models are the most prominent tools. This class of models, however, does not include a formalization of the limited differences theory. Indeed, the standard strategies for incorporating interactions among variables—i.e., as multiplicative terms—in regression models are too crude to represent the more subtle nonlinearities in the limited differences theory.

22. Zuckerman and Merton, *Minerva,* 1971b. The probability of a manuscript being published is largely a function of the effort by the scientist to see the paper through to publication. In fields and specialties with high specific journal rejection rates, the decision to resubmit an article either to the same journal or to a different one almost invariably leads to some form of publication. This may not always be in the journal of first choice, but it will result in publication. Zuckerman and Merton show in their study of *The Physical Review* that eminent scientists not only published more than run-of-the-mill scientists but submitted about twice as many manuscripts for publication over a nine year period: 4.1 for those of the highest rank; 3.5 for the intermediaries; and 2.0 for physicists of the third rank. And the most prolific physicists submitted papers to *The Physical Review* at a rate 12 times that of the rank-and-file. (In Merton 1973: 479.)

23. A full empirical refinement of the publication process outlined here is an important agenda item for future research.

24. A more fine-grained classification involving intensities of kicks and reactions is both possible and meaningful; however, the coarse categories—positive, neutral, and negative—will be utilized to simplify the theoretical formulation herein and focus on the principal concepts.

25. The types of actions and reactions and their sequencing will vary from one historical period to another. Figure 13.1 illustrates an historical pattern of sex differences in scientific publications dating back to the 1930s. Although the aggregate level pattern persists, this does not mean, of course, that the cultural, social structural, or psychological factors that produce the patterns have remained constant. On the contrary, historical evidence suggests that the structure of action (kick)-reaction pairs, and in particular, their intensities, have changed in the past 50 years. The historical changes will be captured in the

transformation of the kick-reaction pairs and, indeed, the replacement of some by others.

26. The analytic division of science careers into three phases is appropriate because there is evidence in the sociology of science literature that early events are critical in shaping subsequent probabilities for publication and rewards. The analytic phasing discussed here may not be appropriate when considering other dynamic processes of differentiation and subsequent fanning out or attenuation of the differences.

27. In a study of physics awards, only one third of scientists report any awards; the most prestigious awards were monopolized by a small subset of scientists (see Cole 1969; Cole and Cole 1973; Zuckerman 1977).

28. Zuckerman 1977, chap. 6.

29. The possible examples here are numerous. To cite only one, Donald A. Glaser, the inventor of the bubble chamber, shifted from physics to biology after receiving the Nobel Prize.

30. This assumption needs testing, since undoubtedly the prestige of the Nobel Prize and other major awards cuts across fields and may in fact increase the initial probabilities that manuscripts produced by major award winners in the new field will meet with a more positive reception than those produced by recent Ph.D.s in the field.

31. In all inequalities in this section, the past histories for the men and women are identical in the conditioning events. Thus comparable histories still yield gender differentiated responses.

32. Empirical evidence supporting these limited differences comes from extensive focused interviews with eminent and rank-and-file men and women scientists conducted by Jonathan R. Cole and Harriet Zuckerman. It should be emphasized that many men and women react in precisely the same way to negative kicks. Indeed, the distributions probably show more similarity than difference; the differences in probabilities are not large.

33. This example suggests that discrimination is one of the fundamental sources for negative kicks in the sequence of action-reaction pairs. Plainly, discrimination need not be rampant for it to have a notable cumulative effect on the publication probability of a subset of women or men who suffer from the initial negative kick and from the negative reactions in terms of motivation and future aspirations.

34. There is actually some empirical support for this assumption. In terms of subsequent publication rate, men scientists are affected more positively by peer recognition in the form of

citations than are women. Conversely, women are more adversely affected than men by a lack of peer recognition. In other terms, positive reinforcement has less of a positive effect on productivity for women than for men; negative reinforcement has a greater negative effect for women than men. Cf. Cole and Zuckerman (1985).

35. There is a growing literature that indicates that marriage and family obligations affect women's careers, but not in terms of published productivity (see Cole 1979; Cole and Zuckerman 1987).

36. The empirical fact is that some eminent women receive more positive and fewer negative kicks than some eminent men. For this subset, the summary of kick-reactions would indicate that these women tend to produce more manuscripts than the men. It is by no means invariably the case that the careers of women show more negative kicks and negative reactions than men. In general, however, this has tended to be the case, and we hypothesize that the cumulation of these micro-level limited differences explain the macro-level disparity in publications.

37. There is, of course, a subset of eminent scientists who accept offers as administrators and public servants after achieving lofty recognition. Their publication rate is usually drastically reduced and, in some instances, virtually eliminated. Thus it is necessary in our formal specifications of career development to allow for the termination of a publication history following the receipt of one or more major awards.

38. For an extended discussion of priority disputes in science, see Merton 1957.

39. Cole and Cole 1981, 1985; Cole, Cole and Dietrich 1978; Cole, Cole and Simon 1981.

40. Merton 1957; Hagstrom 1965; Latour and Woolgar 1979; Knorr-Cetina 1981.

41. The contemporary situation is placed in bold relief by the eminent biochemist Arthur B. Pardee:

> At the heart of current problems [in maintaining high scientific quality and productivity] are the difficulties and uncertainties every scientist faces in obtaining research funds. . . . A scientist perceives now that he has a small probability of getting a grant funded. He cannot afford to be without funds for a year or more if his application fails, because continuity is essential for progress and to retain highly trained, key personnel. So he writes [multiple] proposals in the hope that one of them will be lucky. . . . Fund raising rather than re-

search becomes his major preoccupation. . . . Talents of fine scientists are a rare commodity; wasting them is a very costly proposition. . . . A less evident but also highly important consequence is the diminution of scientists' self-confidence and morale. Rejections by the funding system of one's best ideas are extremely discouraging. We will see scientists in increasing numbers decide that they are in a rat race; they will slow down or get out. Some of the best unfortunately will be among them. (As quoted in Cole and Cole 1985, p. 28).

Similar opinions were voiced by many of the productive men and women scientists that were interviewed over the past four years.

42. Many other less well-known examples could be cited; for example, the competition between the labs of Andrew Schally and Roger Guillemin to identify the structure of TRF(H), Thyrotropin Releasing Factor (Hormone). See Latour and Woolgar 1979.

43. Verifying or refuting this claim is a major research task for the future. The supporting evidence to date is primarily from focused interviews which were not designed a priori to assess these points (see Zuckerman and Cole 1987). A content analysis of the interviews, and 20 years of study of the scientific community, often with various forms of quantitative data, suggested inequalities (5)–(11) and provided informal support for the limited differences explanation of Figure 13.1

44. Of course, even participant observation studies have the weakness of imputation, that is, the observer imputing reactions that are translated through his or her own set of constructs.

Bibliographical References

Introduction

Breiger, Gert. 1980. "Florence Rena Sabin." In B. Sicherman, C. H. Green, I. Kantrov and H. Walker, eds., *Notable American Women: The Modern Period*, pp. 614–617. Cambridge: Harvard University Press.

Committee on the Education and Employment of Women in Science and Engineering, National Research Council. 1983. *Climbing the Academic Ladder: An Update on the State of Doctoral Women Scientists*. Washington, D.C.: National Academy Press.

Harmon, Lindsey R. 1978. *A Century of Doctorates: Data Analyses of Growth and Change*. Washington, D.C.: National Academy of Sciences.

Harmon, Lindsey R. and Herbert Soldz. 1963. *Doctorate Production in United States Universities 1920–62*. Publication no. 1142. Washington, D.C.: National Academy of Sciences–National Research Council.

James, Janet Wilson. 1980. "Ellen Henrietta Swallow Richards." In B. Sicherman, C. H. Green, I. Kantrov and H. Walker, eds., *Notable American Women: The Modern Period*, pp. 514–516. Cambridge: Harvard University Press.

Keller, Evelyn Fox. 1983. *A Feeling for the Organism*. New York: Freeman.

Koblitz, Ann Hibner. 1987. "Career and Home Life in the 1880s: The Choices of Mathematician Sofia Kovalevskaia." In P. Abir-Am and D. Outram, eds., *Uneasy Careers and Intimate Lives*, pp. 172–190. New Brunswick, N.J.: Rutgers University Press.

Kohlstedt, Sally Gregory. 1987. "Maria Mitchell and the Advancement of Women in Science." In P. Abir-Am and D. Outram, eds., *Uneasy Careers and Intimate Lives*, pp. 129–146. New Brunswick, N.J.: Rutgers University Press.

Kramer, Edna E. 1973. "Sonya Kovalevsky." *Dictionary of Scientific Biography*. Vol. 7, pp. 477–480. New York: Scribner.

Levi-Montalcini, Rita. 1988. *In Praise of Imperfection: My Life and Work*. Sloan Foundation Science Series. New York: Basic Books.

Merton, Robert K. 1988. "Le molteplici origini e il carattere epiceno del termini inglese per 'scienziato.'" *Scientia* 123: 279–293.

Pycior, Helena M. 1987. "Marie Curie's 'Anti-natural Path': Time Only for Science and Family." In P. Abir-Am and D. Outram, eds., *Uneasy Careers and Intimate Lives*, pp. 191–214. New Brunswick, N.J.: Rutgers University Press.

Rossiter, Margaret. 1982. *Women Scientists in America: Struggles and Strategies to 1940*. Baltimore: Johns Hopkins University Press.

Sayre, Anne. 1975. *Rosalind Franklin*. New York: Norton.

Vetter, Betty and Eleanor L. Babco, eds. 1989. "1988 Ph.D. Awards." *Manpower Comments* 26:5 (June): 24.

Weill, Adrienne R. 1971. "Marie Curie." *Dictionary of Scientific Biography*. Vol. 3, pp. 497–503. New York: Scribner.

Wilson, David. 1983. *Rutherford: Simple Genius*. London: Hodder and Stoughton.

1. The Careers of Men and Women Scientists

Ahern, Nancy C. and Elizabeth L. Scott, Committee on the Education and Employment of Women in Science and Engineering. 1981. *Career Outcomes in a Matched Sample of Men and Women Ph.D.s: An Analytical Report.* Washington, D.C.: National Research Council, National Academy of Sciences.

Allison, Paul D. and John A. Stewart. 1974. "Productivity Differences Among Scientists: Evidence for Accumulative Advantage." *American Sociological Review* 39: 596–606.

Astin, Helen S. 1969. *The Woman Doctorate in America.* New York: Russell Sage Foundation.

Astin, Helen S. and Alan E. Bayer. 1979. "Pervasive Sex Differences in the Academic Reward System: Scholarship, Marriage, and What Else?" In D. R. Lewis and W. E. Becker, eds., *Academic Rewards in Higher Education,* pp. 211–229. Cambridge, Mass.: Ballinger.

Bakanic, Von, Clark McPhail and Rita J. Simon. 1987. "The Manuscript Review and Decision-Making Process." *American Sociological Review* 52: 631–642.

Bayer, Alan E. 1973. "Teaching Faculty in Academe." *A.C.E. Research Reports* 8, no. 2.

Bayer, Alan E. and Helen S. Astin. 1975. "Sex Differentials in the Academic Reward System." *Science* 188: 796–802.

Bayer, Alan E. and John Folger. 1966. "Some Correlates of a Citation Measure of Productivity in Science." *Sociology of Education* 39: 381–390.

Benbow, C. P. and J. C. Stanley. 1980. "Sex Differences in Mathematical Ability: Factor or Artifact?" *Science* 212: 1262–1264.

———. 1983. "Sex Differences in Mathematical Ability: More Facts." *Science* 222: 1029–1031.

Breiger, G. H. 1980. "Florence Rena Sabin." In B. Sicherman, C. H. Green, I. Kantrov and H. Walker, eds., *Notable American Women: The Modern Period,* pp. 614–617. Cambridge: Belknap Press of Harvard University Press.

Briscoe, Anne M. 1984. "Scientific Sexism: The World of Chemistry." In V. B. Haas and C. C. Perrucci, eds., *Women in Scientific and Engineering Professions,* pp. 147–159. Ann Arbor: University of Michigan Press.

CEEWISE (Committee on the Education and Employment of Women in Science and Engineering). 1979. *Climbing the Academic Ladder; Doctoral Women Scientists in Academe.* Washington, D.C.: National Research Council, National Academy of Sciences.

———. 1980. *Women Scientists in Industry and Government: How Much Progress in the 1970s?* Washington, D.C.: National Research Council, National Academy of Sciences.

———. 1983. *Climbing the Ladder: An Update of the Status of Doctoral Women Scientists and Engineers.* Washington, D.C.: National Research Council, National Academy of Sciences.

Centra, John A. with Nancy M. Kuykendall. 1974. *Women, Men and the Doctorate.* Princeton, N.J.: Educational Testing Service.

Clark, S. M. and M. Corcoran. 1986. "Perspectives on the Professional Socialization of Women Faculty: A Case of Accumulative Disadvantage?" *Journal of Higher Education* 57: 20–43.

Coggeshall, Porter E. 1981. *Postdoctoral Appointments and Disappointments.* Washington, D.C.: National Academy Press.

Cole, Jonathan R. 1979. *Fair Science: Women in the Scientific Community.* New York: Free Press.

Cole, Jonathan R. and Stephen Cole. 1973. *Social Stratification in Science.* Chicago: University of Chicago Press.

Cole, Jonathan R. and Harriet Zuckerman. 1984. "The Productivity Puzzle: Persistence and Change in Patterns of Publication on Men and Women Scientists." In P. Maehr and M. W. Steinkamp, eds., *Advances in Motivation and Achievement,* pp. 217–256. Greenwich, Conn.: JAI Press.

———. 1987. "Marriage and Motherhood and Research Performance in Science." *Scientific American* 256 (2): 119–125.

Cole, Stephen. 1970. "Professional Standing and the Reception of Scientific Discoveries." *American Sociological Review* 76: 286–306.

Douglass, Carl D. and John C. James. 1973. "Support of New Principal Investigators by N.I.H.: 1966–1972." *Science* 181: 241–244.

Edge, D. 1979. "Quantitative Measures of Communication in Science: A Critical Review." *History of Science* 17: 102–134.

Ferber, M. 1986. "Citations: Are They an Objective Measure of Work of Women and Men?" *Signs* 11: 381–389.

Ferber, M. and J. Huber. 1979. "Husbands, Wives, and Careers." *Journal of Marriage and the Family* 41: 315–325.

Fox, Mary Frank. 1985. "Location, Sex-Typing, and Salary Among Academics." *Work and Occupations* 12: 186–205.

Gallese, L. R., 1985. *Women Like Us: What Is Happening to the Women of the Harvard Business School Class of '75—The Women Who Had the First Chance To Make It To the Top.* New York: Morrow.

Garfield, Eugene. 1979. "Is Citation Analysis a Legitimate Evaluation Tool?" *Scientometrics* 1: 359–375.

———. 1982. "The 1,000 Most-Cited Contemporary Authors. Part 2A Details on Authors in the Physical and Chemical Sciences and Some Comments about Nobels and Academy Memberships." *Current Contents* (March 1): 5–13.

Gaston, J. 1978. *The Reward System in British and American Science.* New York: Wiley.

Hargens, Lowell, James McCann and Barbara Reskin. 1978. "Productivity and Reproductivity: Marital Fertility and Professional Achievement Among Research Scientists." *Social Forces* 52: 129–146.

Harmon, L. R. 1978. *A Century of Doctorates: Data Analyses of Growth and Change.* Washington, D.C.: National Academy of Sciences.

Helmreich, R. L. and J. T. Spence. 1982. "Gender Differences in Productivity and Impact." *American Psychologist* 37: 1142.

Helmreich, R. L., J. T. Spence, W. E. Beane, G. W. Lucker and K. A. Matthews. 1980. "Making It in Academic Psychology: Demographic and Personality Correlates of Attainment." *Journal of Personality and Social Psychology* 39: 896–908.

Joas, Hans. 1990. "Das Deutsche Universitätssystem und die Karrieremöglichkeiten junger Wissenschaftler." In H. P. Hofschneider and K. U. Mayer, eds., *Generational Dynamics and Innovation in Science*, pp. 102–113. Munich: Wissenschaftlichen Rat der Max-Planck-Gesellschaft.

Kahle, Jane Butler and Marsha L. Matyas. 1987. "Equitable Science and Mathematics Education: A Discrepancy Model." In Linda S. Dix, ed., *Women: Their Underrepresentation and Career Differentials in Science and Engineering*, pp. 5–41. Washington, D.C.: National Academy Press.

Keller, Evelyn Fox. 1977. "The Anomaly of a Woman in Physics." In S. Ruddick and P. Daniels, eds., *Working It Out*, pp. 77–91. New York: Pantheon Books.

———. 1985. *Reflections on Gender in Science.* New Haven: Yale University Press.

Kyvik, Svein. 1990. "Motherhood and Scientific Productivity." *Social Studies of Science* 20: 149–60.

LeBold, William K., K. W. Linden, C. M. Jagacinski and K. D. Shell. 1983. *National Engineering Career Development Study: Engineers' Profiles of the Eighties.* West Lafayette, In.: Purdue University.

Lefkowitz, Mary R. 1979. "Education for Women in a Man's World." *The Chronicle of Higher Education* (August 6): 56.

Lester, Richard A. 1974. *Antibias Regulation of Universities; Faculty Problems and Their Solutions.* New York: McGraw-Hill.

Lewis, G. L. 1986. "Career Interruptions and Gender Differences in Salaries of Scientists and Engineers." Working paper prepared for the Office of Science and Engineering Personnel, National Research Council, National Academy of Sciences, Washington, D.C.

Long, J. Scott. 1978. "Productivity and Academic Position in the Scientific Career." *American Sociological Review* 43: 889–908.

———. 1987. "Problems and Prospects for Research on Sex Differences in the Scientific Career." In Linda S. Dix, ed., *Women: Their Underrepresentation and Career Differentials in Science and Engineering*, pp. 163–169. Washington, D.C.: National Academy Press.

Long, J. Scott, P. D. Allison and R. McGinnis. 1979. "Entrance Into the Academic Career." *American Sociological Review* 44: 816–830.

Long, J. Scott and R. McGinnis. 1981. "Organizational Context and Scientific Productivity." *American Sociological Review* 46: 422–442.

Luukkonen-Gronow, Terttu and Veronica Stolte-Heiskanen. 1983. "Myths and Realities of Role Incompatibility of Women Scientists." *Acta Sociologica* 26: 267–280.

Marwell, Gerald, Rachel Rosenfeld and Seymour Spilerman. 1979. "Geographic Constraints on Women's Careers in Academia." *Science* 205: 1225–1231.

Merton, Robert K. 1957. "Priorities in Scientific Discovery: A Chapter in the Sociology of Science." *American Sociological Review* 22: 635–659.

———. 1968. "The Matthew Effect in Science." *Science* 159: 156–163.

———. 1972. "The Perspectives of Insiders and Outsiders." *American Journal of Sociology* 77: 9–47.

———. [1942] 1973. "The Normative Structure of Science." In *The Sociology of Science*, pp. 267–278. Chicago: University of Chicago Press.

———. 1987. "Three Fragments from a Sociologist's Notebooks: Establishing the Phenomenon, Specified Ignorance, and Strategic Research Materials (SRMs)." *Annual Review of Sociology* 13: 1–28.

Mittermeir, R. and K. D. Knorr. 1979. "Scientific Productivity and Accumulative Advantage: A Thesis Reassessed in the Light of International Data." *R&D Management* 9: 235–239.

National Institutes of Health, Special Programs Office, Office of Extramural Research and Training. 1981. *Women in Biomedical Research*. Publication No. 81–429. Washington, D.C.: NIH.

National Research Council, Commission on Human Resources. 1980. *Summary Report: 1979 Doctorate Recipients from United States Universities*. Washington, D.C.: National Academy of Sciences.

National Science Board, National Science Foundation. 1977. *Science Indicators 1976*. NSB 77-1. Washington, D.C.: Government Printing Office.

———. 1982. *Science Indicators 1982*. NSB 83-1. Washington, D.C.: Government Printing Office.

———. 1985. *Science Indicators: The 1985 Report*. NSB 85-1. Washington, D.C.: Government Printing Office.

National Science Foundation. 1986. *Women and Minorities in Science and Engineering*. Washington, D.C.: NSF.

Office of Economic Cooperation and Development (OECD). 1984. *Educational Trends in the 1970s: A Quantitative Analysis*. Paris: United Nations.

Perrucci, Carolyn Cummings. 1970. "Minority Status and the Pursuit of Professional Careers: Women in Science and Engineering." *Social Forces* 49: 245–259.

Price, Derek J. de S. 1976. "A General Theory of Bibliometric and Other Cumulative Advantage Processes." *Journal of the American Society for Information Science* 27: 292–306.

Reskin, Barbara F. 1976. "Sex Differences in Status Attainment in Science: The Case of Post-doctoral Fellowships." *American Sociological Review* 41: 597–612.

———. 1977. "Scientific Productivity and the Reward System of Science." *American Sociological Review* 42: 491–504.

———. 1978. "Scientific Productivity, Sex and Location in the Institution of Science." *American Journal of Sociology* 83: 1235–1243.

Reskin, Barbara F. and Lowell L. Hargens. 1979. "Scientific Advancement of Male and Female Chemists." In R. Alvarez and K. Lutterman and associates, eds., *Discrimination in Organizations*, pp. 100–122. San Francisco: Jossey-Bass.

Rose, S. M. 1985. "Professional Networks of Junior Faculty in Psychology." *Psychology of Women Quarterly* 9: 533–547.

Rossiter, Margaret. 1978. "Sexual Segregation in the Sciences: Some Data and a Model." *Signs* 4: 146–151.

———. 1982. *Women Scientists in America: Struggles and Strategies to 1940*. Baltimore: Johns Hopkins Press.

Russo, N. F. and A. N. O'Connell. 1980. "Models from Our Past—Psychology's Foremothers." *Psychology of Women Quarterly* 5: 11–54.

Sigelman, L. and F. P. Scioli, Jr. 1986. "Retreading Familiar Terrain—Bias, Peer Review and the NSF Political Science Program." Unpublished ms.

Simon, R. J., S. M. Clark and L. L. Tifft. 1966. "Of Nepotism, Marriage, and the Pursuit of an Academic Career." *Sociology of Education* 39: 344–358.

Syverson, P. D. 1980. *Summary Report 1979: Doctorate Recipients from United States Universities*. Washington, D.C.: National Academy of Sciences.

Toren, Nina. 1989. "The Nexus Between Family and Work Roles of Academic Women: Reality and Representation." Unpublished ms.

Traweek, Sharon. 1984. "High Energy Physics: A Male Preserve." *Technology Review* 87: 42–43.

Vetter, Betty M. 1981: "Women Scientists and Engineers: Trends in Participation." *Science* 214: 1313–1321.

Vetter, Betty M. and Eleanor R. Babco, eds. 1986. *Manpower Comments* (Commission on Professionals in Science and Technology) 23, no. 7.

Wanner, R. A., L. S. Lewis and D. I. Gregorio. 1981. "Research Productivity in Academia: A

Comparative Study of the Sciences, Social Sciences, and Humanities." *Sociology of Education* 54: 238–253.

White, J. 1967. "Women in the Law." *Michigan Law Review* 65: 1051.

Zuckerman, Harriet. 1970. "Stratification in American Science." *Sociological Inquiry* 40: 235–257.

———. 1977. *Scientific Elite: Nobel Laureates in the United States.* New York: Free Press.

———. 1987. "Citation Analysis and the Complex Problem of Intellectual Influence." *Scientometrics* 12: 329–338.

———. 1989. "Accumulation of Advantage and Disadvantage: The Theory and Its Intellectual Biography." In Carlo Mongardini and Simonetta Tabboni, eds., *L'Opera de R. K. Merton e La Sociologia Contemporanea,* pp. 153–176. Genova: ECIG (Edizioni Culturali Internazionale).

Zuckerman, Harriet and Jonathan R. Cole. 1975. "Women in American Science." *Minerva* 13: 82–102.

2. Citation Classics

Allison, P. D. 1980. *Processes of Social Stratification.* New York: Arno Press.

Astin, H. S. 1979. "Patterns of Women's Occupations." In J. Sherman and F. Denmark, eds., *The Psychology of Women: Future Directions in Research.* New York: Psychological Dimensions.

———. 1983. Unpublished tabulations.

———. 1984. "The Meaning of Work in Women's Lives: A Sociopsychological Model of Career Choice and Work Behavior." *Counseling Psychologist* 12: 117–126.

Astin, H. S. and A. E. Bayer. 1979. "Pervasive Sex Differences in the Academic Reward System: Scholarship, Marriage and What Else?" In D. R. Lewis and W. E. Becker, Jr., eds., *Academic Rewards in Higher Education.* Cambridge, Mass: Ballinger.

Astin, H. S. and M. D. Snyder. 1982. "Affirmative Action 1972–1982: A Decade of Response." *Change* (July–August).

Astin, H. S. and D. Davis. 1985. "Research Productivity Across the Life and Career Cycles: Facilitators and Barriers for Women." In M. F. Fox, ed., *Scholarly Writing and Publishing: Issues, Problems and Solutions.* Boulder, Colo.: Westview Press.

Bayer, A. E. 1973. *Teaching Faculty in Academe: 1972–73.* Vol. 8. Washington, D.C.: American Council on Education.

Cole, J. R. 1979. *Fair Science.* New York: Free Press.

Cole, J. R. and S. Cole. 1973. *Social Stratification in Science.* Chicago: University of Chicago Press.

Cole, J. R. and H. Zuckerman. 1984. "The Productivity Puzzle." In M. L. Maehr and M. W. Steincamp, eds., *Advances in Motivation and Achievement: Women in Science.* Greenwich, Conn.: JAI Press.

Crane, D. 1965. "Scientists at Major and Minor Universities: A Study of Productivity and Recognition." *American Sociological Review* 30: 699–714.

Cunningham, G. K. and N. J. Cunningham. 1986. "Assessing Astin's Sociopsychological Model of Career Choice and Work Behavior." Paper presented at the American Education Research Association meeting, San Francisco.

Davis, D., and H. S. Astin. 1987. "Reputational Standing in Academe." *Journal of Higher Education* (May / June).

Ferber, M. A., J. W. Loeb, and H. M. Lowry. 1978. "The Economic Status of Women Faculty: A Reappraisal." *Journal of Human Resources* 13, 3: 385–401.

Garfield, E. 1979. *Citation Indexing—Its Theory and Application in Sciences, Technology, and Humanities.* New York: Wiley.

Gaston, J. 1973. *Originality and Competition in Science: A Study of the British High Energy Physics Community.* Chicago: University of Chicago Press.

Hagstrom, W. 1971. "Inputs, Outputs and the Prestige of American University Science Departments." *Sociology of Education* 44: 375–379.

Helmreich, R. L., J. T. Spence, W. E. Beane, G. W. Lucker, and K. A. Matthews. 1980. "Making It in Academic Psychology, Demographic and Personality Correlates of Attainment." *Journal of Personality and Social Psychology* 39: 896–908.

Horner, M. S. 1972. "Toward an Understanding of Achievement-related Conflicts in Women." *Journal of Social Issues* 28: 157–176.

Long, J. S. 1978. "Productivity and Academic Position in the Scientific Career." *American Sociological Review* 43: 889–908.

Maslow, A. H. 1954. *Motivation and Personality*. New York: Harper & Row.

McClelland, D. C., J. W. Atkinson, R. A. Clark, and E. L. Lowell. 1953. *The Achievement Motive*. New York: Appleton-Century-Crofts.

Reskin, B. F. 1977. "Scientific Productivity and the Reward Structure of Science." *American Sociological Review* 42: 491–504.

———. 1978. "Sex Differentiation and the Social Organization of Science." In J. Gaston, ed., *The Sociology of Science*. San Francisco: Jossey-Bass.

Simon, J. G. and N. T. Feather. 1973. "Causal Attributions for Success and Failure at University Examinations." *Journal of Educational Psychology* 64: 46–56.

Zuckerman, H. and R. K. Merton. 1971. "Patterns of Evaluation in Science: Institutionalization, Structure and Functions of the Referee System." *Minerva* 9: 66–100.

7. Sex Differences in Careers

Ashmore, Richard D. and Frances K. Del Boca. 1986. "Gender Stereotypes." In R. D. Ashmore and F. K. Del Boca, eds., *The Social Psychology of Female-Male Relations*. New York: Academic.

Association of American Colleges, Project on the Status and Education of Women. 1982. *The Classroom Climate: A Chilly One for Women?* Washington, D.C.: Association of American Colleges.

Baron, James N. and William T. Bielby. 1980. "Bringing the Firms Back In: Stratification, Segmentation, and the Organization of Work." *American Sociological Review* 45: 737–765.

Becker, Gary S. 1974. "A Theory of Marriage." In Theodore W. Schultz, ed., *Economics of the Family: Marriage, Children, and Human Capital*, pp. 299–344. Chicago: University of Chicago Press.

———. 1985. "Human Capital, Effort, and the Sexual Division of Labor." *Journal of Labor Economics* 3: S33–S58.

Beller, Andrea. 1982. "Occupational Sex Segregation: Determinants and Changes." *Journal of Human Resources* 17: 371–372.

———. 1984. "Trends in Occupational Segregation by Sex, 1960–1981." In Barbara F. Reskin, ed., *Sex Segregation in the Workplace: Trends, Explanations, Remedies*, pp. 11–26. Washington, D.C.: National Academy Press.

Berryman, Sue E. 1983. *Who Will Do Science?* New York: Rockefeller Foundation.

Bielby, Denise D. and William T. Bielby. 1988. "She Works Hard for the Money: Household Responsibilities and the Allocation of Effort." *American Journal of Sociology* 93: 1031–1055.

Bielby, William T. and James N. Baron. 1984. "A Women's Place Is with Other Women: Sex Segregation Within Organizations." In Barbara F. Reskin, ed., *Sex Segregation in the Workplace: Trends, Explanations, Remedies*, pp. 27–55. Washington, D.C.: National Academy Press.

———. 1986. "Men and Women at Work: Sex Segregation and Statistical Discrimination." *American Journal of Sociology* 91: 759–799.

Broverman, I. K., S. R. Vogel, D. M. Broverman, F. E. Clarkson and P. S. Rosenkrantz. 1972. "Sex-role Stereotypes: A Current Appraisal." *Journal of Social Issues* 28: 59–78.

Bruer, John T. 1983. "Women in Science: Lack of Full Participation." *Science* 221: 1339.

———. 1985. *Gender Discrimination in Science: Report of a Macy Foundation Symposium*. New York: Josiah Macy, Jr. Foundation.

Clark, Shirley M. and Mary Corcoran. 1986. "Perspectives on the Professional Socialization of Women Faculty: A Case of Accumulative Disadvantage?" *Journal of Higher Education* 57: 20–43.

Cole, Jonathan R. 1979. *Fair Science*. New York: Free Press.

Cole, Jonathan R. and Harriet Zuckerman. 1984. The Productivity Puzzle: Persistence and Change in Patterns of Publication of Men and Women Scientists." In P. Maehr and M. W. Steinkamp, eds., *Advances in Motivation and Achievement*. Greenwich, Conn.: JAI Press.

Cole, Jonathan R. and Harriet Zuckerman. 1987. "Marriage, Motherhood and Research Performance in Science." *Scientific American* 255; 2, 119–125.

Cooper, Joel and Russell H. Fazio. 1979. "The Formation and Persistence of Attributes that Support Intergroup Conflict." In William Austin and Stephen Worchel, eds., *The Social Psychology of Intergroup Relations*, pp. 149–159. Monterey, Calif.: Brooks / Cole.

Cyert, Richard M. and James G. March. 1963. *A Behavioral Theory of the Firm*. Englewood Cliffs, N.J.: Prentice-Hall.

Darley, J. M. and R. H. Fazio. 1980. "Expectancy Confirmation Sequences." *American Psychologist* 35: 867–881.

Deaux, Kay. 1985. "Sex and Gender." *Annual Review of Psychology* 36: 49–81.

DiMaggio, Paul J. and Walter W. Powell. 1983. "The Iron Cage Revisited: Institutional Isomorphism and Collective Rationality in Organizational Fields." *American Sociological Review* 48: 147–160.

Donnell, S. M. and J. Hall. 1980. "Men and Women as Managers: A Significant Case of No Significant Difference." *Organizational Dynamics*, 60–77.

England, Paula. 1984. "Wage Appreciation and Depreciation: A Test of Neoclassical Economic Explanations of Occupational Sex Segregation." *Social Forces* 62: 726–749.

Feldman, S. D. 1974. *Escape from the Doll's House*. New York: McGraw-Hill.

Fox, Mary Frank. 1981. "Sex, Salary, and Achievement: Reward-Dualism in Academia." *Sociology of Education* 54: 71–84.

———. 1984. "Women and Higher Education: Sex Differentials in the Status of Students and Scholars." In J. Freeman, ed., *Women: A Feminist Perspective*, pp. 238–255. 3rd ed. Palo Alto, Calif.: Mayfield.

Hagstrom, Warren O. 1965. *The Scientific Community*. New York: Basic Books.

Hamilton, David L. 1981. *Cognitive Processes in Stereotyping and Intergroup Behavior*. Hillsdale, N.J.: Erlbaum.

Kahle, Jane Butler. 1985. "A View and a Vision: Women in Science Today and Tomorrow." In Jane Butler Kahle, ed., *Women in Science: A Report from the Field*, pp. 193–229. Philadelphia: Falmer Press.

Kanter, Rosabeth M. 1977. "Some Effects of Proportions on Group Life: Skewed Sex Ratios and Responses to Token Women." *American Journal of Sociology* 82: 965–990.

Keller, Evelyn Fox. 1985. *Reflections on Gender and Science*. New Haven: Yale University Press.

Marini, Margaret Mooney and Mary C. Brinton. 1984. "Sex Typing and Occupational Socialization." In Barbara F. Reskin, ed., *Sex Segregation in the Workplace: Trends, Explanations, Remedies*, pp. 192–232. Washington, D.C.: National Academy Press.

Marwell, Gerald R., Rachel A. Rosenfeld and Seymour Spilerman. 1979. "Geographic Constraints on Women's Careers in Academia." *Science* 205: 1225–1231.

Matyas, Marsha Lakes. 1985. "Factors Affecting Female Achievement and Interest in Science and Scientific Careers." In Jane Butler Kahle, ed., *Women in Science: A Report from the Field*, pp. 27–48. Philadelphia: Falmer Press.

Merton, Robert K. [1948] 1968. "The Self-Fulfilling Prophecy." In *Social Theory and Social Structure*, pp. 475–490. Rev. and enl. ed. New York: Free Press.

———. 1968. "The Matthew Effect in Science." *Science* 199: 55–63.

———. [1942] 1973. *The Sociology of Science*, pp. 267–278. Chicago: University of Chicago Press.

Meyer, John W. and Brian Rowan. 1977. "Institutionalized Organizations: Formal Structure as Myth and Ceremony." *American Journal of Sociology* 83: 340–363.

Mincer, Jacob. 1985. "Intercountry Comparisons of Labor Force Trends and of Related Developments: An Overview." *Journal of Labor Economics* 3 (supp.): S1–S32.

Mincer, Jacob and Solomon W. Polachek. 1974. "Family Investments in Human Capital: Earnings of Women." *Journal of Political Economy* 82: S76–S108.

National Center for Education Statistics. 1981. *Faculty, Salary, Tenure, and Benefits, 1980–81*. Washington, D.C.: Government Printing Office.

National Research Council. 1979. *Climbing the Academic Ladder: Doctoral Women Scientists in Academe*. Washington, D.C.: National Academy of Sciences.

———. 1981. *Career Outcomes in a Matched Sample of Men and Women Ph.D.'s*. Washington, D.C.: National Academy of Sciences.

National Science Foundation. 1986. *Women and Minorities in Science and Engineering*. NSF 86–301. Washington, D.C.: National Science Foundation.

O'Neill, June. 1985. "The Trend in the Male-Female Wage Gap in the United States." *Journal of Labor Economics* 3 (supp.): S91–S116.

Pettigrew, Thomas F. and Joanne Martin. 1987. "Shaping the Organizational Context for Black American Inclusion." *Journal of Social Issues*, 43: 41–78.

Pfeffer, Jeffrey. 1981. *Power in Organizations*. Boston: Pitman.

Pfeffer, Jeffrey and Gerald R. Salancik. 1978. *The External Control of Organizations: A Resource Dependence Perspective.* New York: Harper & Row.

Polachek, Solomon W. 1981. "Occupational Self-Selection: A Human Capital Approach to Sex Differences in Occupational Structure." *Review of Economics and Statistics* 63: 60–69.

———. 1984. "Women in the Economy: Perspectives on Gender Inequality." In *Comparable Worth: Issues for the 80's,* pp. 34–53. U.S. Commission on Civil Rights. Washington D.C.: Government Printing Office.

Reagan, B. 1975. "Two Supply Curves for Economists? Implications of Mobility and Career Attachment of Women." *American Economic Review* 65: 100–107.

Reskin, Barbara F. 1978. "Scientific Productivity, Sex, and Location in the Institution of Science." *American Journal of Sociology* 83: 1235–1243.

Reskin, Barbara F. and Lowell L. Hargens. 1978. "Assessing Sex Discrimination in Science." In Rudolfo Alvarez, ed., *Social Indicators of Institutional Discrimination,* pp. 6–37. San Francisco: Jossey-Bass.

Reskin, Barbara F. and Heidi I. Hartmann, eds. 1986. *Women's Work, Men's Work: Sex Segregation on the Job.* Washington, D.C.: National Academy Press.

Roos, Patricia A. and Barbara F. Reskin. 1984. "Institutional Factors Contributing to Sex Segregation in the Workplace." In Barbara F. Reskin, ed., *Sex Segregation in the Workplace: Trends, Explanations, Remedies,* pp. 235–260. Washington, D.C.: National Academy Press.

Rosenfeld, Rachel A. 1984. "Academic Career Mobility for Women and Men Psychologists." In Violet B. Haas and Carolyn F. Perrucci, eds., *Women in Scientific and Engineering Professions,* pp. 89–127. Ann Arbor: University of Michigan Press.

Rossiter, Margaret W. 1982. *Women Scientists in America: Struggles and Strategies to 1940.* Baltimore: Johns Hopkins University Press.

Sanders, G. S. and T. Schmidt. 1980. "Behavioral Discrimination Against Women." *Personality and Social Psychology Bulletin* 6: 484–488.

Schumer, Fran. 1981. "A Question of Sex Bias at Harvard." *New York Times Magazine* (October 18): 96–104.

Skrypnek, B. J. and M. Snyder. 1982. "On the Self-perpetuating Nature of Stereotypes About Women and Men." *Journal of Experimental Social Psychology* 18: 277–291.

Spence, J. T., R. L. Helmreich and J. Stapp. 1975. "Ratings of Self and Peers on Sex Role Attributes and Their Relation to Self-Esteem and Concepts of Masculinity and Femininity." *Journal of Personality and Social Psychology* 32: 29–39.

Szafran, Robert F. 1983. *Universities and Women Faculty: Why Some Organizations Discriminate More Than Others.* New York: Praeger.

Thornton, Arland, Duane F. Alwin and Donald Camburn. 1983. "Causes and Consequences of Sex-Role Attitude Change." *American Sociological Review* 48: 211–227.

Treiman, Donald J. and Heidi Hartmann. 1981. *Women, Work, and Wages: Equal Pay for Jobs of Equal Value.* Washington, D.C.: National Academy Press.

Vetter, Betty M. and Eleanor L. Babco. 1986. *Professional Women and Minorities—A Manpower Data Resources.* Washington, D.C.: Commission on Professionals in Science and Technology.

White, Harrison. 1982. "Fair Science?" *American Journal of Sociology* 87: 951–956.

Word, C. O., M. P. Zanna and J. Cooper. 1974. "The Nonverbal Mediation of Self-Fulfilling Prophecies in Interracial Interaction." *Journal of Experimental Social Psychology* 10: 109–120.

Zald, Mayer N. 1970. "Political Economy: A Framework for Comparative Analysis." In Mayer N. Zald, ed., *Power in Organizations,* pp. 221–261. Nashville: Vanderbilt University Press.

Zuckerman, Harriet and Jonathan R. Cole. 1985. "Sex Discrimination and Career Attainments of American Men and Women Scientists." Paper prepared for presentation at the Josiah Macy, Jr. Symposium on Gender Discrimination in Science, New York, March 1985.

8. Gender, Environmental Milieu, and Productivity in Science

Ahern, Nancy and Elizabeth Scott. 1981. *Career Outcomes in a Matched Sample of Men and Women Ph.D.s.* Washington, D.C.: National Academy Press.

Andrews, Frank. 1976. "Creative Process." In D. Pelz and F. Andrews, *Scientists in Organizations.* Ann Arbor, Mich.: Institute for Social Research.

Astin, Helen. 1978. "Factors Affecting Women's Scholarly Productivity." In H. S. Astin and W. S. Hirsch, eds., *The Higher Education of Women*. New York: Praeger.

Astin, Helen and Diane Davis. 1985. "Research Productivity Across the Life- and Career-Cycles: Facilitators and Predictors for Women." In M. F. Fox, ed., *Scholarly Writing and Publishing: Issues, Problems, and Solutions*. Boulder, Colo.: Westview Press.

Austin, Ann and Zelda Gamson. 1983. *Academic Workplace: New Demands, Heightened Tensions*. Washington, D.C.: Association for the Study of Higher Education.

Bayer, Alan. 1973. *Teaching Faculty in Academe*. ACE Research Reports 8.

———. 1982. "A Bibliometric Analysis of Marriage and Family Literature." *Journal of Marriage and the Family* 44.

Beaver, Donald de B. 1986. "Collaboration and Teamwork in Physics." *Czechoslovak Journal of Physics* 1: 14–18.

Beaver, Donald de B. and R. Rosen. 1978. "The Professional Origins of Scientific Co-authorship." *Scientometrics* 1 (November): 65–84.

Berg, H. M. and Marianne Ferber. 1983. "Women and Women Graduate Students: Who Succeeds and Why." In *Journal of Higher Education* 54 (Nov / Dec): 629–648.

Beyer, Janice. 1978. "Editorial Policies and Practices Among Leading Journals in Four Scientific Fields." *Sociological Quarterly* 19 (Winter): 68–88.

Biglan, Anthony. 1973. "Relationships Between Subject Matter Characteristics and the Structure and Output of University Departments." *Journal of Applied Psychology* 57: 204–213.

Blau, Peter. 1973. *The Organization of Academic Work*. New York: Wiley.

Bowen, Howard and Jack Schuster. 1986. *American Professors*. New York: Oxford University Press.

Broad, William. 1981. "The Publishing Game: Getting More for Less." *Science* 211 (March): 1137–1139.

Cameron, Susan Wilson. 1978. "Women Faculty in Academia: Sponsorship, Informal Networks, and Scholarly Success." Ph.D. dissertation, University of Michigan.

Centra, John A. 1974. *Women, Men, and the Doctorate*. Princeton: Educational Testing Service.

Chubin, Daryl. 1974. "Sociological Manpower and Womanpower: Sex Differences in Career Patterns of Two Cohorts of American Doctorate Sociologists." *American Sociologist* 9 (May): 83–92.

Cole, Jonathan R. 1979. *Fair Science: Women in the Scientific Community*. New York: Free Press.

———. 1981. "Women in Science." *American Scientist* 69 (July–August): 385–391.

Cole, Jonathan R. and Stephen Cole. 1973. *Social Stratification in Science*. Chicago: University of Chicago Press.

Cole, Jonathan R. and Harriet Zuckerman. 1984. "The Productivity Puzzle: Persistence and Change in Patterns of Publication Among Men and Women Scientists." In P. Maehr and M. W. Steinkamp, eds., *Women in Science*. Greenwich, Conn.: JAI Press.

Collins, Randall. 1979. *The Credential Society*. New York: Academic Press.

Deaux, K. and T. Emswiller. 1974. "Explanations of Successful Performance in Sex-Linked Tasks." *Journal of Personality and Social Psychology* 22: 80–85.

Drew, David Eli. 1985. *Strengthening Academic Science*. New York: Praeger.

Epstein, Cynthia F. 1970. *Woman's Place: Options and Limits in Professional Careers*. Berkeley: University of California Press.

———. 1983. *Women in Law*. New York: Anchor Books.

Etzioni, Amitai. 1961. *A Comparative Analysis of Complex Organizations*. New York: Free Press.

Feldt, Barbara. 1985. "An Analysis of Productivity of Nontenured Faculty Women in the Medical and a Related School: Some Preliminary Findings." Ann Arbor, Mich.: Office of Affirmative Action.

———. 1986. "The Faculty Cohort Study: School of Medicine." Ann Arbor, Mich.: Office of Affirmative Action.

Fidell, L. S. 1975. "Empirical Verification of Sex Discrimination in Hiring Practices in Psychology." In R. K. Unger and F. L. Denmark, eds., *Woman: Dependent or Independent Variable?* New York: Psychological Dimensions.

Finkelstein, Martin. 1982. "Faculty Colleagueship Patterns and Research Productivity." Presented at meetings of American Educational Research Association. ERIC Document 216633.

Fox, Mary Frank. 1983. "Publication Productivity Among Scientists: A Critical Review." *Social Studies of Science* 13 (May): 285–305.

———. 1985. "Publication, Performance, and Reward in Science and Scholarship." In J. Smart, ed., *Higher Education: Handbook of Theory and Research*. New York: Agathon Press.

———. 1986. "Mind, Nature, and Masculinity." *Contemporary Sociology* 15 (March): 197–199.

————. 1989. "Women and Higher Education: Gender Differences in the Status of Students and Scholars." In J. Freeman, ed., *Women: A Feminist Perspective*. Palo Alto: Mayfield.

Garvey, William. 1979. *Communication: The Essence of Science*. Oxford: Pergamon Press.

Garvey, William and S. D. Gottfredson. 1979. "Changing the System: Innovations in the Interactive Social System of Scientific Communication." In W. Garvey, ed., *Communication: The Essence of Science*. Oxford: Pergamon Press.

Garvey, William and Belver Griffith. 1967. "Scientific Communication as a Social System." *Science* 157 (September): 1011–1016.

Gordon, M. D. 1980. "A Critical Reassessment of Inferred Relations Between Multiple Authorship, Scientific Collaboration, and the Production of Papers and Their Acceptance for Publication." *Scientometrics* 2: 193–201.

Gornick, Vivian. 1983. *Women in Science: Portraits from a World in Transition*. New York: Simon and Schuster.

Hagstrom, Warren. 1965. *The Scientific Community*. New York: Basic Books.

Hargens, Lowell and Warren Hagstrom. 1982. "Consensus and Status Attainment Patterns in Scientific Disciplines." *Sociology of Education* 55: 183–196.

Heffner, Alan. 1979. "Authorship Recognition of Subordinates in Collaborative Research." *Social Studies of Science* 9 (August): 377–384.

Helmreich, Robert, Janet Spence, William Beane, G. William Lucker, and Karen Matthews. 1980. "Making It in Academic Psychology: Demographic and Personality Correlates of Attainment." *Journal of Personality and Social Psychology* 39 (November): 896–908.

Holmstrom, Engin Inel and Robert Holmstrom. 1974. "The Plight of the Woman Graduate Student." *American Educational Research Journal* 11 (Winter): 1–17.

House, James. 1981. *Work Stress and Social Support*. Reading, Mass.: Addison-Wesley.

Kanter, Rosabeth M. 1977. *Men and Women of the Corporation*. New York: Basic Books.

Kaplan, N. and N. W. Storer. 1968. "Scientific Communication." In D. L. Sills, ed., *International Encyclopedia of the Social Sciences*, vol. 14. New York: Macmillan and Free Press.

Kashket, Eva, Mary Louise Robbins, Loretta Lieve, and Alice Huang. 1974. "Status of Women Microbiologists." *Science* (February): 488–494.

Keller, Evelyn Fox. 1982. "Feminism and Science." *Signs* 7 (Spring): 589–602.

————. 1983. *A Feeling for the Organism: The Life and Work of Barbara McClintock*. New York: Freeman.

————. 1985. *Reflections on Gender and Science*. New Haven: Yale University Press.

Kjerulff, Kristen and Milton Blood. 1973. "A Comparison of Communication Patterns in Male and Female Graduate Students." *Journal of Higher Education* 44 (November): 623–632.

Lawani, S. M. 1986. "Some Bibliometric Correlates of Quality in Scientific Research." *Scientometrics* 9: 13–25.

Lieberman, Marcia L. 1981. "The Most Important Thing for You To Know." In G. De Sole and L. Hoffmann, eds., *Rocking the Boat: Academic Women and Academic Processes*. New York: Modern Language Association of America.

Long, J. Scott. 1978. "Productivity and Academic Position in the Scientific Career." *American Sociological Review* 43: 899–908.

Long, J. Scott and Robert McGinnis. 1981. "Organizational Context and Scientific Productivity." *American Sociological Review* 46: 422–442.

————. 1985. "The Effects of the Mentor on the Academic Career." *Scientometrics* 7: 255–280.

Loring, Katherine. 1985. "Work in Progress: Two GLCA Self-Studies on Equity for Women Faculty." Ann Arbor, Mich.: Great Lakes College Association.

Maanten, A. A. 1970. "Statistical Analysis of a Scientific Discipline: Palynology." *Earth Science Review* 6: 181–218.

Mackie, Marlene. 1977. "Professional Women's Collegial Relations and Productivity." *Sociology and Social Research* 61 (April): 277–293.

Meadows, A. J. 1974. *Communication in Science*. London: Butterworth.

Morlock, Laura. 1973. "Discipline Variation in the Status of Academic Women." In A. Rossi and A. Calderwood, eds., *Academic Women on the Move*. New York: Russell Sage Foundation.

National Research Council, Committee on the Education and Employment of Women in Science and Engineering. 1979. *Climbing the Academic Ladder: Doctoral Women Scientists in Academe*. Washington, D.C.: National Academy Press.

————. 1983. *Climbing the Ladder: An Update on the Status of Doctoral Women Scientists and Engineers*. Washington, D.C.: National Academy Press.

National Science Board. 1985. *Science Indicators: The 1985 Report.* Washington, D.C.: Government Printing Office.

Nieva, Veronica and Barbara Gutek. 1980. "Sex Differences in Evaluation." *Academy of Management Review* 5: 267–276.

O'Connor, J. G. 1969. "Growth of Multiple Authorship." *DRTC Seminar* 7: 463–483.

Pelz, Donald and Frank Andrews. 1976. *Scientists in Organizations: Productive Climates for Research and Development.* Ann Arbor, Mich.: Institute for Social Research.

Pheterson, G. T., S. B. Kiesler and P. A. Goldberg. 1971. "Evaluation of the Performance of Women as a Function of Their Sex, Achievement, and Personal History." 19: 110–114.

Presser, Stanley. 1980. "Collaboration and the Quality of Research." *Social Studies of Science* 10 (February): 95–101.

Price, Derek J. De Solla. 1963. *Little Science, Big Science.* New York: Columbia University Press.

Reskin, Barbara. 1977. "Scientific Productivity and the Reward Structure of Science." *American Sociological Review* 42 (June): 491–504.

———. 1978a. "Sex Differentiation and the Social Organization of Science." *Sociological Inquiry* 48: 6–37.

———. 1978b. "Scientific Productivity, Sex, and Location in the Institution of Science." *American Journal of Sociology* 83 (March): 1235–1243.

Roose, Kenneth and Charles Andersen. 1970. *A Rating of Graduate Programs.* Washington, D.C.: American Council on Education.

Rosen, B. and T. H. Jerdee. 1974. "Influence of Sex-Role Stereotypes on Personnel Decisions." *Journal of Applied Psychology* 59: 9–14.

Rossiter, Margaret. 1982. *Women Scientists in America: Struggles and Strategies to 1940.* Baltimore: Johns Hopkins Press.

Sandler, Bernice with Roberta Hall. 1986. "The Campus Climate Revisited: Chilly for Women Faculty, Administrators, and Graduate Students." Washington, D.C.: Association of American Colleges.

Smart, J. A. and A. E. Bayer. 1986. "Author Collaboration and Impact: A Note on Citation Rates of Single and Multiple Authored Articles." *Scientometrics* 10: 311–319.

Subramanyam, K. and E. M. Stephens. 1982. "Research Collaboration and Funding in Biochemistry and Chemical Engineering." *International Forum on Information and Documentation* 7: 26–29.

Tidball, M. Elizabeth. 1974. "Women Role Models in Higher Education." In *Graduate and Professional Education of Women.* Washington, D.C.: American Association of University Women.

Tuckman, Howard. 1976. *Publication, Teaching, and the Academic Reward Structure.* Lexington, Mass.: Lexington Books.

Yoels, William C. 1974. "The Structure of Scientific Fields and the Allocation of Editorships on Scientific Journals: Some Observations on the Politics of Knowledge." *Sociological Quarterly* 15 (Spring): 264–276.

Ziman, J. M. 1968. *Public Knowledge.* London: Cambridge University Press.

Zuckerman, Harriet. 1965. "Nobel Laureates: Sociological Studies of Scientific Collaboration." Ph.D. dissertation, Columbia University.

———. 1977. *Scientific Elite: Nobel Laureates in the United States.* New York: Free Press.

———. 1987. "Persistence and Change in the Careers of Men and Women Scientists and Engineers: A Review of Current Research." In L. Dix, ed., *Women: Their Underrepresentation and Career Differentials in Science and Engineering.* Washington, D.C.: National Academy Press.

9. Discrimination Against Women in Science

Angrist, Shirley S. and Elizabeth M. Almquist. 1975. *Careers and Contingencies.* New York: Dunellen.

Block, Jeanne H. 1976. "Debatable Conclusions About Sex Differences." *Contemporary Psychology* 21: 517–522.

Cole, Jonathan R. 1979. *Fair Science: Women in the Scientific Community.* New York: Free Press.

Cole, Jonathan R. and Harriet Zuckerman. 1984. "The Productivity Puzzle: Persistence and Change in Patterns of Publication of Men and Women Scientists." In Marjorie W. Steinkamp and

Martin L. Maehr, eds., *Advances in Motivation and Achievement*, pp. 217–258. Vol. 2. Greenwich, Conn.: JAI Press.

Cole, Stephen. 1986. "Sex Discrimination and Admission to Medical School: 1929 to 1984." *American Journal of Sociology* 92: 549–567.

Committee on the Education and Employment of Women in Science and Engineering (CEEWSE). 1979. *Climbing the Academic Ladder: Doctoral Women Scientists in Academia*. Washington, D.C.: Commission on Human Resources, National Research Council, National Academy of Sciences.

Deaux, Kay. 1976. "Sex: A Perspective on the Attribution Process." In J. H. Harvey, W. Ickes and R. F. Kidd, eds., *New Directions in Attribution Research*. Vol. 1. Hillsdale, N.J.: Lawrence Erlbaum.

Douvan, Elizabeth. 1976. "The Role of Models in Women's Professional Development." *Psychology of Women Quarterly* 51: 5–20.

Fiorentine, Robert. 1986. "Sex and Status Attainment: Biological, Cultural and Structural Processes." Ph.D. dissertation, SUNY at Stony Brook.

———. 1987. "Men, Women and the "Pre-Med" Persistence Gap: A Normative Alternatives Approach." *American Journal of Sociology* 92: 1118–1139.

———. 1988. "Increasing Similarity in the Values and Life Plans of Male and Female College Students: Evidence and Implications." *Sex Roles* 18:143–158.

Goldberg, Steven. 1973. *The Inevitability of Patriarchy*. New York: Morrow.

Gould, Stephen Jay. 1981. *The Mismeasurement of Man*. New York: Norton.

Jencks, Christopher et al. 1972. *Inequality*. New York: Basic Books.

Lieberson, Stanley. 1985. *Making it Count: The Improvement of Social Research and Theory*. Berkeley: University of California Press.

Lueptow, Lloyd B. 1981. "Sex-Typing and Change in Occupational Choices of High School Seniors: 1964–1975." *Sociology of Education* 54: 16–24.

Maccoby, Eleanor Emmons and Carol Nagy Jacklin. 1974. *The Psychology of Sex Differences*. Stanford: Stanford University Press.

Marini, Margaret and Mary C. Brinton. 1984. "Sex Stereotyping in Occupational Socialization." In Barbara F. Reskin, ed., *Sex Segregation in the Work Place: Trends, Explanations, Remedies*. Washington, D.C.: National Academy Press.

Medical Education. 1981. *Journal of the American Medical Association* 246: 2913–2929.

Merton, Robert K. 1957. "Reference Group Theory." *In Social Theory and Social Structure*. Glencoe, Ill.: Free Press.

Misztal, Maria. 1986. "Value Systems Among Occupational Groups." In Kazimierz M. Slomczynski and Tadeusz K. Krauze, eds., *Social Stratification in Poland*. New York: M. E. Sharpe.

National Science Board. 1982. *Science Indicators*. Washington, D.C.: National Science Foundation.

Rosenberg, Morris. 1957. *Occupations and Values*. New York: Free Press.

Slomczynski, Kazimierz M. and Tadeusz K. Krauze, eds. 1986. *Social Stratification in Poland*. New York: M. E. Sharpe.

Suter, Larry E. and Herman P. Miller. 1973. "Income Differences Between Men and Career Women." *American Journal of Sociology* 78: 962–974.

Walsh, Mary Roth. 1977. *Doctors Wanted: No Women Need Apply: Sexual Barriers in the Medical Profession, 1835–1975*. New Haven: Yale University Press.

Weitzman, Lenore. 1979. *Sex Role Socialization*. Palo Alto: Mayfield.

White, Harrison. 1982. "Review Essay: Fair Science?" *American Journal of Sociology* 87: 951–956.

Zuckerman, Harriet. 1977. *Scientific Elite*. New York: Free Press.

10. The Wo / Man Scientist

Benbow, Camilla and Julian Stanley. 1980. "Sex Differences in Mathematical Ability: Fact or Artifact?" *Science* (December 12): 1262–1264.

Cott, Nancy. 1987. *The Grounding of Modern Feminism*. New Haven: Yale University Press.

Keller, Evelyn Fox. 1981. "New Faces in Science and Technology: A Study of Women Students at M.I.T." Unpublished manuscript.

———. 1983. *A Feeling for the Organism: The Life and Work of Barbara McClintock*. New York: Freeman.

———. 1985. *Reflections on Gender and Science*. New Haven: Yale University Press.

Rossiter, Margaret. 1982. *Women Scientists in America*. Baltimore: Johns Hopkins University Press.

Traweek, Sharon. 1984. "High Energy Physics: A Male Preserve." *Technology Review* (November / December): 42.

11. Constraints on Excellence

American Medical Association. 1986. *Physicians' Characteristics Distribution in the U.S.* Chicago: AMA.

Ben-David, Joseph. 1971. *The Scientist's Role in Society: A Comparative Study*. Englewood Cliffs, N.J.: Prentice-Hall.

Barber, Bernard. 1952. *Science and the Social Order*. New York: Free Press.

Berryman, Sue E. 1983. *Who Will Do Science?* New York: Rockefeller Foundation.

Bielby, Denise Del Vento. 1982. "Career Commitment of Female College Graduates: Conceptualization and Measurement of Issues." In P. J. Perun, ed., *The Undergraduate Woman: Issues in Educational Equity*. Lexington, Mass.: D. C. Heath.

Brim, Orville G., Jr. 1976. "Theories of the Male Mid-Life Crisis." *The Counseling Psychologist* 6 (1): 2–9.

Brim, Orville G., Jr., and Jerome Kagan. 1980. "Constancy and Change: A View of the Issues." In Brim and Kagan, eds., *Constancy and Change in Human Development*, pp. 1–25. Cambridge: Harvard University Press.

Cole, Jonathan. 1979. *Fair Science: Women in the Scientific Community*. New York: Free Press.

Cole, Jonathan and Harriet Zuckerman. 1987. "Marriage, Motherhood and Research Performance in Science." *Scientific American* 256: 119–125.

Coleman, James. 1957. *Community Conflict*. Glencoe, Ill.: Free Press.

Cyert, R. M. and J. G. March. 1963. *A Behavioral Theory of the Firm*. Englewood Cliffs, N.J.: Prentice-Hall.

"Doctoral Recipients." *Newsbriefs* (National Research Council) 37 (3): 18.

Durant, Ariel and Will Durant. 1961. *The Story of Civilization, vol. 7, The Age of Reason Begins: A History of European Civilization in the Period of Shakespeare, Bacon, Montaigne, Rembrandt, Galileo and Descartes: 1558–1648*. New York: Simon and Schuster.

Eagly, Alice, 1987. *Sex Differences in Social Behavior: A Social Role Interpretation*. Hillsdale, N.J.: Erlbaum.

Ehrenreich, Barbara and Deirdre English. 1978. *For Her Own Good: 150 Years of Experts' Advice to Women*. New York: Doubleday.

Epstein, Cynthia Fuchs. 1968. "Women and Professional Careers: The Case of the Woman Lawyer." Ph.D. dissertation of sociology, Columbia University.

———. 1969. "Women Lawyers and Their Profession: Inconsistency of Social Controls and Their Consequences for Professional Performance." In Athena Theodore, ed., *The Professional Woman*, pp. 669–684. Cambridge: Schenkman.

———. 1970. *Woman's Place: Options and Limits in Professional Careers*. Berkeley: University of California Press.

———. 1976. "Sex Role Stereotyping, Occupations and Social Exchange." Paper presented at Radcliffe Institute Conference, "Women: Resource for a Changing World," Radcliffe College, April 18, 1972. Reprinted in *Women's Studies* 3: 183–194.

———. 1981. *Women in Law*. New York: Basic Books.

———. 1982. "Ambiguity as Social Control: Consequences for the Integration of Women in Professional Elites." In Vivian Stewart and Muriel Cantor, eds., *Varieties of Work Experiences*, pp. 61–72. Cambridge: Schenkman, 1974; Beverly Hills, Calif.: Sage.

———. 1985. "Ideal Roles and Real Roles or the Fallacy of the Misplaced Dichotomy." *Research in Social Stratification and Mobility* 4: 29–51.

———. 1988. *Deceptive Distinctions: Sex, Gender and the Social Order*. New Haven: Yale University Press; New York: Russell Sage Foundation.

Friedan, Betty. 1963. *The Feminine Mystique*. New York: Norton.

"Gender Stratification in Higher Education." 1987. *Women's Studies International Forum* 10 (2): 157–164.

Gilligan, Carol. 1982. *In a Different Voice: Psychological Theory and Women's Development*. Cambridge: Harvard University Press.

Goode, William J. 1957. "Community Within a Community: The Professions." *American Sociological Review* 22: 194–200.

Hansot, Elisabeth and David Tyack. 1988. "Gender in Public Schools: Thinking Institutionally." *Signs* 13 (4).

Harris, Joseph. 1987. "The Newest Quilt Fad Seems To Be Going Like Crazy." *Smithsonian* (May).

Homans, George C. 1961. "Social Behavior as Exchange." *American Journal of Sociology* 62 (May): 595–606.

Hyde, Janet S. 1990. "Meta-analysis and the Psychology of Gender Differences." *Signs* 16(1): 55–73.

Hyde, Janet S. and Marcia C. Linn, eds. 1986. *The Psychology of Gender*. Baltimore: Johns Hopkins University Press.

Kanter, Rosabeth Moss. 1977. *Men and Women of the Corporation*. New York: Basic Books.

Kolko, Gabriel and Joyce Kolko. 1962. *Wealth and Power in America*. New York: Praeger.

Larson, Magali Sarfatti. 1977. *The Rise of Professionalism: A Sociological Analysis*. Berkeley: University of California Press.

Linn, Marcia and Anne C. Petersen. 1986. "A Meta-analysis of Gender Differences in Spatial Ability: Implications for Mathematics and Science Achievement." In J. S. Hyde and M. C. Linn, eds., *The Psychology of Gender*, pp. 67–101. Baltimore: Johns Hopkins University Press.

Lorber, Judith. 1985. *Women Physicians: Careers, Status and Power*. New York: Methuen.

March, James and Johan Olsen. 1984. "The New Institutionalism: Organizational Factors in Political Life." *American Political Science Review* 78: 734–749.

Marek, George. 1972. *Gentle Genius: The Story of Felix Mendelsohn*. New York: Funk and Wagnall's.

Merton, Robert K. 1957. *Social Theory and Social Structure: Toward the Codification of Theory and Research*. Glencoe, Ill.: Free Press. Originally published 1949.

Milford, Nancy. 1983. *Zelda*. New York: Harper & Row.

Project on the Status and Education of Women. 1986. *On Campus with Women* 16 (2).

Rosenberg, Rosalind. 1982. *Beyond Separate Spheres: Intellectual Roots of Modern Feminism*. New Haven: Yale University Press.

Rossiter, Margaret W. 1982. *Women Scientists in America: Struggles and Strategies to 1940*. Baltimore: Johns Hopkins University Press.

Sachs, Albie and Joan Hoff Wilson. 1978. *Sexism and the Law: Male Beliefs and Legal Bias in Britain and the United States*. New York: Free Press.

Sayre, Anne. 1975. *Rosalind Franklin and DNA*. New York: Norton.

Sokoloff, Alice Hunt. 1969. *Cosima Wagner: Extraordinary Daughter of Franz Liszt*. New York: Dodd, Mead.

Stiehm, Judith H. 1981. *Bring Me Men and Women: Mandated Change at the United States Air Force Academy*. Berkeley: University of California Press.

Tangri, S. 1972. "Determinants of Occupational Role Innovation Among College Women." *Journal of Social Issues* 28: 177–200.

U.S. Women Engineering (November / December 1985), as cited in Project on the Status and Education of Women, *On Campus with Women* 16 (1).

Walsh, Mary Roth. 1977. *Doctors Wanted—No Women Need Apply: Sexual Barriers in the Medical Profession*. New Haven, Conn.: Yale University Press.

Watson, J. D. 1968. *The Double Helix*. New York: Atheneum.

Who's Who of American Women. 1958. 1st ed. Chicago: Marquis Who's Who.

Zuckerman, Harriet. 1977. *Scientific Elite: Nobel Laureates in the United States*. New York: Free Press.

———. 1987. "Persistence and Change in the Careers of Men and Women Scientists and Engineers: A Review of Current Research." In Linda S. Dix, ed., *Women: Their Underrepresentation and Career Differentials in Science and Engineering*. Washington, D.C.: National Academy Press.

Zuckerman, Harriet and Jonathan Cole. 1987. Personal communication.

13. A Theory of Limited Differences

Allison, P. D. 1980. "Inequality and Scientific Productivity." *Social Studies of Science* 10: 163–179.

Allison, P. D. and J. A. Stewart. 1974. "Productivity Differences Among Scientists: Evidence for Accumulative Advantage." *American Sociological Review* 39: 596–606.

Allison, P. D., J. S. Long, and T. Kraus. 1982. "Cumulative Advantage and Inequality in Science." *American Sociological Review* 47: 615–625.

Andrews, F. M. 1979. *Scientific Productivity*. Cambridge: Cambridge University Press.

Astin, H. S. 1969. *The Woman Doctorate in America*. New York: Russell Sage Foundation.

Astin, H. S. and A. E. Bayer. 1975. "Sex Discrimination in Academe." In M. T. S. Mednick, S. S. Tangri, and L. W. Hoffman, eds., *Women and Achievement*. Washington, D.C.: Hemisphere.

Bandura, A. 1977. *Social Learning Theory*. Pp. 11–12. Englewood Cliffs, N.J.: Prentice-Hall.

Bayer, A. E. and H. S. Astin. 1972. "Sex Differentials in the Academic Reward System." *Science* 188: 796–802.

Bayer, A. E. and J. Folger. 1966. "Some Correlates of a Citation Measure of Productivity in Science." *Sociology of Education* 39: 381–390.

Berryman, S. E. 1983. *Who Will Do Science?* New York: The Rockefeller Foundation.

Bielby, W. 1991. "Sex Differences in Careers: Is Science a Special Case." In Harriet Zuckerman, Jonathan R. Cole, and John Bruer, eds., *The Outer Circle*. New York: Norton.

Bielby, W. and J. Baron. 1984. "Work Commitment, Sex-Role Attitudes, and Women's Employment." *American Sociological Review* 49: 234–247.

Centra, J. A. 1974. *Women, Men, and the Doctorate*. Princeton, N.J.: Educational Testing Service.

Clemente, F. 1973. "Early Career Determinants of Research Productivity." *American Journal of Sociology* 79: 409–419.

Cohen, J. E. 1980. "Publication Rate as a Function of Laboratory Size in a Biomedical Research Institution." *Scientometrics* 2: 35–52.

Cole, J. R. 1969. *The Social System of Science*. Ph.D. dissertation, Columbia University.

———. 1979. *Fair Science: Women in the Scientific Community*. New York: The Free Press.

———. 1987. "The Paradox of Individual Particularism and Institutional Universalism." In Jon Elster, ed., *Justice and the Lottery*. Cambridge: Cambridge University Press.

Cole, J. R. and S. Cole. 1973. *Social Stratification in Science*. Chicago: Chicago University Press.

———. 1976. "The Reward System of the Social Sciences." In C. Frankel, ed., *Controversies and Decisions: The Social Sciences and Public Policy*, pp. 55–88. New York: Russell Sage Foundation.

———. 1985. "Experts' 'Consensus' and Decision-Making at the National Science Foundation." In Kenneth Warren, ed., *Selectivity in Information Systems—Survival of the Fittest*. pp. 27–63. New York: Praeger.

Cole, J. R., S. Cole, and the Committee on Science and Public Policy. 1981. *Peer Review at the National Science Foundation: Phase II of a Study*. Washington, D.C.: National Academy of Sciences Press.

Cole, J. R., and H. Zuckerman. 1984. "The Productivity Puzzle: Persistence and Change in Patterns of Publication of Men and Women Scientists." In Marjorie W. Steinkamp and Martin L. Maehr, eds., *Advances in Motivation and Achievement*, vol. 2, pp. 217–258. Greenwich, Ct.: JAI Press.

———. 1987. "Marriage, Motherhood, and Research Performance in Science." *Scientific American* 255, 2: 119–125.

Cole, S. 1970. "Professional Standing and the Reception of Scientific Discoveries." *American Sociological Review* 76:286–306.

———. 1975. "The Growth of Scientific Knowledge: Theories of Deviance as a Case Study." In L. A. Coser, ed., *The Idea of Social Structure: Papers in Honor of Robert K. Merton*, pp. 175–220. New York: Harcourt Brace Jovanovich.

Cole, S., J. R. Cole, and L. Dietrich. 1978. "Measuring the Cognitive State of Scientific Disciplines." In Y. Elkana et al., eds., *Toward a Metric of Science*, pp. 209–251. New York: Wiley.

Cole, S., L. Rubin, and J. R. Cole. 1978. *Peer Review at the National Science Foundation: Phase I of a Study*. Washington, D.C.: National Academy of Sciences Press.

Cole, S., J. R. Cole, and G. A. Simon. 1981. "Chance and Consensus in Peer Review." *Science* 214: 881–886.

Crane, D. 1969. "Social Class Origin and Academic Success: The Influence of Two Stratification Systems on Academic Careers." *Sociology of Education* 42: 1–17.

Darwin, C. 1859. *The Origin of Species*. New York: Modern Library.

Duncan, B. and O. D. Duncan. 1978. *Sex Typing and Social Roles: A Research Report*. New York: Academic Press.

Feller, W. 1968. *Introduction to Probability Theory and Applications*, vol. 1. 3rd ed. New York: Wiley.

Foulkes, M. A. and C. E. Davis. 1981. "An Index of Tracking for Longitudinal Data." *Biometrics* 37: 439–446.

Fox, M. F. 1981a. "Sex, Salary, and Achievement: Reward-Dualism in Academia." *Sociology of Education* 54: 71–84.

———. 1981b. "Patterns and Determinants of Research Productivity Among Male and Female Social Scientists." University of Michigan, Preprint.

———. 1983. "Published Productivity Among Scientists: A Critical Review." *Social Studies of Science* 13: 285–305.

———. 1985. "Publication, Performance, and Reward in Science and Scholarship." In J. Smart, ed., *Higher Education: Handbook of Theory and Research*. New York: Agathon Press.

Gaston, J. 1973. *Originality and Competition in Science*. Chicago: University of Chicago Press.

———. 1978. *The Reward System in British and American Science*. New York: Wiley Interscience.

Gilbert, G. N. and M. Mulkay. 1984. *Opening Pandora's Box*. Cambridge: Cambridge University Press.

Hagstrom, W. O. 1965. *The Scientific Community*. New York: Basic Books.

———. 1971. "Inputs, Outputs and the Prestige of American University Science Departments." *Sociology of Education* 44: 375–379.

Hargens, L. L., J. C. McCann, and B. R. Reskin. 1978. "Productivity and Reproductivity: Fertility and Professional Achievement Among Research Scientists." *Social Forces* 52: 129–146.

Harmon, L. R. 1963. "The Development of a Criterion of Scientific Competence." In C. W. Taylor and Frank Baron, eds., *Scientific Creativity: Its Recognition and Development*, pp. 44–53. New York: Wiley.

———. 1965. *High School Ability Patterns: A Backward Look from the Doctorate*. Scientific Manpower Report, no. 6. Washington, D.C.: National Research Council.

Helmreich, R. L., J. T. Spence, W. E. Beane, G. W. Lucker, and K. A. Matthews. 1980. "Making It in Academic Psychology: Demographic and Personality Correlates of Attainment." *Journal of Personality and Social Psychology* 39: 896–908.

Kahle, J. B. and M. L. Matyas. 1985. "Equitable Science and Mathematical Education: A Discrepancy Model." In L. S. Dix, ed., *Women: Their Underrepresentation and Career Differentials in Science and Engineering*, pp. 5–41. Washington, D.C.: National Academy Press.

Knorr-Cetina, K. D. 1981. *The Manufacture of Knowledge*. Oxford: Pergamon Press.

Latour, B. and S. Woolgar. 1979. *Laboratory Life: The Social Construction of Scientific Facts*. Beverly Hills: Sage.

Long, J. S. 1978. "Productivity and Academic Position in the Scientific Career." *American Sociological Review* 43: 889–908.

Long, S. and R. McGinnis. 1981. "Organizational Context and Scientific Productivity." *American Sociological Review* 46: 422–442.

Maccoby, E. E. and C. N. Jacklin. 1975. *The Psychology of Sex Differences*. Stanford: Stanford University Press.

Marini, M. M. 1987. "Sex and Gender." In E. F. Borgotta and K. S. Cook, eds., *The Future of Sociology*. Beverly Hills: Sage.

Marini, M. M. and M. C. Brinton. 1984. "Sex Typing in Occupational Socialization." In B. Reskin, ed., *Sex Segregation in the Workplace: Trends, Explanations, Remedies*, pp. 192–232. Washington, D.C.: National Academy Press.

Marwell, G., R. Rosenberg and S. Spilerman. 1979. "Geographic Constraints on Women's Careers in Academia." *Science* 205: 1225–1231.

Merton, R. K. 1957. "Priorities in Scientific Discovery." *American Sociological Review* 22: 635–659.

———. 1968. "The Matthew Effect in Science." *Science* 159: 56–63.

———. 1973. *The Sociology of Science*. Chicago: University of Chicago Press.

———. 1987. "Three Fragments from a Sociologist's Notebooks: Establishing the Phenomenon,

Specified Ignorance, and Strategic Research Materials." *Annual Review of Sociology* 13: 1–28.

Mittermeir, R. and K. D. Knorr. 1979. "Scientific Productivity and Accumulative Advantage: A Thesis Reassessed in the Light of International Data." *R&D Management* 9: 235–239.

Over, R. 1982. "Research Productivity and Impact of Male and Female Psychologists." *American Psychologist* 37: 24–31.

Over, R. and D. Moore. 1980. "Research Productivity and Impact of Men and Women in Psychology Departments of Australian Universities, 1975–1977." *Australian Psychologist* 15: 413–418.

Pfeffer, J. 1982. *Organizations and Organization Theory*. Boston: Pitman.

Price, D. J. deSolla. 1963. *Little Science, Big Science*. New York: Columbia University Press.

Reskin, B. 1977. "Scientific Productivity and the Reward Structure of Science." *American Sociological Review* 42: 491–504.

———. 1978a. "Scientific Productivity, Sex, and Location in the Institution of Science." *American Journal of Sociology* 83: 1235–1243.

———. 1978b. "Sex Differentiation and the Social Organization of Science." In J. Gaston, ed., *The Sociology of Science*. San Francisco: Jossey-Bass.

———. 1979. "Review of the Literature on the Relationship Between Age and Scientific Productivity. In *Research Excellence Through the Year 2000*. Washington, D.C.: Commission on Human Resources, National Academy of Sciences Press.

Rossi, P. 1979. "Vignette Analysis: Uncovering the Normative Structure of Complex Judgments." In R. K. Merton, J. S. Coleman and P. S. Rossi, eds., *Qualitative and Quantitative Social Research*, pp. 176–186. New York: The Free Press.

Spence, J. T. and R. L. Helmreich. 1978. *Masculinity and Femininity: Their Psychological Dimensions, Correlates and Antecedents*. Austin: University of Texas Press.

———. 1979. "Comparison of Masculine and Feminine Personality Attributes and Sex-Role Attitudes Across Age Groups." *Developmental Psychology* 15: 583–584.

Spence, J. T., R. L. Helmreich, and J. Stapp. 1975. "Ratings of Self and Peers on Sex Role Attributes and Their Relation to Self-Esteem and Conceptions of Masculinity and Femininity." *Journal of Personality and Social Psychology* 32: 29–39.

Waddington, C. H. 1940. *Organization and Genes*. Cambridge: Cambridge University Press.

Ware, J. H. and M. C. Wu. 1981. "Tracking: Prediction of Future Values from Serial Measurement." *Biometrics* 37: 427–437.

Weiner, N. 1956. *I Am a Mathematician*. New York: Doubleday.

Wilson, E. O. 1975. *Sociobiology*. Cambridge: The Belknap Press.

Zuckerman, H. 1970. "Stratification in American Science." *Sociological Inquiry* 40: 235–257.

———. 1977. *Scientific Elite: Nobel Laureates in the United States*. New York: The Free Press.

———. 1978. "Theory Choice and Problem Choice in Science." *Sociological Inquiry* 48:65–95.

———. 1989. "Accumulation of Advantage and Disadvantage: The Theory and Its Intellectual Biography." In C. Mongardini and S. Tabboni, eds., *L'Opera di R. K. Merton e La Sociologia Contemporanea*, pp. 153–176. Genova: Edizioni Culturali Internazionali (ECIG).

Zuckerman, H. and J. R. Cole. 1975. "Women in American Science." *Minerva* 13: 84 ff.

Zuckerman, H. and R. K. Merton. 1971a. "Age, Aging, and Age Structure in Science." In M. W. Riley, M. Johnson and A. Foner, eds., *Aging and Society: A Sociology of Age Stratification*, vol. 3. New York: Russell Sage Foundation.

———. 1971b. "Patterns of Evaluation in Science: Institutionalization, Structure and Functions of the Referee System." *Minerva* 9: 66–100. Reprinted in R. K. Merton, *The Sociology of Science*, p. 479.

Contributors

HELEN S. ASTIN is Professor and Associate Director of the Higher Education Research Institute at the University of California at Los Angeles. The author of many studies of the careers of men and women academics, Astin recently served as Associate Provost of Letters and Science at UCLA.

WILLIAM T. BIELBY is Professor of Sociology at the University of California at Santa Barbara. A widely known specialist on the nature of work in different kinds of organizations, Bielby is also interested in the connections between economy and society, more generally.

JOHN T. BRUER is President of the James S. McDonnell Foundation in St. Louis. He holds a degree in the Philosophy of Science from the Rockefeller University and served in various administrative posts at the Rockefeller and Josiah Macy, Jr. Foundations.

JONATHAN R. COLE is Provost and Quetelet Professor of the Social Sciences at Columbia University. His current research, in collaboration with Zuckerman, compares the careers of men and women in science. He is also studying the processes by which provisional claims to knowledge are solidified into scientific "facts."

STEPHEN COLE is Professor of Sociology at the State University of New York at Stony Brook. A long-time student of the social organization of science, he is the author of the forthcoming volume, *Social Influences on the Growth of Knowledge*.

ANDREA DUPREE serves as Astrophysicist at the Harvard Smithsonian Center for Astrophysics, where she led its Solar and Stellar Physics Division. Active in a variety of professional activities, Dupree has won awards for her work, including the Lifetime Achievement Award, given by the Women in Aerospace Association.

CYNTHIA F. EPSTEIN is Professor of Sociology at the Graduate Center of the City University of New York. One of the first to study the careers of professional women, Epstein's most recent book is *Deceptive Distinctions: Sex, Gender and the Social Order*.

ROBERT FIORENTINE holds a Ph.D. in Sociology from the State University of New York at Stony Brook and is now associated with the U.S. General Accounting Office in Washington, D.C.

OWEN M. FISS is Alexander M. Bickel Professor of Public Law at the Yale University Law School. He has written on legal procedure and is especially interested in the social and ethical aspects of the law, including questions of equity in American society. Fiss clerked for Associate Justice Brennan of the U.S. Supreme Court.

MARY FRANK FOX holds an Associate Professorship at the Pennsylvania State University. Her research has focused both on the Sociology of Science and the Sociology of Gender.

EVELYN FOX KELLER is Professor of Rhetoric and Women's Studies at the University of California at Berkeley. The author of a critically acclaimed biography of Barbara McClintock and numerous works on women and science, her essays are collected in *Reflections on Gender and Science*. She was trained in theoretical physics and molecular biology.

SANDRA PANEM holds a Ph.D. in the biological sciences from the University of Chicago and is now a venture capitalist specializing in biotechnology with Salomon Brothers, Inc., in New York.

BURTON SINGER is Professor of Biostatistics and Department Chairman at Yale University School of Medicine. His research has ranged widely in the sciences and social sciences and includes studies of the epidemiology of stroke and malaria and processes of occupational mobility in organizations.

SALOME WAELSCH, an eminent developmental biologist, is Professor of Molecular Genetics at the Albert Einstein College of Medicine in New York and is also a member of the National Academy of Sciences.

HARRIET ZUCKERMAN is Professor of Sociology at Columbia University. She is president of the Society for Social Studies of Science and a member of the American Academy of Arts and Sciences. She continues her studies of the Nobel prizes and is also doing research, with Jonathan Cole, on the careers of men and women in science.

Index

	DATE DUE		
DEC 13 1996 S			